AF755101

INTRODUCTION TO
FOOD CHEMISTRY

Leandro Oviedo

Toronto Academic Press

INTRODUCTION TO FOOD CHEMISTRY

Leandro Oviedo

Toronto Academic Press
4164 Lakeshore Road
Burlington ON L7L 1A4
Canada
www.tap-books.com

Email: orders@arclereducation.com

© 2025 Toronto Academic Press

ISBN: 978-1-77956-342-2

This book contains information obtained from highly regarded resources. Reprinted material sources are indicated and copyright remains with the original owners. Copyright for images and other graphics remains with the original owners as indicated. A Wide variety of references are listed. Reasonable efforts have been made to publish reliable data. Authors or Editors or Publishers are not responsible for the accuracy of the information in the published chapters or consequences of their use. The publisher assumes no responsibility for any damage or grievance to the persons or property arising out of the use of any materials, instructions, methods or thoughts in the book. The authors or editors and the publisher have attempted to trace the copyright holders of all material reproduced in this publication and apologize to copyright holders if permission has not been obtained. If any copyright holder has not been acknowledged, please write to us so we may rectify.

Notice: Registered trademark of products or corporate names are used only for explanation and identification without intent of infringement.

Toronto Academic Press publishes wide variety of books and eBooks. For more information about Toronto Academic Press and its products, visit our website at www.tap-books.com.

ABOUT THE AUTHOR

Leandro Rodrigues Oviedo is a chemical engineer with a solid background in machine/deep learning, chemical kinetics, reactor design, and water/wastewater treatment (especially Advanced Oxidation Processes - AOPs). Master in Nanosciences (Nanoscience Post-Graduation Program at Franciscan University - UFN, Brazil), with expertise in nanozeolite synthesis from solid (agro)industrial wastes for adsorption and photocatalytic degradation of organic pollutants.

Table of Contents

List of Figures

List of Tables

PREFACE

The production and distribution of an adequate supply of nutritious and appealing food are fundamental to human survival. Historically, food preparation and preservation were tasks managed primarily within individual households and small-scale trade. However, industrialization has progressively shifted these activities to a larger, industrial scale. This transformation has not only increased food production efficiency but also influenced lifestyles, resulting in the widespread availability of convenience foods and the rise of catering services to meet the growing demand for out-of-home dining.

Industrialization has also fostered a global network of food trade and exchange, yet challenges such as food shortages persist, particularly in regions with unbalanced access to resources. This disparity underscores food scarcity as a critical demographic, economic, and political issue worldwide.

In addressing these complexities, food science, particularly food chemistry, plays a pivotal role. The primary aim of this book is to acquaint readers with the structures and properties of three essential food components: carbohydrates, fats, and proteins. Understanding these compounds is crucial for comprehending the chemical reactions and transformations that occur during food production, storage, and handling. These processes can significantly impact the texture and quality of food products, either enhancing or diminishing their appeal and nutritional value.

Food chemistry, as a branch of both chemistry and food science, investigates the chemical processes and interactions within foods. It encompasses the study of biological and non-biological components of food, their chemical structures, and how these evolve throughout the stages of food processing, preparation, distribution, consumption, and storage. By addressing these issues, food chemists contribute vital insights and innovations to enhance food safety, quality, and sustainability, thereby tackling the multifaceted challenges of present and future food supply.

Content/Structure of the book

This book is an indispensable guide structured into 10 comprehensive chapters, offering a detailed exploration of the molecular composition, properties, and interactions that define the chemistry of food.

Chapter 1, Introduction to Food Chemistry, lays the groundwork by outlining the fundamental principles and significance of studying food at a molecular level. It introduces key concepts such as food components, composition, and the role of chemistry in food quality and safety.

Water and Ice are the focal points of Chapter 2, focusing into their crucial roles as solvents and structural elements in food matrices. This chapter explores water activity, phase transitions, and their impact on food texture and stability.

Chapter 3, Carbohydrates and Lipids, elucidates the diverse structures and functions of these essential macronutrients in foods. It covers carbohydrate types, molecular structures, digestion, and lipid classes, highlighting their roles in flavor, texture, and nutritional value.

Amino Acids, Peptides, and Proteins take center stage in Chapter 4, providing a comprehensive overview of protein structures, functions, and enzymatic activities crucial for food processing and nutrition.

Chapter 5, Enzymes, explores the pivotal role of enzymatic reactions in food chemistry, including enzyme functionality, kinetics, and their applications in food production and preservation.

Vitamins (Chapter 6) and Minerals (Chapter 7) are extensively covered, focusing on their biochemical roles, sources, deficiencies, and bioavailability in food systems essential for human health and metabolism.

Colorants (Chapter 8) and Flavors (Chapter 9) cover the sensory aspects of food chemistry, discussing natural and synthetic compounds responsible for color perception and flavor profiles in foods.

Chapter 10, Food Additives, examines the diverse roles of additives in food processing, preservation, and enhancement of sensory attributes, addressing safety regulations and consumer concerns.

Collectively, these chapters provide a comprehensive foundation in food chemistry, equipping readers with essential knowledge to understand the molecular basis of food composition, quality, and safety, essential for professionals and enthusiasts alike in the food science and technology field.

Audience

This book is designed for a diverse readership encompassing students, professionals, and researchers in the fields of food science, nutrition, and culinary arts. It serves as a fundamental resource for undergraduate and graduate students pursuing degrees in food science, nutrition, dietetics, and related disciplines, providing them with a solid foundation in the chemical principles underlying food composition and processing. Additionally, it is an invaluable reference for food industry professionals, including food technologists, quality assurance specialists, and product developers, who seek to deepen their understanding of the chemical interactions and transformations that influence food quality and safety. Researchers engaged in food chemistry, biochemistry, and related fields will find the comprehensive coverage of topics and the latest advancements in the field beneficial for their investigative endeavors. Culinary professionals and chefs interested in the scientific aspects of cooking and food preparation can also gain valuable insights from the book, enhancing their culinary techniques with a robust understanding of the chemistry behind food ingredients and processes. Overall, this book is a versatile resource that caters to a wide audience invested in the scientific, practical, and innovative aspects of food chemistry.

Learning Features

- Start-of-chapter learning objectives
- Key terms in each chapter
- Summary Points at the end of each chapter
- Review Questions at the end of each chapter
- Case Studies and Activities in each chapter for solving real problems
- References at the end of each chapter

CHAPTER

1 Introduction to Food Chemistry

LEARNING OBJECTIVES

After studying this chapter, you will be able to:

- Explain what food chemistry is
- State the history of food chemistry
- Discuss the functions of food chemistry
- Review the gross chemical composition of foods
- Understand the approach to the study of food chemistry
- Describe the societal role of food chemists

KEY TERMS FROM THIS CHAPTER

Biochemical reactions
Biogas
Chemistry
Food chemistry
Food science
Major nutrients
Molecular biology
Biochemistry
Botany
Environment protection act
Food quality
Hidden quality
Minor nutrients
Nutraceuticals

INTRODUCTION

Food is any material that an organism consumes for nutritional support, mostly solid or liquid. Food is typically derived from plants, animals, or fungi and includes vital nutrients like proteins, fats, carbohydrates, and vitamins and minerals. An organism consumes the substance, which is then absorbed by the organism's cells to supply energy, sustain life, or promote growth. Different animal species have developed distinct eating habits to meet the demands of their respective metabolisms and to occupy particular ecological niches in particular geographic settings.

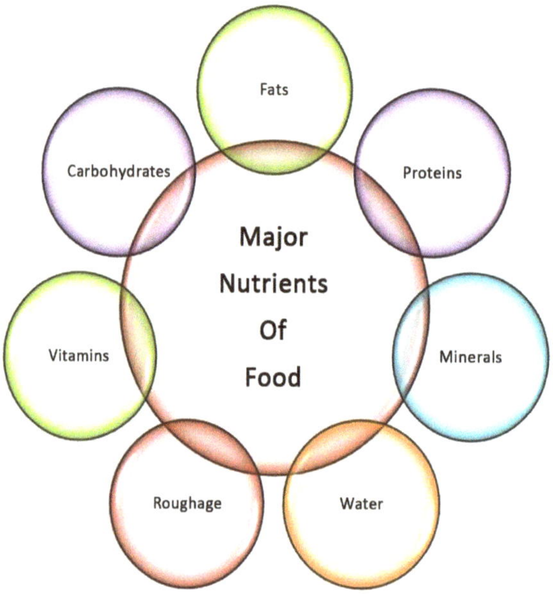

Figure 1.1. *Major components of food.*

Source: https://i.pinimg.com/736x/b2/b1/54/b2b15456151bd7f73f83d42eee8b152c.jpg.

The study of food chemistry focuses on the chemical makeup of foods, the characteristics and structures of their constituent ingredients, and the alterations in food composition that occur during processing and storage. The food chemist heavily depends on his understanding of chemistry in his work. However, food chemistry is also linked to microbiology (food safety, spoilage and preservation), nutritional science (nutritive value of foods, contaminants), and biochemistry (especially in the study of chemical changes of food constituents).

The content included in the food chemistry theme is separated into the following major categories of food components: minerals and other micro components, additives and contaminants, proteins and enzymes, lipids, carbohydrates, vitamins, flavors and colorants, and other micro components.

1.1. WHAT IS FOOD CHEMISTRY?

Food is a global issue, but the nature of the concerns varies depending on where you live. The majority of people work in food production in developing nations. However, obtaining sufficient quantities of the appropriate essential nutrients remains a persistent challenge.. Merely a small portion of the population is engaged in the highly mechanized food production found in developed nations. Food is widely accessible, processed, and frequently contains chemical additives. Concerns about food in these lucky places are mostly about price, quantity, variety, ease of use, and the impact of added chemicals and processing on food's nutritional value and wholesomeness. Food science addresses the physical, chemical, and biological aspects of food in relation to stability, cost, quality, processing, safety, nutritional value, wholesomeness, and convenience. All of these issues are covered by this field of study.

Food science is an interdisciplinary field that mainly involves engineering, biology, chemistry, and bacteriology. A significant area of food science is food chemistry, which examines the characteristics and makeup of food as well as the chemical alterations it goes through during handling, processing, and storage. Chemistry, biochemistry, physiological chemistry, botany, zoology, and molecular biology are all closely linked to food chemistry (John M. de Man, 1999). The understanding of the aforementioned sciences is crucial for food chemists to effectively study and manipulate biological substances used as food sources for humans. Food chemists and biological scientists share an interest in understanding the fundamental characteristics of biological materials and becoming proficient in their manipulation. Reproduction, growth, and changes that biological substances go through in environments that are either marginally or completely compatible with life are the main areas of interest for biologists. On the other hand, postharvest physiology of plants and postmortem physiology of muscle are biological substances that are dead or dying, and changes they undergo when exposed to a very wide range of environmental conditions are the primary concerns of food chemists. For instance, when fresh fruits and vegetables are being marketed, food chemists are concerned with conditions that support residual life processes, but when long-term food preservation is being attempted, conditions that are incompatible with life processes are of great interest. Food chemists also study the chemical characteristics of single-cell food sources (eggs and microorganisms), one major biological fluid (milk), and disrupted food tissues (flour, fruit and vegetable juices, isolated and modified constituents, and manufactured foods). In conclusion, food chemists and biological scientists share many similarities, but they also pursue very different and vital fields for humankind.

1.2. HISTORY OF FOOD CHEMISTRY

Food chemistry has unclear beginnings, and its past has not yet been thoroughly investigated and documented. This is not unexpected given that food chemistry did not establish a distinct identity until the 20th century and its history is closely linked to that of agricultural chemistry, for which historical records are not thought to be complete (Browne, 1944; Ihde, 1964). As such, the brief historical overview of food chemistry that follows is biased and incomplete. However, the information currently available is sufficient to pinpoint the exact time, location, and purpose of a few significant food chemistry events and to link some of these events to significant shifts in the quality of the food supply that have occurred since the early 1800s.

While food chemistry has roots in antiquity, the most important discoveries as we understand them today started in the late 1700s. The information presented here has primarily been sourced from Filby (Filby, 1934) and Browne (Browne, 1944), whose accounts of developments during this period are considered the best.

Modern food chemistry has its roots in the significant discoveries made by several well-known chemists between 1780 and 1850, many of which had some connection to food. One of history's greatest chemists, Carl Wilhelm Scheele (1742–1786) was a Swedish pharmacist who made significant contributions to the field of food chemistry. Apart from his well-known discoveries of oxygen, glycerol, and chlorine (three years prior to Priestly, but unpublished), he also isolated and investigated the characteristics of lactose (1780), made music acid by oxidizing lactic acid (1780), developed a method of preserving vinegar through heat application (1782, well before Appert's "discovery"), separated malic acid from apples (1785), isolated citric acid from lemon juice (1784), and tested 20 common fruits for the presence of citric, tartaric, and malic acids (1785). His discovery of numerous novel chemical compounds derived from plant and animal materials is regarded as the start of precise analytical studies in the fields of food and agricultural chemistry.

The modern chemistry principles were formulated and the phlogiston theory was ultimately rejected thanks in large part to the work of the French chemist Antoine Laurent Lavoisier (1743–1794). In terms of food chemistry, he created the foundational ideas of combustion organic analysis, demonstrated for the first time that fermentation could be described as a balanced equation, attempted to identify the elemental makeup of alcohol for the first time (1784), and gave one of the earliest papers (1786) on the organic acids of different fruits.

French chemist Nicolas-Théodore de Saussure (1767–1845) made significant contributions to the formalization and elucidation of Lavoisier's agricultural and food chemistry principles. Along with researching the changes in CO_2 and O_2 during plant respiration (1804), he also used ashing to determine the mineral contents of plants and produced the first precise elemental analysis of alcohol (1807).

The first technique to calculate the percentages of carbon, hydrogen, and nitrogen in dry vegetable substances was developed in 1811 by Joseph Louis Gay-Lussac (1778–1850) and Louis-Jacques Thenard (1777–1857).

The elements K, Na, Ba, Sr, Ca, and Mg were isolated in 1807 and 1808 by the English chemist Sir Humphrey Davy (1778–1829). His books on agricultural chemistry, the first of which was published in 1813 and was titled Elements of Agriculture Chemistry, in a Course of Lectures for the Board of Agriculture, were largely responsible for his contributions to agricultural and food chemistry. His books helped to arrange and make sense of the knowledge that was available at the time. He said in the first edition.

Plants have various parts that can all break down into a few different elements. Compound arrangements of these elements are required for their uses as food or in the arts. These arrangements can be made from the organized parts of the elements or from the juices they contain. An important aspect of agricultural chemistry is the study of these substances.

He said that in the fifth edition, "the most essential vegetable substances consist of hydrogen, carbon, and oxygen in different proportion, generally alone, but in some few cases combined with azote [nitrogen]" and that "plants are usually composed of only seven or eight elements" (Davy, 1936). The chemists Jons Jacob Berzelius (1779–1848) from Sweden and Thomas Thomson (1773–1852) from Scotland are credited with developing the first organic formulas, without which "food analysis would be an endless task and organic analysis would be a trackless desert." By identifying the fundamental elements of roughly 2,000 compounds, Berzelius confirmed the validity of the law of definite proportions. Additionally, he created a technique to precisely ascertain the water content of organic materials, which was lacking in Gay-Lussac and Thenard's approach. Furthermore, and this is a crucial point, Thomson demonstrated that the laws governing the composition of inorganic substances also apply to organic substances.

In a book entitled Considérations générales sur l'analyze organique et sur ses applications, Michel Eugene Chevreul (1786–1889), a French chemist, listed the elements known to exist at that time in organic substances (O, Cl, I, N, S, P, C, Si, H, Al, Mg, Ca, Na, K, Mn, and Fe) and cited the processes then available for organic analysis:

1. extraction with a neutral solvent, such as water, alcohol, or ether-water mixture;
2. slow distillation or fractional distillation;
3. steam distillation;
4. passing the substance through a tube heated to incandescence; and
5. analysis with oxygen.

Chevreul was a pioneer in the analysis of organic substances, and his classic research on the composition of animal fat led to the discovery and naming of stearic and oleic acids.

William Beaumont (1785–1853), an American Army surgeon stationed at Fort Mackinac, Michigan, conducted groundbreaking research on gastric digestion that disproved the notion that food was composed of only one nutrient, as held since Hippocrates. Between 1825 and 1833, he conducted experiments on Alexis St., a Canadian. Martin, whose musket wound allowed for direct entry into the stomach's interior, allowing food to be added and then checked for digestive alterations (Beaumont, 1833).

Among his many noteworthy achievements, Justus von Liebig (1803–1873) demonstrated in 1837 that during the fermentation of vinegar, acetaldehyde occurs as a transition between alcohol and acetic acid. He divided food into two categories in 1842: nonnitrogenous (fats, carbohydrates, and alcoholic beverages) and nitrogenous (vegetable fibrin, albumin, casein, and animal flesh and blood). This classification helped to identify significant differences between different foods, even though it is incorrect in a few areas. In addition, he developed techniques for the quantitative analysis of organic materials, especially through combustion, and published his work titled 'Researches on the Chemistry of Food, arguably the first book on food chemistry, in 1847.

It is interesting to note that the emergence of serious and widespread food adulteration coincided with the developments just reviewed. It is also true that the development of analytical chemistry in general and analytical food chemistry in particular was greatly influenced by the need to identify food impurities. Unfortunately, it is also true that chemistry's advancements played a part in food adulteration, as dishonest food vendors profited from the availability of chemical literature, which included recipes for contaminated food, and were able to replace less successful empirical methods of food adulteration with more effective methods based on scientific principles. Hence, it is appropriate to approach the topic of food adulteration from a historical perspective because the history of food chemistry and food adulteration are intricately entwined by the threads of multiple causal relationships.

Three distinct phases can be identified in the history of food adulteration in the world's more developed countries today. Food adulteration was not a significant issue from antiquity until roughly 1820, at which point detection techniques were not really necessary. The most obvious explanation for this scenario is that a significant amount of interpersonal accountability was involved in the transactions and the food was purchased from small businesses or individuals. The deliberate adulteration of food increased significantly in frequency and severity during the second phase, which started in the early 1800s. This development can be partially attributed to the previously mentioned rise of modern chemistry, as well as to increased centralization of food processing and distribution, which has resulted in a decline in interpersonal accountability. Food adulteration on purpose continued to be a significant issue until roughly 1920, when phase two came to an end and phase three began. At this point, the frequency and severity of intentional food adulteration were brought down to acceptable levels by regulatory pressures and efficient detection techniques, and things have progressively improved up until the present.

Some claim that around 1950, a fourth phase of food adulteration began, marked by the rise in foods containing permissible chemical additives, the use of highly processed foods that eventually made up a large portion of people's diets in most industrialized nations, and the growing public and regulatory concern over certain foods being contaminated with unwanted industrial byproducts like pesticides, mercury, and lead. Disagreement continues to this day regarding the veracity of this claim. However, it appears obvious what needs to be done over the next few years. More voluntary and involuntary changes to the production, handling, and processing of food are unavoidable as knowledge about safe food handling practices expands and as estimates of the maximum tolerable intake of undesirable constituents get more precise. Public concerns about the safety and nutritional adequacy of the food supply continue to motivate these changes. There was a particularly strong public concern about the safety and quality of the food supply in the early 1800s. A Treatise on Adulterations of Food (Accum, 1966) by Frederick Accum and Death in the Pot, an anonymous publication, were the sources of this worry—or more accurately, indignation—in England. As Accum put it, "There are some substances which are scarcely ever to be procured genuine; and it would be difficult to mention a single article of food which is not to be met with in an adulterated state." "That the widespread application of chemistry to the useful purposes of life, should have been perverted into an auxiliary to this nefarious traffic [adulteration]" was his additional statement.

It was true—as cited by Accum and Filby—that intentional adulteration of a number of foods and ingredients was common in the 1800s. These included annatto, black pepper, cayenne pepper, essential oils, vinegar, lemon juice, coffee, tea, milk, beer, wine, sugar, butter, chocolate, bread, and confectionary products. Filby claimed that Accum's accusations were a little exaggerated.

Remedial efforts grew over time after the public was made aware of how serious food adulteration was in the early 1800s. These came in the form of new laws that outlawed adulteration and significantly increased chemists' efforts to discover the natural ingredients of food, the substances that are frequently used as adulterants, and how to detect them. As a result, chemistry and food chemistry started to gain prominence in Europe between 1820 and 1850. The creation of new journals dedicated to chemical research as well as the establishment of chemical research laboratories for undergrads at various universities made this possible, thanks in large part to the efforts of the scientists previously mentioned. Food chemistry has since advanced at an accelerated rate; some of these developments and their contributing variables are listed in the text that follows.

The first agricultural experiment station funded by the public was founded in 1860 in Weede, Germany, and W. Hanneberg and F. Stohmann was named chemist and director, respectively. They created a crucial method for the regular identification of food's primary ingredients, heavily drawing on the work of past chemists. They were able to ascertain the moisture content, "crude fat," ash, and nitrogen by splitting a sample into multiple parts. They then calculated the protein content by multiplying the nitrogen value by 6.25. The result of sequential digestion using diluted alkali and acid was a residue known as "crude fiber." The part that was left over after crude fiber, ash, fat, protein, and fat were removed was called "nitrogen-free extract," and it was thought to be usable carbohydrates. Regretfully, regardless of the type of food, chemists and physiologists mistakenly believed for a long time that values obtained by this process indicated a nutritional value similar to that of the food. (McCollum, 1959).

According to Jean Baptiste Duman (1800–1884), a diet limited to fat, carbohydrates, and protein was insufficient to sustain life in 1871.

The Land-Grant College Act, written by Justin Smith Morrill, was approved by the US Congress in 1862. This act significantly boosted the training of agricultural and food chemists and contributed to the founding of agricultural colleges in the United States. The United States Department of Agriculture was founded in 1862 as well, with Isaac Newton serving as its first commissioner.

Harvey Washington Wiley was appointed chief chemist of the US Department of Agriculture in 1863. From this position, he spearheaded the campaign against adulterated and mislabeled food, which resulted in the first Pure Food and Drug Act of the US being passed (1906).

Most of the important nutrients found in food were found and identified in the first part of the 20th century. These include vitamins, minerals, fatty acids, and certain amino acids. The invention and widespread application of chemicals to support food production, distribution, and manufacturing was a particularly significant and controversial development in the mid-1900s.

This brief historical overview makes the food supply of today seem almost perfect when compared to that of the 1800s. In terms of what the food science community needs to address to further promote the wholesomeness and nutritional value of foods while mitigating the real or perceived threats to the safety of the food supply, however, as of this writing, a number of contemporary issues have replaced the historical ones. These concerns include molecular engineering of crops (genetically modified organisms [GMOs]) and the benefits weighed against the perceived risks to safety and human health, the nature, effectiveness, and impact of non-nutrient components in foods, dietary supplements, and botanicals that can promote human health beyond simple nutrition, and the relative nutritive value of crops raised by organic vs. traditional farming techniques.

1.3. FUNCTIONS OF FOOD CHEMISTRY

Food chemistry, food science, and nutrition are important fields that work together to provide humans with wholesome, safe food. It goes without saying that an in-depth understanding of the precise chemical (nutritional) makeup of foods is essential to fulfilling these roles.

1.3.1. Chemical Composition of Foods

Food chemistry has made it possible for us to determine the nutritional makeup of the majority of everyday foods. Additionally, information about their significance and role in nutrition has accumulated. Studies on nutrition have revealed that while the human body needs large amounts of proteins, fats, and carbohydrates, it needs much smaller amounts of vitamins and minerals. Consequently, the term "major nutrients" was applied to the former group of nutrients and "minor nutrients" to the latter. Foods were divided into categories based on their composition: "carbohydrate rich (starchy) foods" (e.g., Root crops, cereals, and foods high in protein (e.g., fatty foods (oil seeds, fatty meat, fish, poultry, and legumes), etc. Fruits and vegetables are generally good sources of the minor nutrients, i.e., minerals and vitamins that play a protective role against certain diseases caused by deficiencies. Fruits and vegetables were therefore categorized as protective foods.

Understanding nutrition and food composition has also made it possible to plan and create balanced meals that are appropriate for a variety of age groups, sexes, recovering from illness, etc. A balanced diet is one that contains all the nutrients in the necessary amounts. This knowledge allowed fortification to supplement nutrients where there is still a deficiency in food formulation.

Consumers who are knowledgeable about nutrition today are requesting details on the nutrients included in the foods they eat. As a result, food products now have to include nutrition labels, which is required in some nations. A food product's nutrition label tells us which nutrients are present in it and what proportion of each nutrient's Recommended Dietary Allowance are in a typical serving size. The serving size is stated in grams or milliliters. It should go without saying that accurate chemical analysis of the products is necessary for nutrition labeling.

Numerous bioactive compounds, particularly from fruits, vegetables, and herbs, have been isolated from food in addition to the major and minor nutrients already discussed. Together, they are referred to as "Nutraceuticals" or "Phytonutrients."

Carotenoids, flavonoids, sulfides, thiols, and phenolic cyclic compounds are a few of them. Numerous of them have demonstrated antioxidative defense against the human body, cancer growth suppression, vascular health improvement, osteoporosis retardation, and cataract control. The idea of "food as medicine" has been brought back to life by these advancements.

Modification of foods and food components has also been made possible by our growing understanding of the chemistry of food ingredients. One instance of food modification is the process of making fermented foods. Numerous traditional fermented foods have been created in various nations. Because of their understanding of chemical (biochemical) reactions, they have been able to enhance their processing procedures and guarantee consistent quality. Examples of how food ingredients are modified include the production of glucose syrup and high fructose syrup from starch and protein hydrolysates from proteins.

1.3.2. Quality Changes in Foods

A food's quality is a multifaceted amalgam of its hidden (intrinsic) qualities as well as various sensory aspects. Some of the sensory characteristics, such as color and appearance, firmness, taste, and mouth feel, are known to you. Changes to these characteristics in a food item, such as fruits, for instance, cause the buyer to reject the item or reduce its value.

Chemical (or biochemical) processes are the root cause of all these alterations in sensory quality. For instance, the breakdown of chlorophyll is the reason spinach loses its green color when it is cooked or stored. Similarly, the enzymatic oxidation of phenolic substances is what causes the browning of cut apples. For instance, the breakdown of pectins causes fruits to soften, while post-mortem chemical changes that cause tissue hardening and a drop in pH cause meat to toughen. Chemical reactions are also the cause of flavor variations.

For instance, the oxidation of unsaturated fatty acids causes a change in flavor in fatty foods known as rancidity. It is crucial to realize that methods to control quality deterioration can be developed once the chemical cause is identified.

The amount of nutrients in food and its lack of adulterants and toxicants are its hidden quality factors. Sensations are incapable of perceiving hidden qualities. Only chemical methods can be used to evaluate them. Changes in nutritional quality during food processing and storage are more significant than other hidden quality traits. Among the nutrients, certain vitamins are conditionally sensitive during processing. Ascorbic acid, a form of vitamin C, is particularly heat labile. Understanding the chemical kinetics of reactions has facilitated the development of non-thermal and high temperature short time processing techniques. Thiamine's (vitamin B1) breakdown is widely known. Consequently, foods high in thiamine are preserved by avoiding sulphites. The assessment of processed foods' nutritional value has been the focus of extensive research recently.

Food processing and storage can alter the quality of food because of two main factors: product factors and environmental factors. Product characteristics include a food's pH, available water content, and chemical makeup. Temperature and duration, light, vulnerability to microbial and insect attack, and the gas composition of the storage atmosphere are all significant environmental factors. Although it has been done in a few instances, changing the composition of food products to regulate quality changes is difficult, with the exception of drying (or removing water). For instance, egg powder is enzymatically oxidized to remove glucose in order to stop browning.

Storage at low temperatures regulates the temperature effect on quality to a large extent. The prevention or reduction of light's influence on food quality is one of packaging's primary purposes.

The gas atmosphere has an equally significant impact on quality. Several oxidative reactions occur when a food product is exposed to the atmosphere, which is primarily composed of oxygen (21%) and nitrogen (79%). Ascorbic acid oxidation, fatty acid oxidation, oxidative flavor changes, and browning reactions are a few examples. Fruit juices are deaerated, antioxidants are added to fatty foods, cans are exhausted (steam heated of cans filled with the product before sealing) to remove air, and other methods are used to prevent the effect of oxygen in canned foods. Altering the gas atmosphere is another technique, particularly if the goal is to reduce or eliminate the amount of oxygen in the gas atmosphere inside a package. You have undoubtedly seen potato chip pillow packs, or bloating pouches. They are packed full of nitrogen to keep the chips from physically deteriorating and from browning. Complete oxygen deprivation is detrimental to fresh fruits and vegetables. Understanding the biochemistry of plant respiration reveals that these commodities can have their storage lives prolonged by decreasing the concentration of oxygen and raising the concentration of carbon dioxide. As a result, fresh fruit and vegetable storage methods such as Modified Atmosphere Packaging (MAP) and Controlled Atmosphere (CA) have been developed. While the respiring commodities in MAP primarily alter the gas composition, the corresponding gases in CA storage physically alter the gas composition. Toxins and adulterants in food must be closely watched to guarantee food safety. The food chemist is facing an increasingly difficult task.

1.3.3. Safety Evaluation of Foods

The majority of scientific advancements have both positive and negative effects. Chemistry in food is not an exception. The newfound information provided dishonest vendors with lots of opportunities to adulterate food. Almost every food item has the potential to be tampered with. Below are some instances of food adulterants.

Table 1.1. List the Few Examples of Food Adulterants

Food produtcts	Common adulterantsProtein
Milk	Cane sugar, starch, water
Species and condiments	Sand, colouring matter, paddy husk, lead chromate, sawdust, argemone seed
Oils and fat	Mineral oil, argemone oil, ti-ortho-cresyl phosphate (TOCP), animal fat (tallow) in vegetable fat

Cereals and cereal products	Kesari dhal, colouring matter, talc, inorganic pigments
Beverages	Artificial sweeteners, unpermitted colouring matter, excessive preservatives
Coffee and tea	Cashew nut endocarp, date seed, tamarind seed/ powder, sawdust, added colouring
Ice cream	Artifical sweeteners, unpermitted colours such as metanil yellow
Alcoholic drinks	Methanol, chloral hydrate

Source: http://www.uprtou.ac.in/other_pdf/dvapfv_block_3.pdf.

In addition to the deliberate addition of toxic substances to food (adulteration), food contaminants such as heavy metals, agricultural chemicals, and pesticide residues can also find their way into food supplies. also require observation.

This threat to the public's health led to the development of analytical food chemistry, a new field of food chemistry whose primary goal is the detection of contaminants and adulteration. New laws prohibiting adulteration also came into being at the same time, which significantly increased chemists' efforts to study the natural ingredients of food, the substances that are frequently used as adulterants, and how to detect them.

1.3.4. Food Waste Management

The food processing industry generates enormous amounts of both liquid and solid waste. Sugars and other organic materials are highly concentrated in liquid wastes, or effluents. They are frequently dumped into adjacent streams. Environmentalists have good reason to be outraged. The Environment Protection Act (EPA) has established a number of guidelines for the safe release of wastewater. This indicates that in order to reduce the components of the effluent to safe levels, the effluent must be appropriately treated. Therefore, developing the effluent treatment protocols requires an understanding of the effluent's chemical composition.

When food raw materials are processed, solid wastes range from 10% to 50%. They can take the shape of rice husk and bran, fruit and vegetable peels and pomace, milling wastes from other grains and legumes, wastes from slaughterhouses, and wastes from fish processing. The simplest ways to get rid of them are to burn them for fuel, compost them, or dump them in the closest land. A thorough understanding of the wastes' chemical makeup has made it possible to separate their byproducts, some of which are more valuable than the main product. For instance, the peel and pomace of citrus fruits can yield more than 20 by-products, which generate greater profits for the industry than the primary product, citrus juice concentrate. Pectin, peel oil, seed meal and oil, candied juice sacks, etc., are a few of the byproducts made.

Utilizing the organic compounds in the wastes through microbial (also biochemical) conversion to produce biogas (mostly methane) is another way to make use of the wastes from food processing. Fuel can be made from biogas.

1.3.5. Societal Roles

As time went on, food chemists were forced to take on more duties, the most significant of which was becoming involved in social concerns. The phenomenon known as "food adulteration" was brought about by advancements in food chemistry, as was previously mentioned. The food chemist bears the responsibility of keeping it in check. The understanding of food chemistry has also made it possible to alter or enhance the functional qualities of food by adding a variety of chemicals known as food additives. Among them are coloring and flavoring agents, emulsifying, stabilizing, and antioxidant agents. Without the inclusion of certain of these additives, many of the new food products won't have their intended uses. Their population is constantly growing. There is a great deal of fear, which is frequently caused by misuse ignorance. Food chemists can serve as public educators and advisors regarding their use. Together with physiologists, microbiologists, nutritionists, food scientists, and technologists, food chemists today play a very complementary role in giving people safe foods in the format and location of their choice.

1.4. GROSS CHEMICAL COMPOSITION OF FOODS

Many books, data banks, and other resources are currently available for obtaining information about the chemical composition of foods as a result of investigations conducted by food quality control stations and numerous other institutions. Considering that the structure of the topic of food chemistry is based on the kind of constituents rather than the various food types, a brief overview of the gross chemical composition and characteristics of these foods will be provided in this section.

1.4.1. Cereals and Cereal Products

In most nations, cereals are a staple diet and a major source of nutrients. Cereal grains have a high starch content, a substance with a comparatively significant protein content, and a relatively low lipid content. These characteristics define their chemical composition. The primary minor constituents are magnesium, potassium, phosphorus, and B-group vitamins. Table 1.2 summarizes the gross chemical composition of cereals that are significant to the commercial industry.

Table 1.2. Gross Chemical Composition of Cereal Grains

Cereal Grain	Protein	Lipid	Starch	Fiber	Ash
Wheat	12.2	1.9	71.9	1.9	1.7
Rye	11.6	1.7	71.9	1.9	2.0
Triticale	11.9	1.8	71.9	1.9	1.8
Barley	10.9	2.3	73.5	4.3	2.4
Oats	10.2	5.8	55.5	10.9	3.2
Maize	10.2	4.6	79.5	2.3	1.3
Millet	10.3	4.5	58.9	8.7	4.7
Sorghum	11.0	3.5	65.0	4.9	2.6
Rice (brown)	8.1	2.3	75.8	1.8	1.4

Source: https://www.eolss.net/sample-chapters/c10/e5–08–07–00.pdf.

The ratio of constituents in the various botanical parts of the kernel (germ or embryo, pericarp and hull, aleurone layer, and endosperm) varies significantly because the various components of grain are not evenly distributed within the kernel.

Protein (20–35%) and lipids (12–35%) are abundant in embryos. As such, they have less carbohydrates in them. Additionally, minerals and some vitamins are more abundant in embryos (for specifics see Vitamins and Minerals). The endosperm has the highest starch content and a somewhat lower protein content than the kernel as a whole. The least amount of fat, vitamin, and mineral content is found in endosperm. The primary constituent of the pericarp (and hull in covered cereals) is cellulose; significant quantities of hemicelluloses, pentosans, and other non-starch polysaccharides are also present, as is a small amount of lignin. Hulls have very little protein and fat content and no starch. Phosphorus and potassium are the two primary mineral components. Silicon is the primary mineral found in rice husk. The remaining ingredients are coloring materials and waxes. The endosperm's outermost layer is called the aleurone layer. From a nutritional perspective, it is possible to highlight the high contents of protein (24% on average), fat (8%), and minerals (11%). The B-group vitamins are abundant in aleurone. Phytic acid, which is present as granules containing the compound's potassium and calcium salts, is an important ingredient. Phytic acid, which is the primary phosphorus store of seeds, makes up about 70–75% of the phosphorus found in cereals.

Cereal grains are widely used in a wide range of products across the globe because of their low moisture content and unique kernel structure, which also contributes to their good storability. The type of milling product used determines the chemical composition of cereal products (products made from white flour have less vitamins, dietary fiber, and minerals than products made from whole grain meal), and the type and quantity of other ingredients naturally affect this as well.

1.4.2. Legumes and Oilseeds

When it comes to food raw materials, oilseeds and grain legumes have the highest protein content (Table 1.3). Humans have long eaten the seeds of leguminous plants, which provide a significant protein boost to cereals and other starchy foods.

Table 1.3. Chemical Composition of Some Legumes and Oilseeds

Legume (Oilseed)	Moisture	Protein	Lipid	Carbohydrate
Pea	12.0	23.3	2.0	69.2
Field bean	12.0	30.1	2.2	61.0
Peanut	9.1	25.0	49.9	20.0
Sunflower seed	6.0	30.0	43.5	26.2
Rapeseed	5.2	26.0	33.8	34.9
Soybean	7.0	45.2	25.3	16.0

Source: https://www.eolss.net/sample-chapters/c10/e5–08–07–00.pdf.

The search for new sources of protein for food was spurred by the increasing issues related to the world population's food supply of protein. Plant protein isolates and concentrates are being produced on a commercial scale and are being used as ingredients in many novel and traditional food products as a result of the efforts to increase the production of food protein. Because of its advantageous chemical makeup, soy protein concentrate production has increased in particular. Following mild oil extraction, non-protein components are eliminated, allowing for the extraction of concentrates with a protein content of roughly 60–70% and isolates with a protein content of roughly 90%.

1.4.3. Fruits and Vegetables

A wide variety of plants, or more specifically, the edible portions (tissues) of plants, are included in the terms fruits and vegetables. From a botanical perspective, there are leaf crops like lettuce and spinach, roots like sugar beet, sweet potato, yams, and cassava, and stems like potato, onion, asparagus, rhubarb, and kohlrabi. The edible plant parts are made of living, metabolically active tissues that continue to be so even after harvest. This indicates that their composition is more or less continuously changing based on factors such as the growing conditions before harvest, the physiological function of the plant part, the cell's genetic pool, and the post-harvest environment. Usually, between 70 and 90% of the fresh weight is made up of water. Carbohydrates are the primary components of plant tissues' solid matter (Table 1.4). The main source of carbohydrates in fruits is sugar, whereas the main source in potatoes and other starchy vegetables is starch. Apart from the easily digested sugars and starch found in plant tissues, fruits and vegetables also contain a variety of substances referred to as cell wall constituents. Pectin, lignin, hemicelluloses, and cellulose are the main components of cell walls. The proportion of these compounds varies depending on the species, level of maturity at harvest, and storage conditions following harvest. There is strong evidence supporting the protective effects of dietary fiber in both health and disease, despite the fact that it's main components—the cell wall constituents—are not digestible.

Table 1.4. Chemical Composition of Some Fruits and Vegetables (%)

Fruit/Vegetable	Water	Protein	Carbohydrate	Ash
Apple	90.5	0.4	7.0	0.4
Apricot	86.1	0.9	10.2	0.7
Sour cherry	85.7	0.8	11.0	0.6
Orange	89.6	0.9	7.0	0.5
Banana	74.1	1.3	22.8	0.9
Walnut	8.1	18.6	11.7	1.7
Hazelnut	3.0	15.6	8.7	2.0
Potato	73.0	2.6	21.1	0.9
Green pea	75.1	2.5	13.5	1.0
Tomato	94.1	0.8	3.8	0.6
Carrot	90.1	1.1	7.2	0.8
Spinach	91.5	2.5	1.7	1.6

Lettuce	94.6	1.8	1.2	0.9
Cucumber	96.8	0.6	1.8	0.6
Green pepper	89.5	1.1	5.2	0.6

Source: https://www.eolss.net/sample-chapters/c10/e5–08–07–00.pdf.

Although the amount of protein in fruits and vegetables varies depending on the species, it usually makes up a small portion of their fresh weight and, with the exception of potatoes, has no appreciable impact on the population's supply of protein. Nitrogen that is not found in proteins, such as free amino acids and amines, can be found in significant amounts in plant tissues. Vitamins like provitamins of vitamin A, thiamine, niacin, riboflavin, and vitamin C are all found in significant amounts in fruits and vegetables. As ash (residue left over after incineration), the total mineral content of fruits and vegetables varies widely (0.1–5%). The main minerals are phosphorus, calcium, magnesium, and potassium. Fruit organic acids (malic, citric, and tartaric acid), fragrance compounds, and pigments (carotenoids, chlorophylls, and flavonoids) can be listed as additional components.

1.4.4. Meat, Fish and Their Products

The world's population gets most of its high-quality protein from meat, fish, and their byproducts. From a quantitative perspective, the majority of meat and fish are made up of water (Table 1.5). Its proportion is mostly determined by the amount of fat. In general, it can be computed that the water and fat contents are constant, ranging from 70–80%. Lean meat and fish portions have a protein content of 19–23%.

Table 1.5. Chemical Composition of Meat (%)

Type of Animal	Protein	Fat	Water
Cattle (lean)	20.6	3.5	74.2
Cattle (fatty)	18.9	24.5	58.3
Pig (lean)	20.1	6.3	78.7
Pig (fatty)	15.1	35.0	49.0
Lamb (lean)	19.0	6.4	72.1
Lamb (fatty)	17.6	26.4	53.5
Broiler	19.8	11.5	67.5
Turkey	19.9	8.2	68.4
Duck	13.0	26.8	51.0
Goose	12.2	32.1	52.1
Carp	20.5	6.1	73.3
Trout	23.1	8.8	67.2
Herring	18.6	3.5	78.0
Codfish	23.9	1.0	76.1

Source: https://www.eolss.net/sample-chapters/c10/e5–08–07–00.pdf.

Contractile proteins, or main constituent myosin, make up the majority of proteins. Soluble proteins, which include muscle enzymes and red-colored myoglobin, and connective tissue proteins like collagen, keratin, and elastine, are other categories. Fat can be found partially in adipose and partially in muscle tissue in mammals and birds. Generally speaking, chicken muscles are lower in fat than those of mammals. The liver stores fat in lean fish. Depot fat appears to exist in the muscle tissue of fatty fish as extracellular droplets. Vitamins, in particular those of the B-group (thiamin, riboflavin, niacin, and vitamin B_{12}), are abundant in muscle tissue. Meat has extremely low levels of vitamin C. While meat is a fairly poor source of calcium, it is a good source of iron and phosphorus.

1.4.5. Milk and Dairy Products

The complex mixture of proteins, lipids, carbohydrates, vitamins, and minerals that makes up 12–15% of fresh milk is called the solid matter (Table 1.6). 3.0–3.5% high-quality protein can be found in cow's milk. The proteins are categorized as whey and casein proteins. Approximately 80% of the total protein is made up of casein. In the processes used to produce cheese and yogurt, respectively, it can be precipitated with rennet or acid. Milk contains globules with a diameter of 2–3 μm that contain an average fat content, which is approximately 4%.

Triacylglycerols make up its composition, along with trace amounts of phospholipids and sterols (0.1 g/l). The primary carbohydrate in milk is lactose, or 50 mg/l of milk sugar. Some people rarely exhibit lactose intolerance, also known as lactose malabsorption. This is related to the breakdown of lactose into simple sugars by the digestive enzyme lactase, which is absent or inactive. Digestion, bloating, and cramping in the abdomen are signs of lactose malabsorption. People who are lactose intolerant can freely consume acid milk products like yogurt and kefir because the lactose in them is converted to lactic acid. Milk is a great source of water- and fat-soluble vitamins and minerals, especially iron and calcium. Table 1.6 displays the chemical makeup of a few milk products.

Table 1.6. Chemical Composition of Milk and Some Dairy Products

Product	Water	Protein	Fat	Carbohydrate
Cow milk	87.3	3.6	3.8	4.7
Sheep milk	80.1	5.5	7.4	4.8
Goat milk	87.4	2.9	4.5	4.1
Hard cheese	38–48	22–27	28–36	-
Processed cheese	45–58	18–20	20–25	-
Milk powder	5–6	19	26	36
Butter	16–20	1	78–82	1

Source: https://www.eolss.net/sample-chapters/c10/e5–08–07–00.pdf.

1.5. APPROACH TO THE STUDY OF FOOD CHEMISTRY

In order to enhance food formulation, processing, and storage stability, food chemists usually focus on determining the molecular determinants of material properties and the chemical reactivity of food matrices. Finding cause-and-effect and structure-function relationships between various classes of chemical components is one of the main goals. It is possible to apply the knowledge gained from studying one food or model system to the comprehension of other food items. Four elements comprise an analytical approach to food chemistry, which are as follows:

1. Determining those properties that are important characteristics of safe, high-quality foods;
2. Determining those chemical and biochemical reactions that have important influences on the loss of quality and/or wholesomeness of foods;
3. Integrating the first two points so that one understands how the key chemical and biochemical reactions influence the quality and safety; and
4. Applying this understanding to various situations encountered during the formulation, processing, and storage of food.

The primary requirement for any food is safety. This basically means that at the time of consumption, food must be free of any hazardous chemicals or microorganisms. This definition assumes a more applied form when used for operational purposes. When it comes to low-acid foods, "commercial" sterility in the canning industry refers to the lack of viable Clostridium botulinum spores. This can then be converted into a particular set of heating parameters for a particular product in a particular container. One can then choose particular time-temperature conditions that will maximize the retention of quality attributes based on these heating requirements.

Table 1.7. Classification of Alterations That Can Occur during Handling, Processing, or Storage

Attribute	Alteration
Texture	Loss of solubility Loss of water-holding capacity Toughening Softening
Flavor	Development of Rancidity (hydrolytic or oxidative) Cooked or caramel flavors Other off-flavors Desirable flavors
Color	Darkening Bleaching Development of desirable colors (e.g., browning of baked goods)
Nutritive value	Loss, degradation, or altered bioavailability of proteins, lipids, vitamins, minerals, and other health-promoting components
Safety	Generation of toxic substances Development of substances that are protective to health Inactivation of toxic substances

Source: Srinivasan Damodaran and Kirk L. Parkin. Fennema's Food Chemistry, Fifth Edition. CRC Press. ISBN: 9781482208122.

Similar to this, operational safety in a product like peanut butter can be defined as mostly being free of aflatoxins, which are substances that cause cancer and are produced by specific types of mold. Although actions taken to stop the mold from growing may or may not affect the preservation of another quality feature, safe production conditions must be followed.

Table 1.7 provides a list of food quality attributes along with some changes they may experience during processing and storage. The consumer can easily observe the potential changes, aside from those related to safety and nutritional value.

Food safety and quality can be changed by a variety of chemical and biological processes. Table 1.8 contains a list of some of the more significant classes of these reactions. Depending on the particular food and the specific handling, processing, or storage conditions, each reaction class may involve a different combination of reactants or substrates. Since the general characteristics of the substrates and reactants are the same for all foods, they are categorized as reaction classes. As a result, carbonyl compounds are produced during nonenzymic browning.

Table 1.8. Chemical and Biochemical Reactions That Can Lead to Alteration of Food Quality or Safety

Types of Reaction	Examples
Nonenzymic browning	Baked goods, dry and intermediate moisture foods
Enzymic browning	Cut fruits and some vegetables
Oxidation	Lipids (off-flavors), vitamin degradation, pigment decoloration, proteins (loss of nutritive value)
Hydrolysis	Lipids, proteins, vitamins, carbohydrates, pigments
Metal interactions	Complexation (anthocyanins), loss of Mg from chlorophyll, catalysis of oxidation
Lipid isomerization	cis → trans isomerization, nonconjugated → conjugated
Lipid cyclization	Monocyclic fatty acids
Lipid oxidation–polymerization	Foaming during deep-fat frying
Protein denaturation	Egg white coagulation, enzyme inactivation
Protein cross-linking	Loss of nutritive value during alkali processing
Polysaccharide synthesis and degradation	In plants postharvest
Glycolytic changes	Animal postmortem, plant tissue postharvest

Source: Srinivasan Damodaran and Kirk L. Parkin. Fennema's Food Chemistry, Fifth Edition. CRC Press. ISBN: 9781482208122.

These compounds can originate from various reactions, including the oxidation of ascorbic acid, the hydrolysis of starch, or the oxidation of lipids, as well as from already-existing reducing sugars. Triacylglycerols in one food or phospholipids in another are examples of how lipids can oxidize. Proteins, vitamins, and pigments can also oxidize.

The changes indicated in Table 1.7 are brought about by the reactions indicated in Table 1.9. Understanding the reasons behind food deterioration can result from integrating the data in both tables. Food deterioration typically happens as a result of a sequence of primary events followed by secondary events, which show up as changed quality attributes (Table 1.7). Table 1.9 displays examples of these kinds of sequences. Take particular note of the fact that multiple primary events can change a particular quality attribute.

Table 1.9. Examples of Cause-and-Effect Relationships Pertaining to Food Alteration During Handling, Storage, and Processing

Primary Causative Event	Secondary Event	Attribute Influenced
Hydrolysis of lipids	Free fatty acids react with protein.	Texture, flavor, nutritive value
Hydrolysis of polysaccharides	Sugars react with protein.	Texture, flavor, color, nutritive value
Oxidation of lipids	Oxidation products react with many other constituents.	Texture, flavor, color, nutritive value; toxic substances can be generated
Bruising of fruit	Cells break, enzymes are released, and oxygen is accessible.	Texture, flavor, color, nutritive value
Heating of horticultural products	Cell walls and membranes lose integrity, acids are released, and enzymes become inactive.	Texture, flavor, color, nutritive value
Heating of muscle tissue	Proteins denature and aggregate, and enzymes become inactive.	Texture, flavor, color, nutritive value
cis → trans conversion in lipids	Enhanced rate of polymerization during deep-fat frying.	Excessive foaming during deep-fat frying, diminished nutritive value and bioavailability of lipids, solidification of frying oil

Source: Srinivasan Damodaran and Kirk L. Parkin. Fennema's Food Chemistry, Fifth Edition. CRC Press. ISBN: 9781482208122.

There are two ways to apply the sequences in Table 1.9. One can think about a specific primary event, related secondary events, and the impact on a quality attribute by working from left to right. As an alternative, one can identify the likely cause(s) of an observed quality change (column 3, Table 1.9) by taking into account every primary event that might be relevant and then identifying the primary event that is most important by means of suitable chemical testing. Building these kinds of sequences has the benefit of encouraging an analytical approach to food alteration problems. Processing inevitably modifies the physical and chemical characteristics of the main food ingredients—proteins, carbohydrates, and fats. Both intra- and intercomponent interactions and reactions are involved in these modifications.

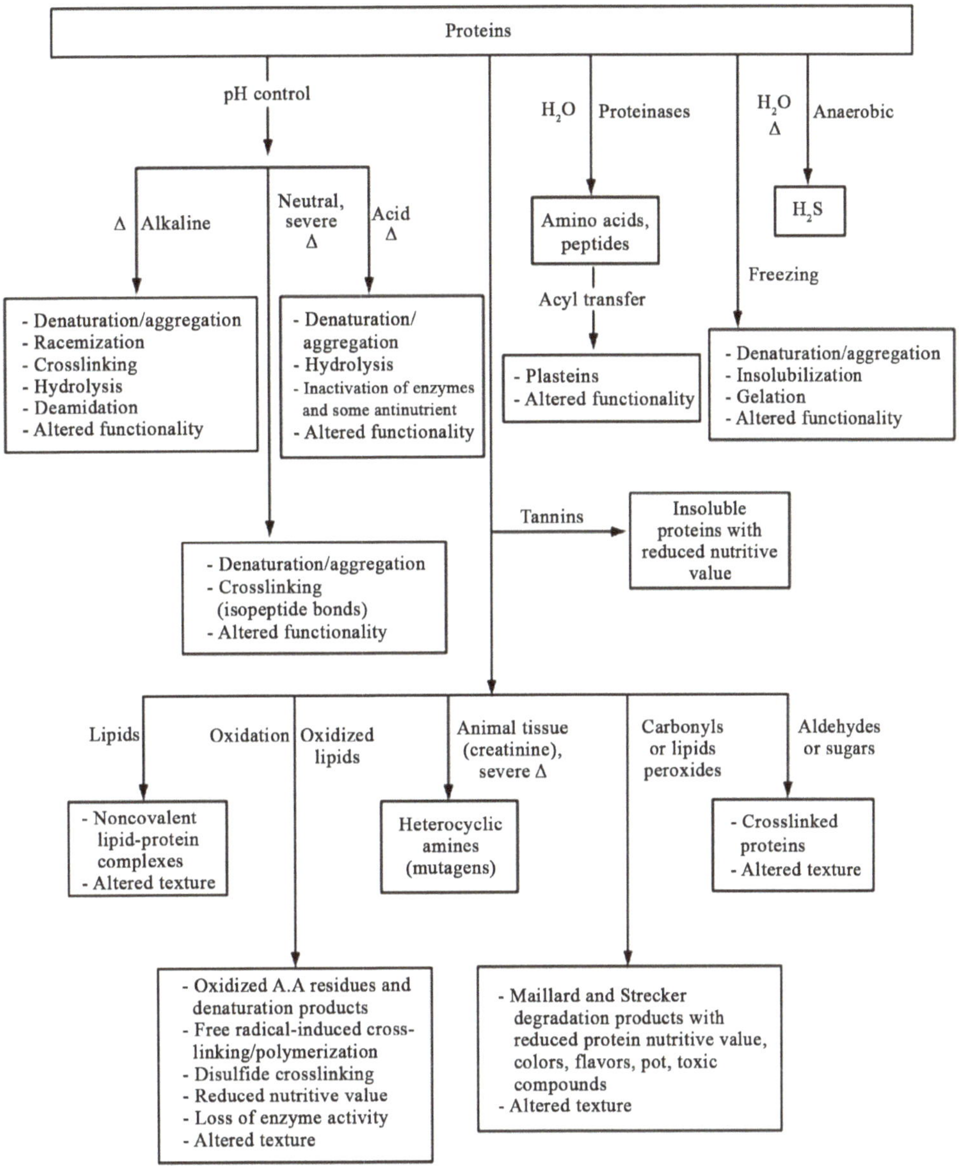

Figure 1.2. *Major reactions that proteins can undergo during the processing and handling of foods.*

Source: Srinivasan Damodaran and Kirk L. Parkin. Fennema's Food Chemistry, Fifth Edition. CRC Press. ISBN: 9781482208122.

The main reactions that occur when proteins, carbs, and fats are handled and processed are enumerated in Figures 1.2 through 1.4. Both desirable and undesirable sensory and nutritional qualities of foods are developed through these intricate networks of interactions and reactions.

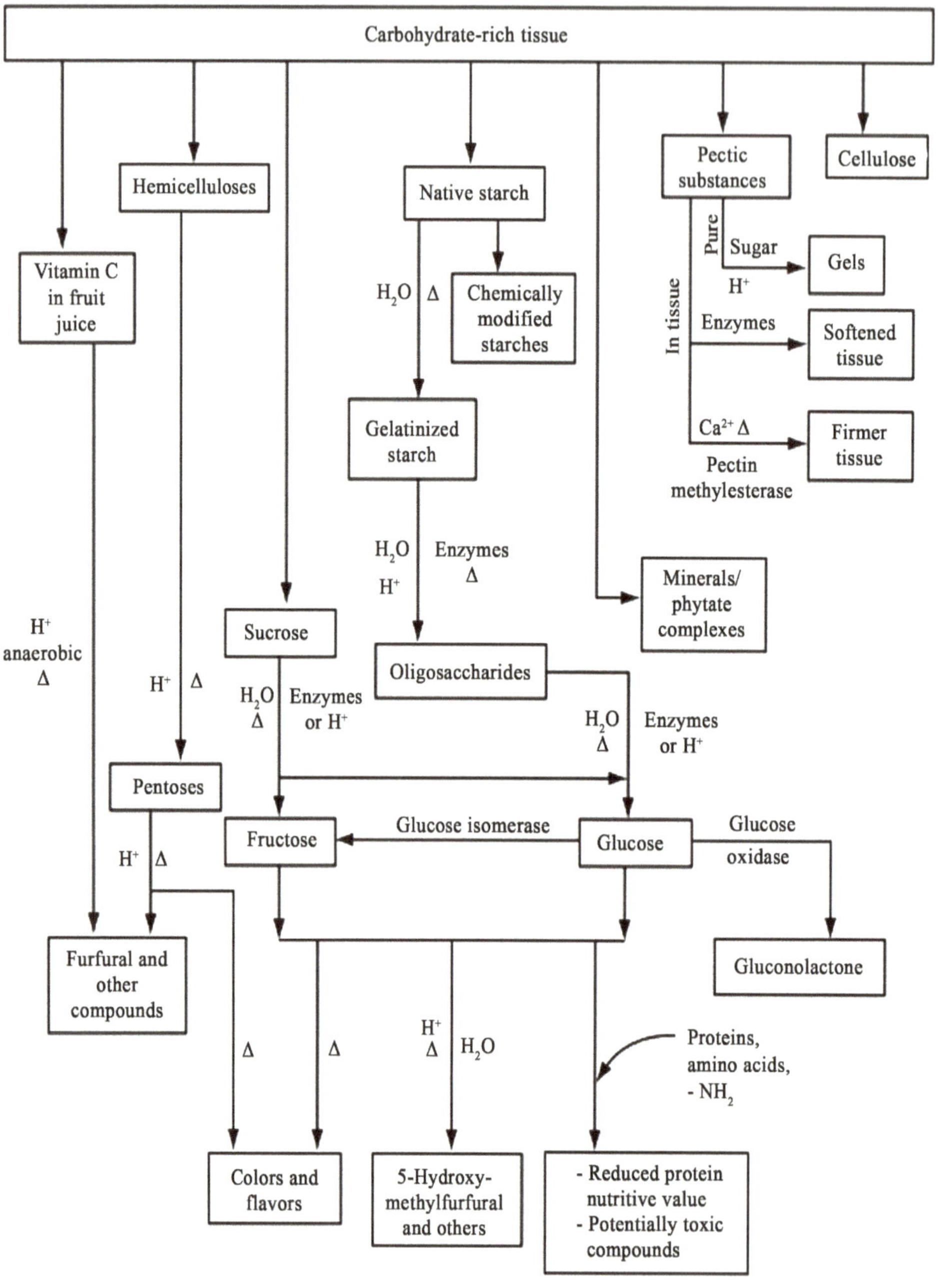

Figure 1.3. *Major reactions that carbohydrates can undergo during the processing and handling of foods.*

Source: Srinivasan Damodaran and Kirk L. Parkin. Fennema's Food Chemistry, Fifth Edition. CRC Press. ISBN: 9781482208122.

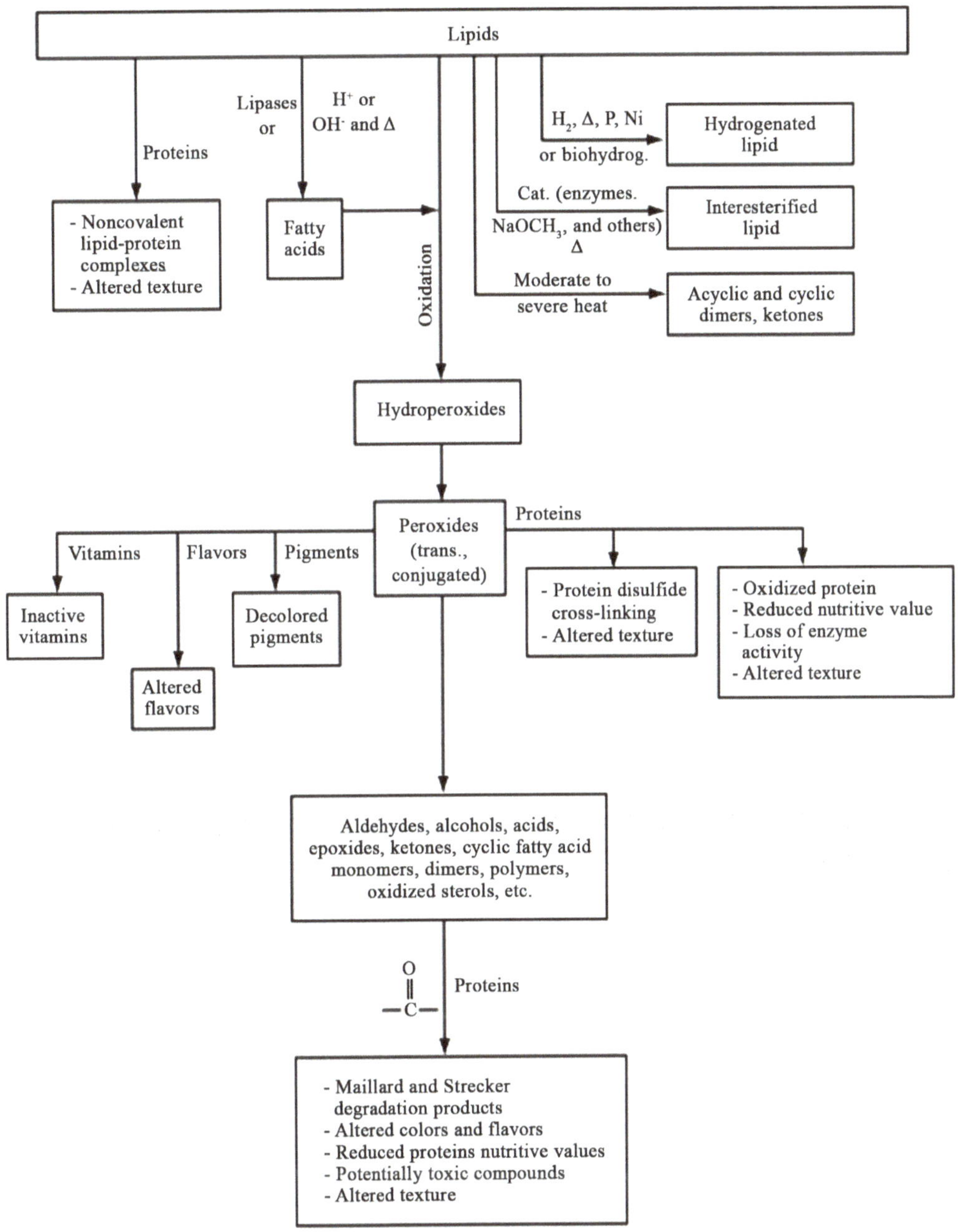

Figure 1.4. *Major reactions that lipids can undergo during the processing and handling of foods.*

Source: Srinivasan Damodaran and Kirk L. Parkin. Fennema's Food Chemistry, Fifth Edition. CRC Press. ISBN: 9781482208122.

A basic synopsis of the interactions and reactions between the main food ingredients that cause food quality to decline can be found in Figure 1.5. Every class of compound is capable of experiencing a unique kind of deterioration. The part that carbonyl compounds play in numerous deterioration processes is noteworthy. They primarily result from the oxidation of lipids and the breakdown of carbohydrates, and they can cause the loss of nutrients as well as odd colors and flavors. Naturally, when many foods are cooked, these same reactions produce flavors and colors that are desired.

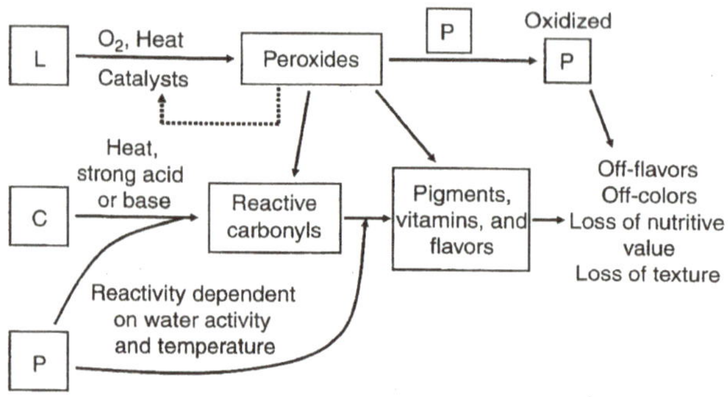

Figure 1.5. *Summary of chemical interactions among major food constituents: L, lipid pool (triacylglycerols, fatty acids, and phospholipids); C, carbohydrate pool (polysaccharides, sugars, organic acids, etc.); P, protein pool (proteins, peptides, amino acids, and other N-containing substances).*

Source: Srinivasan Damodaran and Kirk L. Parkin. Fennema's Food Chemistry, Fifth Edition. CRC Press. ISBN: 9781482208122.

1.5.1. Analysis of Situations Encountered during the Storage and Processing of Food

Now that we have a clear understanding of the characteristics of safe, high-quality foods, the important chemical processes that cause food to deteriorate, and the connection between the two, we can start thinking about how to apply this knowledge to scenarios that arise during the processing and storage of food.

Table 1.10 contains a list of the variables that are crucial during food processing and storage. Given its extensive impact on all kinds of chemical reactions, temperature is arguably the most significant of these factors. The Arrhenius equation, $k = Ae^{-\Delta E/RT}$, can be used to estimate the impact of temperature on a single reaction. Plotting log k vs 1/T yields a straight line for data that follows the Arrhenius equation. The activation energy, denoted by the parameter ΔE, is the free energy change necessary to move a chemical entity from its ground state to its transition state, where a reaction can take place. Reactions crucial to food deterioration are represented by Arrhenius plots in Figure 1.6. It is clear that, within a narrow range of intermediate temperatures, food reactions primarily follow the Arrhenius relationship; however, at high or low temperatures, deviations from this relationship may occur. As a result, it's critical to keep in mind that the Arrhenius relationship for food systems is valid only within an experimentally confirmed temperature range. The following events, the majority of which are brought on by either high or low temperatures, can cause deviations from the Arrhenius relationship: (1) loss of enzyme activity; (2) modification of the reaction pathway or rate-limiting step or influence from competing reaction(s); and (3) change in the system's physical state (e.g., by freezing), or (4) the reactants might run out of resources.

Table 1.10. Important Factors Governing the Stability of Foods during Handling, Processing, and Storage

Product Factors	Environmental Factors
Chemical properties of individual constituents (including catalysts), oxygen content, pH water activity, T_g, and W_g	Temperature (T); time (t); composition of the atmosphere; chemical, physical, or biological treatments imposed; exposure to light; contamination; physical abuse
Note: Water activity = p/p_o, where p is the partial pressure of water vapor above the food and p_o is the vapor pressure of pure water; T_g is the glass transition temperature; W_g is the product water content at T_g.	

Source: Srinivasan Damodaran and Kirk L. Parkin. Fennema's Food Chemistry, Fifth Edition. CRC Press. ISBN: 9781482208122.

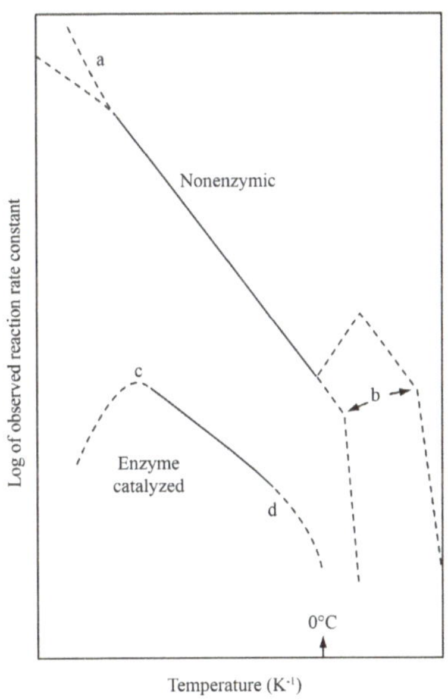

Figure 1.6. *Conformity of important deteriorative reactions in food to the Arrhenius relationship.*

Source: Srinivasan Damodaran and Kirk L. Parkin. Fennema's Food Chemistry, Fifth Edition. CRC Press. ISBN: 9781482208122.

Note: (a) Above a certain value of T, there may be deviations from linearity due to a change in the path of the reaction. (b) As the temperature is lowered below the freezing point of the system, the ice phase (essentially pure) enlarges, and the fluid phase, which contains all the solutes, diminishes. This concentration of solutes in the unfrozen phase can decrease reaction rates (supplement the effect of decreasing temperature) or increase reaction rates (oppose the effect of declining temperature), depending on the nature of the system. (c) For an enzymic reaction there is a high temperature at which the enzyme is denatured, resulting in loss of its activity, and (d) in the vicinity of the freezing point of water where subtle changes, such as the dissociation of an enzyme complex, can lead to a sharp decline in reaction rate.

Time is another crucial component in Table 1.10. When storing food products, people often want to know how long the food will stay at a certain quality level. As a result, time is important in terms of the integral chemical and/or microbiological changes that take place during a given storage period and how these changes work together to establish the product's specified storage life. The length of time it takes to inactivate a specific population of microorganisms or for a reaction to proceed to a predetermined degree are common processing-related questions. For instance, it could be interesting to know how long it takes for potato chips to fry to the appropriate brown color. The temperature change with time, or dT/dt, must be taken into consideration in order to achieve this. This relationship is significant because it makes it possible to calculate how much the reaction rate alters as the food matrix's temperature changes during processing. An integrative analysis provides a prediction of the net accumulation of a reaction product if the food's temperature profile and the reaction's ΔE are known. Foods that deteriorate through multiple processes, such as lipid oxidation and nonenzymic browning, should also be aware of this. It is crucial to determine whether there will be a significant interaction between the products of the browning reaction, which are antioxidants, and the relative rates of these reactions.

The pH variable is another factor that affects the rates of numerous chemical and enzymatic reactions. Severe inhibition of microbial growth or enzymatic processes typically requires extremely high pH values, which can speed up reactions that are catalyzed by acids or bases. However, some foods, like muscle, can undergo significant changes in quality even with a slight pH shift.

The product's composition is significant because it establishes the reactants that can be used for chemical transformation. Another crucial factor is how cellular vs. homogenous and noncellular versus. Diverse food systems have an impact on reactants' reactivity and disposition. The relationship between the composition of the raw material and the finished product is particularly significant in terms of quality. Examples include: (1) how postharvest handling of fruits and vegetables affects sugar content, which affects the level of browning achieved during dehydration or deep-fat frying; (2) how postmortem handling of animal tissues affects the amounts and rates of glycolysis and ATP degradation, which affect storage life, water-holding capacity, toughness, flavor, and color; and (3) how blending raw materials may result in unexpected interactions, such as how salt content can either accelerate or inhibit the rate of oxidation.

Water activity (a_w) is another significant compositional determinant of food reaction rates. The rate of enzyme-catalyzed reactions, lipid oxidation, nonenzymic browning, sucrose hydrolysis, chlorophyll degradation, anthocyanin degradation, and other processes have all been demonstrated by numerous researchers to be significantly influenced by aw. The majority of reactions tend to slow down below an a_w that corresponds to the range of foods with intermediate moisture content (0.75–0.85). This is not always the case; for example, oxidation of lipids and related secondary effects like carotenoid discoloration accelerate at the lower end of the a_w scale. In recent times, it has been observed that there is a causal relationship between the rates of diffusion-limited events in food and the glass transition temperature (T_g) of the food and the corresponding water content (W_g) at T_g. The physical characteristics of frozen and dried foods, the ideal conditions for freeze drying, the physical changes associated with crystallization, recrystallization, gelatinization, and starch retrogradation, and the diffusion-limited chemical reactions are all affected by T_g and W_g. The composition of manufactured foods can be altered by adding approved substances like flavors, chelating agents, acidulants, or antioxidants, or by eliminating unwanted reactants like glucose from dehydrated egg albumen. Although ethylene and CO_2 are also significant during the storage of living plant foods, relative humidity and oxygen content make up the majority of the atmosphere's composition. Regretfully, it is nearly impossible to fully exclude oxygen in circumstances where this is desired. Sometimes during product storage, the negative effects of a tiny amount of residual oxygen become evident. For instance, Maillard browning during storage may result from the early production of a tiny quantity of dehydroascorbic acid (from the oxidation of ascorbic acid). When light exposure can be harmful to a product, it is appropriate to package it in a material that doesn't absorb light or, if that isn't possible, control the light's intensity and wavelengths. Food chemists solve problems about food formulation, processing, and storage stability by integrating knowledge about food quality attributes, deteriorative reactions to which foods are susceptible, and the factors governing kinds and rates of these deteriorative reactions.

1.6. SOCIETAL ROLE OF FOOD CHEMISTS

Food chemistry is a significant branch of food science that studies the composition, characteristics, and chemical changes that foods undergo. Food chemists are frequently very important in ensuring that the food produced is safe and of high quality.

1.6.1. Why Should Food Chemists Become Involved in Societal Issues?

For the following reasons, food chemists ought to feel compelled to get involved in socioeconomic issues that have relevant technological components (technosocietal issues):

- Food chemists have the advantage of having received advanced training and developing specialized scientific skills; however, these advantages and skills come with a high degree of responsibility.
- The work of food chemists affects a number of factors, including population health, food costs, waste generation and disposal, water and energy consumption, and the structure of food laws. It is understandable that food chemists would feel obligated to focus their work on advancing society since these issues affect the public's overall well-being.
- The views of others—scientists from other fields, professional lobbyists, news media personnel, consumer activists, charlatans, and antitechnology zealots—will win out if food chemists stay out of the technosocietal issues. Many of these people are plainly unqualified to speak on food-related topics, and many are even less qualified than food chemists.
- Food chemists have a responsibility and a chance to assist in resolving disputes that affect public health and public perceptions of scientific and technological advancements. The safety of genetically modified organisms (GMOs) and cloned organisms, the use of animal growth hormones in agricultural production, and the relative nutritional value of crops grown using conventional and organic methods are a few examples of current debates.

1.6.2. Types of Involvement

Food chemists have social responsibilities as well, such as doing a good job on the job, being a good citizen, and protecting the scientific community's ethics. However, these vital roles alone are insufficient. Contributing to the interpretation and application of scientific knowledge by society is another crucial role that food chemists frequently fail to fulfill. To make informed decisions, food chemists and other food scientists should contribute to these discussions, but they shouldn't be the only ones. Their opinions should be heard and taken into consideration. When one accepts this position—which is undeniably true—one must naturally ask, "What exactly should food chemists do to properly discharge their responsibilities in this regard?" Several activities are appropriate:

1. Participate in pertinent professional societies.
2. Serve on governmental advisory committees, when invited.
3. Undertake personal initiatives of a public service nature.

The third point can involve speeches or dialog with civic groups, including sessions with K–12 students and all other stakeholders, as well as letters to newspapers, journals, legislators, government regulators, business executives, university administrators, and others.

Enlightening and educating the public about food and dietary practices are the main goals of these initiatives. Enhancing the public's capacity to assess information on these subjects sensibly is one way to do this. This will not be easy to accomplish because food has connotations for many people that go far beyond the narrow perspective of chemists, and because a large segment of the public has deeply ingrained false beliefs about food and appropriate dietary practices. For these people, eating is a necessary component of ritual, religious practice, cultural heritage, social symbolism, or a means of achieving bodily well-being. These viewpoints are generally incompatible with learning how to evaluate foods and dietary habits objectively and scientifically.

The use of chemicals to modify food is one of the most controversial food issues that has escaped public evaluation in a systematic, scientific way. Food additives are perceived by many as risks that are not supported by evidence because of chemophobia, or the fear of chemicals, which affects a sizable segment of the population. Articles in popular literature that assert that the American food supply is poisoned to the point where it is at best unwholesome and at worst potentially fatal can be found with unsettling ease. The way that avaricious industrialists contaminate our food for financial gain, with a helpless Food and Drug Administration looking on indifferently, is said to be highly startling. How reliable and knowledgeable the author is about the scientific matter at hand will determine how much weight to give to their opinions. Should authors who espouse this position be taken seriously? Scholarly contributions to the body of knowledge that a given dispute is linked to, along with formal education, training, and practical experience, are the foundations of credibility. Investigating new information, reviewing and interpreting existing knowledge, and conducting research are all examples of scholarly activity.

Furthermore, the author's credibility is based on their efforts to remain objective, which means that rather than focusing solely on facts and interpretations that support their preferred position, they must take into account other points of view and as much of the body of knowledge as is practical. As a result of peer review and adherence to strict professional guidelines regarding documentation, protocol, and ethics, research findings published in the scientific literature contribute to the body of knowledge by being more reputable than those found in the popular press.

In closer proximity to the everyday life of the student or aspiring food science professional, a modern concern about the veracity of information relates to the volume of information (including scientific information) that is easily and quickly available via the World Wide Web. Certain information is rarely credited to a specific author, and the website might not have the necessary credentials to be trusted as a reliable source. Certain information may be posted as part of a marketing campaign, to promote a cause, or to sway the viewer's opinion or purchasing behavior.

The student is urged to carefully consider the source of information obtained from the World Wide Web and not just give in to the ease of access, even though some information on the web is as authoritative as media released by qualified scientists and scientific publishers.

Disagreement regarding food safety and other food science matters persists despite the present and expanding body of knowledge in the field. The vast majority of informed people agree that our food supply is generally safe and nutritious and that legally permitted food additives do not present unnecessary risks, though ongoing monitoring for negative effects is necessary. But a small but informed minority feels that our food supply is needlessly dangerous, especially when it comes to some of the legally allowed food additives.

The safety of genetically modified organisms for humans and the environment, the nutritional value of crops grown conventionally versus organically, and the propriety of marketing-driven claims that the public may mistake for health claims that accompany dietary supplements are just a few of the topics that have recently been the subject of scientific debate in public forums. The gradual and rapid advancement of scientific knowledge seldom adequately equips us for the next controversy. In order to help fully informed policymakers come to the right decision, scientists should be involved in the process and encourage all parties to concentrate objectively on science and knowledge.

In conclusion, people with formal scientific education have lower societal obligations than scientists. It is not sufficient for scientists to produce knowledge in an ethical and productive manner. They ought to take on the duty of making sure that scientific knowledge is applied in a way that will most benefit society. To fulfill this duty, scientists must cultivate a sincere concern for the public's scientific education and well-being in addition to their daily professional activities, which should involve striving for excellence and adhering to high ethical standards.

CASE STUDY

STRUCTURAL ANALYSIS OF A NOVEL ENZYME TO PRODUCE SOY ISOFLAVONES

- **Adapting β-Glucosidases to Function in Extreme Production Processes**

A group of enzymes called β-Glucosidases are used in the food industry to hydrolyze glycosidic bonds in complex sugars. Since many food industrial processes involve harsh conditions like high concentrations of solvents and sugars, low pH, and high temperatures, there is a need for these kinds of enzymes to be adapted to function in extreme environmental conditions. These enzymes can be employed for the cleavage of isoflavone glucosides from soy flour, to produce soy isoflavones, a valuable food supplement.

- **The Challenge**

Researchers from the Universities of Nottingham, Bern, and Milano identified a novel β-Glucosidase from the extremophilic organism Alicyclobacillus herbarius that can hydrolyze isoflavone glucosides under extreme conditions.

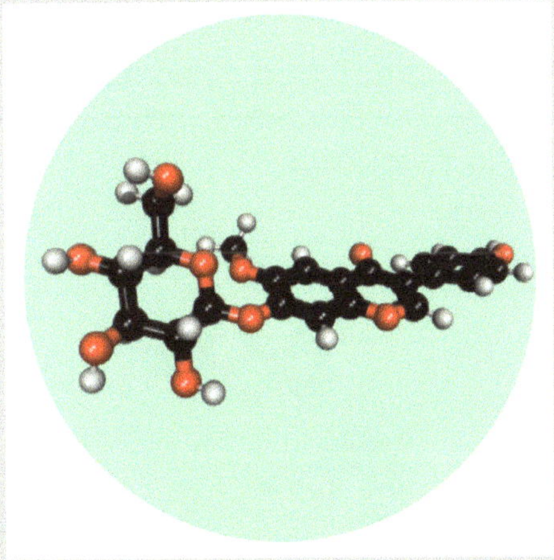

Source: https://www.diamond.ac.uk/industry/Case-Studies/Case-study-novel-enzyme-for-soy-isoflavones.html.

A high-resolution structure was needed to understand how the enzyme works and what makes it tolerate these extreme conditions.

- **The Solution**

The enzyme was expressed and purified in sufficient quantities that permitted the researchers to set up crystallization experiments and obtain crystals.

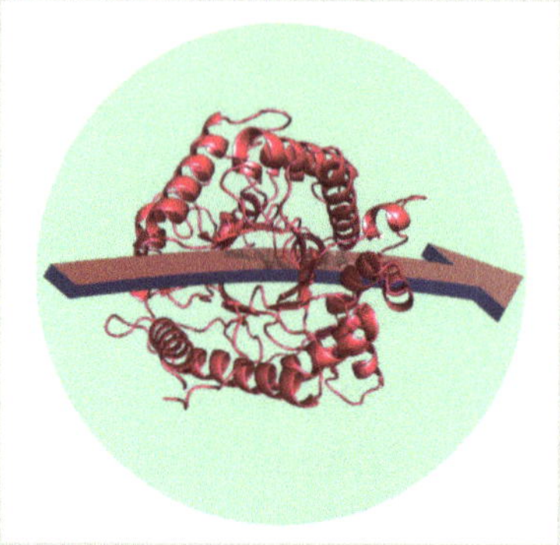

Source: https://www.diamond.ac.uk/industry/Case-Studies/Case-study-novel-enzyme-for-soy-isoflavones.html.

X-ray diffraction data were collected using beamline I04 at Diamond. The high quality and resolution of the data made it possible to determine the atomic resolution structure of the enzyme.

Structural analysis showed that even though this β-Glucosidase shared low sequence similarity with related enzymes, the structural similarity was high, and conserved amino acid residues important for enzyme stability and activity could be identified.

- **The Benefits**

The advanced data collection capabilities of beamline I04 combined with the highly intense X-ray beam generated by the Diamond synchrotron made it possible to model the enzyme of interest at atomic resolution.

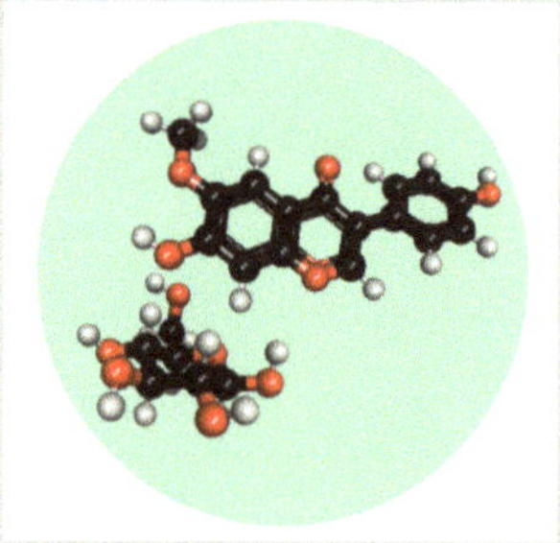

Source: https://www.diamond.ac.uk/industry/Case-Studies/Case-study-novel-enzyme-for-soy-isoflavones.html.

An understanding of how these types of enzymes work and why they can retain activity in extreme conditions is important for current and future applications within the food industry.

CLASS ACTIVITY

ROAD DESIGN

Students will use pectin in conjunction with acid and sugar to form jelly. By varying the sugar concentrations, the students will observe that there is an optimum ratio for the creation of this spreadable gel.

SUMMARY

- Food is any substance (mainly solid or liquid) consumed by an organism for nutritional support. Food is usually of plant, animal, or fungal origin and contains essential nutrients such as carbohydrates, fats, proteins, vitamins, or minerals.
- Food chemistry is the science that deals with chemical composition of foods, with chemical structure and properties of food constituents, and with chemical changes food undergoes during processing and storage.
- The origins of food chemistry are obscure, and details of its history have not yet been rigorously studied and recorded. This is not surprising since food chemistry did not acquire a clear identity until the twentieth century, and its history is deeply entangled with that of agricultural chemistry for which historical documentation is not considered exhaustive.
- Food chemistry has enabled us to know the nutrient composition of most of the common foods. This knowledge on their nutritional role and importance also accumulated. Nutrition studies showed that the human system require certain nutrients like carbohydrates, proteins and fats in large quantities and some others like vitamins and minerals in much smaller quantities.
- Quality of a food is a complex combination of several sensory and hidden (intrinsic) attributes. You are familiar with some of the sensory qualities like color and appearance, firmness, mouth feel, flavor, taste etc. Alterations in these qualities of a food material, like for example, fruits results in value reduction and even rejection by the buyer. All these sensory quality changes are caused by chemical (or biochemical) reactions.
- Most scientific developments have both benefits and harmful effects. Food chemistry is no exception. The new knowledge gave ample opportunity to unscrupulous purveyors for food adulteration. Practically all foods are liable for adulteration.
- Food processing industries produce huge quantities of wastes – solid and liquid. The liquid wastes (effluents) are loaded with high concentration of sugars and

other organic substances. They are quite often discharged into nearby streams. Environmentalists cry foul for justifiable reasons. The Environment Protection Act (EPA) has stipulated various parameters for safe discharge of effluents. This means that the effluent has to be treated suitably to degrade the constituents of the effluents to the safe limits.

- Cereals are staple foods in majority of countries and play a central role in nutrient supply. The chemical composition of cereals grains is characterized by a high content of starch, a substance with a relatively significant protein content, and a relatively low lipid content. From minor constituents the vitamins of B-group, potassium and phosphorus and magnesium are the main ones.
- Grain legumes and oilseeds are the richest in protein among food raw materials. Seeds of leguminous plants have been traditionally consumed by humans, and they constitute an important protein complement to cereals and starchy foods.
- The terms fruits and vegetables cover a big number of plants or more precisely edible parts (tissues) of plants. From botanical point of view there are roots such as sugar beet, sweet potato, yams, and cassava; stems such as potato, onion, asparagus, rhubarb, and kohlrabi; leaf crops such as lettuce and spinach; and fruits. The edible plant parts are composed from living tissues that are even after harvesting metabolically active. It means that they are more or less constantly changing in composition depending on growing conditions prior to harvest, the physiological role in the plant part, the genetic pool of the cell, and the post-harvest environment.
- Meat, fish and their products play a decisive role in providing high quality protein to the world population. From quantitative point of view water is the major component of meat and fish. Its proportion is primarily dependent on fat content. Generally it may be calculated that the fat + water content is constant (about 70–80%). The protein content of lean portion of meat and fish is 19–23%.

REVIEW QUESTIONS

1. What is food?
2. What is food chemistry?
3. Name three scientists who have done pioneering work in the development of food chemistry.
4. What is a balanced diet?
5. Explain nutrition labeling.
6. What is meant by nutraceuticals?
7. Explain the chemical basis of two sensory quality changes.
8. List the factors responsible for food quality changes.
9. Explain how the effect of oxygen on quality change can be overcome.
10. List a few examples of food adulteration.

REFERENCES

1. Accum, F. (1966). *A Treatise on Adulteration of Food, and Culinary Poisons, 1920, Facsimile Reprint by Mallinckrodt Chemical Works, St. Louis, MO.*
2. Acker, L. W. (1969). Water activity and enzyme activity. *Food Technology, 23*(11), 1257–1270.
3. Beaumont, W. (1833). *Experiments and Observations on the Gastric Juice and the Physiology of Digestion*. F. P. Allen. Plattsburgh, NY.
4. Braverman, J. B. S. (1963). *Introduction to the Biochemistry of Foods*. Elsevier Publishing Company.
5. Browne, C. A. (1944). *A Source Book of Agricultural Chemistry*. Chronica Botanica Co. Waltham, MA.
6. Clydesdale, F. M., & Francis, F. J. (1977). *Food, Nutrition and You*. Prentice-Hall.
7. Davy, H. (1936). *Elements of Agricultural Chemistry* (5th ed.). Longman, Rees, Orme, Brown, Green and Longman.
8. de Man, J. M. (1999). *Principles of Food Chemistry* (Food Science Text Series). Springer Science.
9. Eichner, K., & Karel, M. (1972). The influence of water content and water activity on the sugar-amino browning reaction in model systems under various conditions. *Journal of Agricultural and Food Chemistry, 20*(2), 218–223.
10. Erlandson, J. A., & Wrolstad, R. E. (1972). Degradation of anthocyanins at limited water concentration. *Journal of Food Science, 37*(4), 592–595.
11. Fennema, O. R. (1976). *Principles of Food Science, Part I–Food Chemistry*. Marcel Dekker Inc.
12. Filby, F. A. (1934). *A History of Food Adulteration and Analysis*. George Allen & Unwin.
13. Hall, R. L. (1982). Food additives. In D. Kirk & I. K. Eliason (Eds.), *Food and People* (pp. 148–156). Boyd & Fraser.
14. Ihde, A. J. (1964). *The Development of Modern Chemistry*. Harper & Row.
15. Jukes, T. H. (1978). How safe is our food supply? *Archives of Internal Medicine, 138*(6), 772–774.
16. Labuza, T. P., Tannenbaum, S. R., & Karel, M. (1970). Water content and stability of low-moisture and intermediate-moisture foods. *Food Technology, 24*(10), 543–550.
17. LaJollo, F., Tannenbaum, S. R., & Labuza, T. P. (1971). Reaction at limited water concentration. 2. Chlorophyll degradation. *Journal of Food Science, 36*(5), 850–853.
18. Mayer, J. (1975). *A Diet for Living*. David McKay, Inc.
19. McCollum, E. V. (1959). The history of nutrition. *World Review of Nutrition and Dietetics, 1*, 1–27.
20. McWeeny, D. J. (1968). Reactions in food systems: Negative temperature coefficients and other abnormal temperature effects. *Journal of Food Technology, 3*(1), 15–30.

21. Meyer, L. H. (1969). *Food Chemistry*. Van Nostrand Reinhold Company.
22. Quast, D. G., & Karel, M. (1972). Effects of environmental factors on the oxidation of potato chips. *Journal of Food Science, 37*(4), 584–588.
23. Riviere, J. H. (2002). *Chemical Food Safety—A Scientist's Perspective*. Iowa State Press.
24. Roberts, C. A. (2001). *The Food Safety Information Handbook*. Oryx Press.
25. Schoebel, T., Tannenbaum, S. R., & Labuza, T. P. (1969). Reaction at limited water concentration. 1. Sucrose hydrolysis. *Journal of Food Science, 34*(4), 324–329.
26. Stare, F. J., & Whelan, E. M. (1978). *Eat OK—Feel OK*. Christopher Publishing House.
27. Taoukis, P., & Labuza, T. P. (1996). Summary: Integrative concepts (shelf-life testing and modeling). In O. Fennema (Ed.), *Food Chemistry, 3rd ed.* (pp. 1013–1042). Marcel Dekker.
28. Taylor, R. J. (1980). *Food Additives*. John Wiley & Sons.
29. Watson, D. H. (2001). *Food Chemical Safety* (Vol. 1: Contaminants; Vol. 2: Additives). Woodhead Publishing Ltd.
30. Whelan, E. M. (1993). *Toxic Terror*. Prometheus Books.
31. Wilcock, A., Pun, M., Khanona, J., & Aung, M. (2004). Consumer attitudes, knowledge and behavior: A review of food safety issues. *Trends in Food Science & Technology, 15*(2), 56–66.

CHAPTER

2 Water and Ice

LEARNING OBJECTIVES

After studying this chapter, you will be able to:

- State the physical properties of water and ice
- Discuss the water molecule
- Identify the water in the presence of solutes
- Describe the water activity and relative vapor pressure

KEY TERMS FROM THIS CHAPTER

Biochemical reactions	Biochemistry	Biogas
Botany	Chemistry	Environment protection act
Food chemistry	Food quality	Food science
Hidden quality	Major nutrients	Minor nutrients
Molecular biology	Nutraceuticals	Nutrition labeling
Physiological chemistry	Zoology	

INTRODUCTION

Water turns out to be a significant component of most foods when we look at their composition. Water is the main solvent used in our own metabolic processes, which we should also take into consideration. In order to comprehend the pivotal role that water plays in food chemistry, it is appropriate to examine the nature and characteristics of water and aqueous solutions, as well as the various functions that water performs in food systems. Water and ice play significant roles in food chemistry, impacting the texture, stability, and preservation of various food products. In its liquid form, water is a solvent for many ingredients, facilitating chemical reactions and the mixing of flavors. It helps in the dispersion of salts, sugars, and proteins, which is essential for processes like baking, brewing, and cooking.Ice, on the other hand, is crucial in food preservation and texture.

Freezing food slows down the activity of enzymes and microorganisms that cause spoilage, extending the shelf life of perishable items. The formation of ice crystals during freezing can affect the texture of food. Small, uniform ice crystals help preserve the food's texture, while large ice crystals, often formed during slow freezing, can damage cell structures and lead to a mushy texture upon thawing.

Additionally, ice is used in the preparation of certain food items, such as ice cream and frozen desserts, where the control of ice crystal size is critical to the final product's quality.

The interaction of water and ice also plays a role in the stability of emulsions, foams, and gels in food products, affecting their consistency and mouthfeel. Understanding the behavior of water and ice in food chemistry is essential for developing and maintaining the quality of a wide range of food products.

2.1. THE PHYSICAL PROPERTIES OF WATER AND ICE

It is appropriate to start by thinking about the physical characteristics of water, as Table 2.1 illustrates. One can ascertain whether the behavior of water is normal or abnormal by contrasting its properties with those of molecules with similar molecular weights and atomic compositions (Table 2.2).

These comparisons show that water has somewhat lower than expected density, unusual properties such as expansion upon solidification, unusually high melting and boiling point temperatures, unusually large values for surface energy, permittivity, heat capacity, and heats of phase transformation (fusion, vaporization, and sublimation), and yet a relatively normal viscosity despite these unusual properties. We'll get into why this liquid, which is obviously anomalous, appears normal later.

There are other amazing qualities of water. When compared to most other liquids, water has a high thermal conductivity, and ice has a higher thermal conductivity than one might anticipate for a nonmetallic solid. Notably, at 0°C, the thermal conductivity of ice is roughly four times higher than that of liquid water, suggesting that ice will conduct thermal energy far more quickly than immobilized (e.g., water, tissue).

The thermal diffusivities of water and ice differ by a factor of nine because water has a heat capacity that is roughly twice that of ice. Since thermal diffusivity is a good indicator of how quickly a material will change in temperature, we would anticipate that ice will change in temperature nine times faster than liquid water in a given thermal environment. Under symmetrically applied temperature differentials, tissues freeze more quickly than they thaw. This can be explained in part by the differences in thermal conductivity and diffusivity values for water and ice (Fennema, 1973).

Table 2.1. Physical Properties of Water and Ice

Property	Value
Molecular weight	18.0153
Melting point (at 101.3 kPa)	0.00°C
Boiling point (at 101.3 kPa)	100.00°C
Critical temperature	373.99°C
Critical pressure	22.064 Mpa
Triple point temperature	0.01°C
Triple point pressure	611.73 Pa
ΔH_{vap} at 100°C	40.647 kJ/mol
ΔH_{sub} at 0°C	50.91 kJ/mol
ΔH_{fus} at 0°C	6.002 kJ/mol

Other Temperature-Dependent Properties	Temperature (°C)			
	Ice		Water	
	−20	0	0	+20
Density (g/cm^3)	0.9193	0.9168	0.99984	0.99821
Vapor pressure (kPa)	0.103	0.6113	0.6113	2.3388
Heat capacity (J/g/K)	1.9544	2.1009	4.2176	4.1818
Thermal conductivity (W/m/K)	2.433	2.240	0.561	0.5984
Thermal diffusivity (m^2/s)	11.8×10^{-7}	11.7×10^{-7}	1.3×10^{-7}	1.4×10^{-7}
Compressibility (Pa^{-1})		2	4.9	
Permittivity	98	90	87.9	80.2

Please try to enhance the quality of this figure. Try to copy the original figure and Paste Special, in the enhanced metafile.Source: Lide, D.R. (Ed.) (1993/1994) Handbook of Chemistry and Physics, 74 edn. CRC Press: Boca Raton, FL.

Table 2.2. Properties of Related Small Molecules

	CH_4	NH_3	H_2O	H_2S	H_2Se	HF
MW	16.04	17.0	18.01	34.08	80.9	20.01
mp (°C)	−182.6	−77.7	0	−86	−60	−83.1
bp (°C)	−161.4	−33.3	100	−61	−41	19.5
ΔH_V (kJ/mol)	8.16	23.26	40.71	18.66		

Source: Lide, D. R. (1993/1994) Handbook of Chemistry and Physics, 74 edn. CRC Press: Boca Raton, FL.

2.2. THE WATER MOLECULE

The peculiar characteristics of water imply that there are strong forces of attraction between its molecules and that the structures of ice and water may also be peculiar. It is best to think about the nature of a single water molecule, followed by the features of clusters of increasing size, and then the nature of the bulk system in order to explain the properties and peculiar behavior of water and ice. Frequently, the water molecule is explained as consisting of two hydrogen atoms interacting with the two oxygen sp^3 bonding orbitals to form two covalent sigma (σ) bonds with 40% ionic character. The dissociation energy of each bond is 4.6 ×102 kJ/mol. It is assumed that the localized molecular orbitals will maintain their symmetric orientation with respect to the initial orbital axes, thereby preserving an approximate tetrahedral structure. Figure 2.1a depicts a schematic model, and Figure 2.1b shows the van der Waals radii. Although this model fits the geometric behavior of water molecules bound by hydrogen bonds, there has been growing doubt about the widespread sp3 hybridization of the lone pairs. An isolated water molecule's bond angle in the vapor state is 104.5◦, which is quite near to the ideal tetrahedral angle of 109.5◦. The van der Waals radii for hydrogen and oxygen are 1.2 and 1.40 Å, respectively.

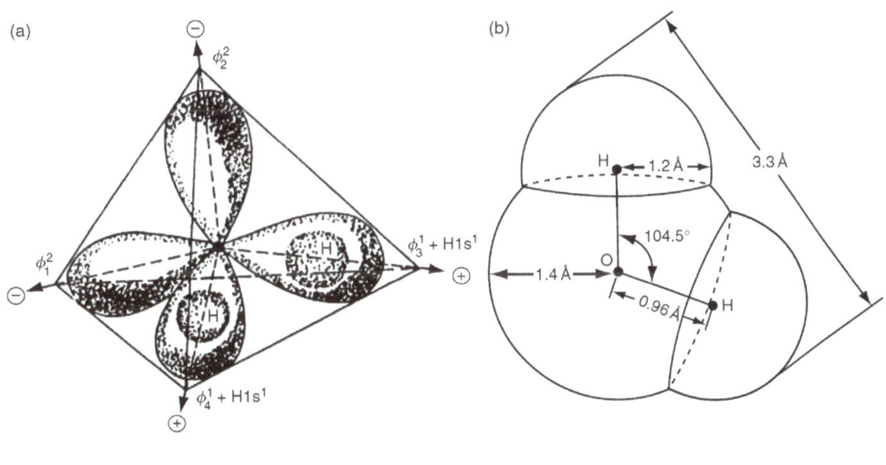

Figure 2.1. *Schematic model of a single HOH molecule. (a) Possible sp^3 configuration and (b) van der Waals radii for a HOH molecule in vapor state.*

Source: https://edisciplinas.usp.br/pluginfile.php/7989549/mod_resource/content/1/%C3%81gua%20nos%20Alimentos.pdf.

It is crucial to recognize at this point that the current representation, which solely describes the HOH molecule, is simplistic. Pure water is actually a mixture of HOH molecules and numerous other related substances. There are 18 isotopic variations of molecular HOH resulting from the presence of ^{17}O, ^{18}O, ^{2}H (D), and ^{3}H (T) in addition to the common isotopes of hydrogen and oxygen, ^{16}O and 1H.

Ionic species found in water also include hydroxyl ions and hydrogen ions, both of which have isotopic variations. Hydrogen ions can exist in forms such as H_3O^+ and $H_9O^+_4$. Thus, "pure" water is made up of over 33 different chemical forms of HOH; however, because these forms are only found in trace amounts, the HOH species dominates the water's characteristics.

2.2.1. Association of Water Molecules

Pure water has a dipole moment in its vapor state of 1.84 D due to the polarized O-H bond and the V-shaped structure of the HOH molecule, which causes an asymmetric charge distribution within the molecule. This level of polarity creates strong intermolecular attraction forces, which is why water molecules bind together so firmly. However, take note that the large molecular dipole moment alone cannot account for the entire observed unusually large intermolecular attractive force of water. This makes sense because dipole moments are a property of the entire molecule and do not reveal information about the geometry of the molecule or the extent to which individual charges are exposed—aspects that significantly impact the strength of the intermolecular association.

The ability of water molecules to form multiple three-dimensional hydrogen bonding associations provides a satisfactory explanation for the significant intermolecular attractive forces between them. Hydrogen bonds, which are weaker than covalent bonds, have an average bond energy of roughly 335 kJ/mol. They also tend to have longer and more variable lengths. The dissociation energy of the oxygen–hydrogen bond is approximately 11–25 kJ/mol, and its length varies from approximately 1.7 to 2.0 Å, whereas the oxygen–hydrogen covalent bond has a length of about 1.0 Å.

Further discussion of the geometric patterns formed by associating HOH molecules will emphasize electrostatic effects, as these forces play a major role in the energy of the hydrogen bond. An electrostatic model of water is straightforward and yields a geometric picture of HOH molecules that is essentially accurate. Though perfectly adequate for this purpose, this simplified approach will prove insufficient for explaining other behavioral characteristics of water, such as the influence of apolar solutes.

Each hydrogen atom has a partial positive charge and a minimal electron shield as a result of the highly electronegative oxygen in the water molecule partially removing the single electrons from the two covalently bonded hydrogen atoms. In other words, each hydrogen atom takes on some of the properties of a bare proton. Two of the axes of an imaginary tetrahedron (Figure 2.1a) contain orbitals that form hydrogen–oxygen bonds, hence these axes can be thought of as lines of positive force (donor sites for hydrogen bonds). The two lone pair orbitals of oxygen can be thought of as lines of negative force (hydrogen bond acceptor sites) that are located along the other two axes of the tetrahedron. Each water molecule has the capacity to form hydrogen bonds with a maximum of four other molecules thanks to these four lines of force arranged in a tetrahedral orientation. Figure 2.2 shows the resulting tetrahedral arrangement. Water molecules exhibit unusually large attractive forces, even when compared to other small molecules that participate in hydrogen bonding associations. This is because each molecule has an equal number of hydrogen bond donor and acceptor sites, arranged in a way that allows for three-dimensional hydrogen bonding, NH_3, HF. Both hydrogen fluoride and ammonia, with their tetrahedral arrangements of one donor and three acceptor sites and three donor and one acceptor site, respectively, do not have an equal number of donor and acceptor sites, and as a result, neither can form three-dimensional hydrogen bonded networks similar to those found in water. Both can only create large, two-dimensional networks with fewer hydrogen bonds between molecules than in water.

Figure 2.2. *Hydrogen bonding of water molecules in a tetrahedral configuration. Open circles are oxygen atoms, closed circles are hydrogen atoms. Hydrogen bonds are represented by dashed lines.*

Source: https://edisciplinas.usp.br/pluginfile.php/7989549/mod_resource/content/1/%C3%81gua%20nos%20Alimentos.pdf.

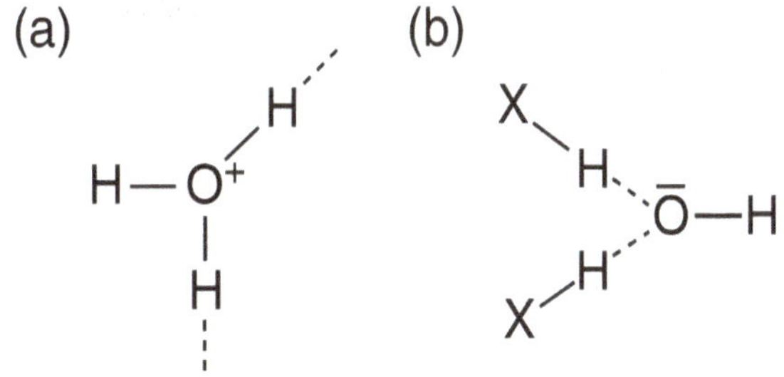

Figure 2.3. *Structure and hydrogen bond possibilities: (a) for a hydronium ion and (b) for a hydroxyl ion. Dashed lines represent hydrogen bonds, X--H represents a solute or another water molecule.*

Source: https://edisciplinas.usp.br/pluginfile.php/7989549/mod_resource/content/1/%C3%81gua %20nos%20Alimentos.pdf.

Considering isotopic variations as well as hydronium and hydroxyl ions makes conceptualizing the association of a few water molecules much more difficult. Due to its positive charge, the hydronium ion is predicted to show a higher potential for donating hydrogen bonds than non-ionized water (Figure 2.3a), while the hydroxyl ion is predicted to show a higher potential for accepting hydrogen bonds than non-ionized water (Figure 2.3b). The enthalpies of phase transition, surface tension, melting and boiling points, large heat capacities, and surface tensions that have been observed all make sense given that water is capable of extensive three-dimensional hydrogen bonding. The extra energy required to break a lot of intermolecular hydrogen bonds can be linked to all of these. Hydrogen bonding also affects water's permittivity, or dielectric constant. Water is a dipole, but this does not explain its high permittivity on its own. Multimolecular dipoles are thought to be produced by hydrogen-bonded molecular clusters, thereby raising the permittivity.

2.2.2. Dissociation of Water Molecules

The ions created by the molecule's self-dissociation are two of the species found in pure water, as previously mentioned. These ions are simplest known as the hydrogen ion (H^+) and the hydroxyl ion (OH^-), though in actuality they exist in a hydrated form. These come from the process of self-dissociation, so they will be present in equimolar amounts in pure water.

$$H_2O \rightleftharpoons H^+ + OH^- \qquad (1)$$

At 298 K, the equilibrium constant for this dissociation is $K_w = 10^{-14}$ and the pH is 7. It is important to realize that this dissociation is enhanced at higher temperatures, and in consequence, the pH of pure water is temperature dependent. K_w approaches 10^{-12} at 373 K, leading to a pH close to 6 at this temperature. Note that, while a pH of 6 at 298 K implies a concentration of OH^- of 10^{-8} M, at 373 K a pH of 6 implies a concentration of OH^- close to 10^{-6} M.

2.2.3. Structures in Pure Water Systems

2.2.3.1. The Structure of Ice

The structure of ice is better understood than that of liquid water, so it makes sense to talk about it before discussing the structure of liquid water. Water crystallizes in an open, low-density structure that has been precisely determined due to its tetrahedrally directed forces. In ice, the O–H–O bond angle is approximately 109°, which is very near to the perfect tetrahedral angle of 109.28° (Figure 2.4).

The internuclear nearest neighbor distance between O atoms is 2.76 Å. Considering molecule W and its four nearest neighbors, 1, 2, 3, and W,' makes it easy to visualize in the unit cell of Figure 2.4 the way in which each HOH bond can associate with four others (coordination number of 4).

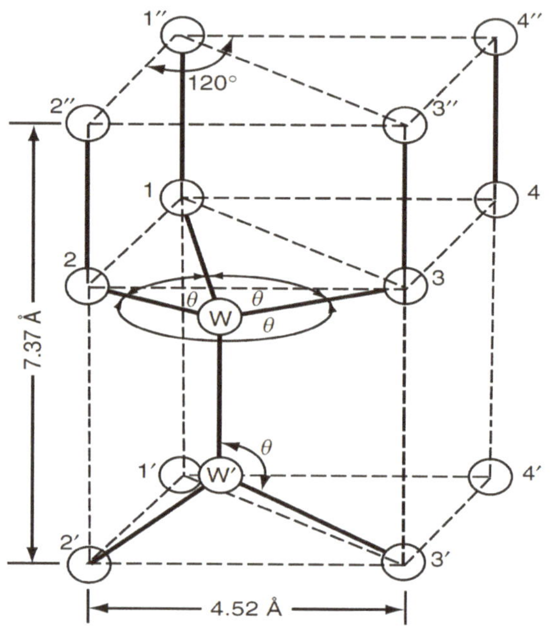

Figure 2.4. *Unit cell of ordinary ice at 0°C. Circles represent oxygen atoms of water molecules. The nearest neighbor internuclear O--O distance is 2.76 Å. θ is 109°.*

Source: https://edisciplinas.usp.br/pluginfile.php/7989549/mod_resource/content/1/%C3%81gua%20nos%20Alimentos.pdf.

The hexagonal symmetry of ice is visible when multiple unit cells are joined and viewed from above (down the c-axis) (Figure 2.5). Molecule W and its four closest neighbors, 1, 2, and 3, all exhibit the tetrahedral substructure; the fourth neighbor is directly beneath molecule W and lies below the paper's plane.

It is clear that two planes of molecules are involved when Figure 2.5a is viewed in three dimensions, as shown in Figure 2.5b (open and filled circles). The "slip" or flow of ice under pressure, as in a glacier, causes these two parallel, extremely close to one anotherplanes to move together as a single unit. The basal planes of ice are pairs of these kinds of planes.

An extended structure of ice can be created by stacking multiple basal planes. The structure shown in Figure 2.6 is the result of combining three basal planes. The appearance is identical to that seen in Figure 2.5a when viewed along the c-axis, demonstrating the perfect alignment of the basal planes.

In this direction, ice is monorefringent, while in all other directions, it is birefringent. Thus, the optical axis of ice is the c-axis. Interestingly, the c-axis is frequently found to be perpendicular to the main plane of large sheets of ice (Hobbs, 1974). Although a completely satisfactory explanation has not yet been put forth, it might be due to the varying velocities at which ice grows along the various symmetrical axes.

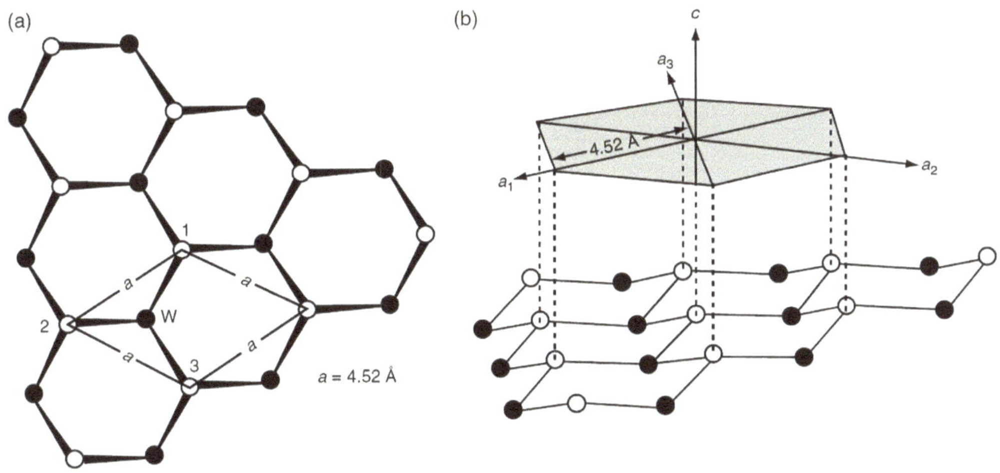

Figure 2.5. *The basal plane of ice (a combination of two layers of slightly different elevation). Each circle represents the oxygen atom of a water molecule. Open and shaded circles represent, respectively, oxygen atoms in the upper and lower layers of the basal plane. (a) Hexagonal structure viewed down the c-axis.*

Source: https://edisciplinas.usp.br/pluginfile.php/7989549/mod_resource/content/1/%C3%81gua%20nos%20Alimentos.pdf.

With regard to the location of hydrogen atoms in ice, there is general agreement regarding the following:

- One hydrogen atom, centered 1 ± 0.01 Å from the covalently bonded oxygen and 1.76 ± 0.01 Å from the hydrogen bonded oxygen, occupies each line that joins the two nearest neighbor oxygen atoms. Figure 2.7a depicts this configuration.
- However, a slightly different picture than the one above is obtained if the locations of hydrogen atoms are viewed over time instead of as a snapshot in time. One of two possible positions for a hydrogen atom on a line connecting

the two nearest neighbor oxygen atoms, X and Y, is 1 Å from X or 1 Å from Y. It is thought that each of these two positions is occupied on average half of the time because they have an equal chance of being occupied. This is made possible by the cooperative rotation of HOH molecules, which permits hydrogen atoms to "jump" between neighboring oxygen atoms, with the exception of very low temperatures. Figure 2.7b depicts the resulting mean structure, also referred to as the half hydrogen, Pauling, or statistical structure.

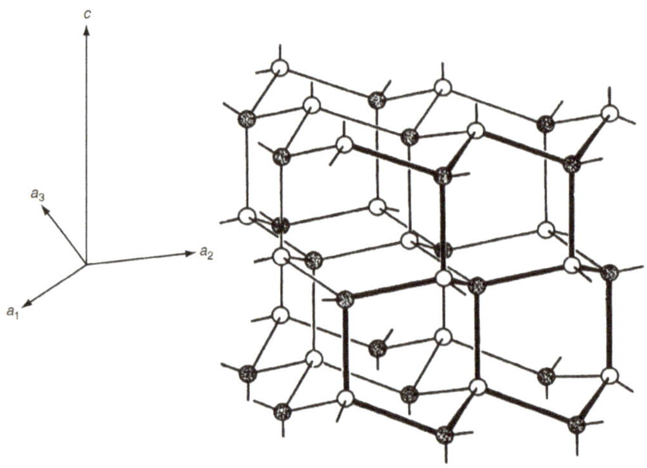

Figure 2.6. *The extended structure of ordinary ice. Only oxygen atoms are shown. Open and shaded circles represent, respectively, oxygen atoms in upper and lower layers of a basal plane.*

Source: https://edisciplinas.usp.br/pluginfile.php/7989549/mod_resource/content/1/%C3%81gua%20nos%20Alimentos.pdf.

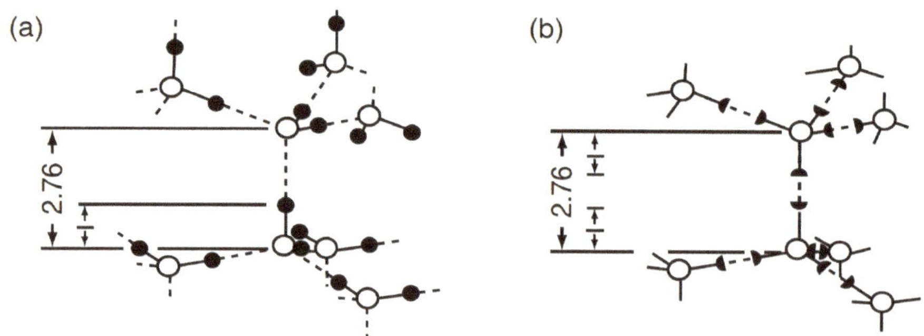

Figure 2.7. *The location of hydrogen atoms (•) in the structure of ice: (a) instantaneous structure and (b) mean structure (also known as half hydrogen (•), Pauling, or statistical structure). Open circles are oxygen atoms.*

Source: https://edisciplinas.usp.br/pluginfile.php/7989549/mod_resource/content/1/%C3 %81gua%20nos%20Alimentos.pdf.

Ordinary ice is classified as a dihexagonal bipyramidal class within the hexagonal system based on crystal symmetry. Amorphous or vitreous states of uncertain, but primarily noncrystalline structure, are also possible states for ice, as are nine other crystalline polymorphic structures. Only regular hexagonal ice is stable at 0°C and normal pressure out of the 11 total structures. The previous discussion may lead one to believe that the true structure of ice is simple. First of all, in addition to regular HOH molecules, pure ice also includes isotopic and ionic forms of HOH, which have been identified as trace amounts of water. Fortunately, because the isotopic variants are present in such small quantities, we can usually ignore their structural influence. From a structural perspective, only the contributions from HOH, H^+ (H_3O^+), and OH^- need to be considered.

As real ice crystals are never perfect, the structural defects that occur are usually of the ionic type (caused by proton dislocation with formation of H_3O^+ and OH^-) or orientational type (caused by proton dislocation with neutralizing orientational adjustments) (*see* Figure 2.8). These structural flaws explain the surprisingly high mobility of protons in ice as well as the relatively minor electrical conductivity loss that happens when water freezes, despite the intuitive expectation of a significant conductivity loss during solidification. Apart from the atomic mobilities found in crystal defects (lattices), ice exhibits other forms of motional activity. At −10°C, it is estimated that the root mean amplitude of vibration for each HOH molecule in ice is approximately 0.4 Å, assuming that each molecule vibrates as a unit. Furthermore, rather than being trapped in a specific interstitial space, the individual HOH molecules that presumably occupy some of the ice's interstitial spaces appear to be able to diffuse slowly through the lattice. As a result, ice is not a static or homogenous molecular assembly, and its properties vary with temperature. Even though HOH molecules in ice are four coordinated at all temperatures, the hydrogen atoms must be constrained to only one of the numerous possible configurations by lowering the temperature to roughly −180°C or lower. Therefore, all hydrogen bonds will only be intact at temperatures close to −180°C or lower, and the mean number of intact (fixed) hydrogen bonds will progressively decrease as the temperature rises.

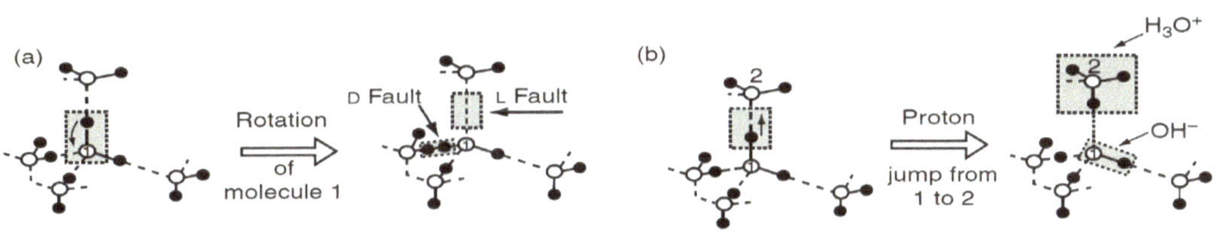

Figure 2.8. *Schematic representation of proton defects in ice.*

Source: https://edisciplinas.usp.br/pluginfile.php/7989549/mod_resource/content/1/%C3 %81gua%20 nos%20Alimentos.pdf.

Note: (a) Formation of orientational defects; and (b) formation of ionic defects. Open and filled circles represent, respectively, oxygen and hydrogen atoms. Solid and dashed lines represent, respectively, chemical bonds and hydrogen bonds.

2.2.3.2. The Structure of Water (Liquid)

The idea of structure in a liquid may seem strange at first because fluidity is the essence of the liquid state. However, it is a long-standing and widely accepted theory that liquid water has some structure, not enough to produce long-range rigidity, but far more organized than the vapor state, and enough to cause neighboring water molecules to affect the orientation and mobility of any given water molecule. Thinking of the structure in the liquid as a collection of short-term structured associations that constantly change quickly but always retain an average level of structure within the liquid has proven to be a helpful conceptual approach. There is a wealth of convincing evidence to support the idea that water is a structured liquid. Water, for instance, is an "open" liquid because its density is only 60% of what would be predicted for a liquid with closely spaced molecules. Liquid water's low density can be easily explained by partial retention of the tetrahedral arrangement of ice, which is open and hydrogen-bonded. Moreover, even though ice has an exceptionally high enthalpy of fusion for a solid, this energy would only be needed to break 15% of the hydrogen bonds that are thought to be present in ice. While this does not necessarily mean that liquid water retains 85% of the hydrogen bonds found in ice (e.g., the results of numerous independent investigations strongly support the theory that many water-water hydrogen bonds continue to exist in the liquid, with the extent of hydrogen bonding decreasing as the liquid temperature increases. Alternatively, more bonds could be broken but the resulting change in energy could be hidden by a simultaneous increase in van der Waals interactions.

Clarifying the structure or structures of pure liquid water is a difficult and complex task. Numerous theories have been put forth, but they are all flawed, simplistic, and prone to flaws that opponents of competing theories can easily point out. This is a favorable circumstance that should ultimately lead to a precise structural description of liquid water. Computer simulations of the molecular dynamics of water, which are governed by the equations of motion, and molecular potential functions, which aim to approximate the significant interactive modes of the water molecule, have become possible in recent years due to advances in computer power. Even though these simulations are constrained by the potential function's errors and approximations, it is discovered that they exhibit many of the characteristics of water, offering compelling new insights into the realities of liquid water. Though difficult to depict on paper, visual displays of the movements of the molecules represented in the simulation are highly instructive. Despite the growing complexity of these simulations and the insightful information they offer, it is still worthwhile to look at models created before widespread access to the kind of raw computational power that is available today.

Mixture models, interstitial models, and continuum models—also known as homogeneous or uniformist models—are the three general categories of models for liquid water that have been proposed. The idea of intermolecular hydrogen bonds being momentarily concentrated in bulky clusters of water molecules that are in dynamic equilibrium with more dense species is embodied in mixture models; the word "momentarily" denotes a timescale of roughly 10–11 s. (Luck, 1981). This kind of methodology is frequently embodied in molecular dynamic computer simulations, which offer a time series of snapshots of the location (and frequently orientation) of the constituent molecules represented in the model. The interaction potential function assigned to water determines the precise properties displayed by the model. A variety of potential functions, each with unique advantages and disadvantages, have been suggested and used. According to continuum models, many of the bonds found in ice simply distort when it melts rather than breaking. These bonds are thought to be uniformly distributed throughout the sample. According to some theories, this makes it possible for a continuous, dynamic network of water molecules to exist, with the distortions able to move across the network and relocate in space.

The idea behind the interstitial model is that water can retain, with minimal distortion, either an ice-like or clathrate-type hydrogen-bonded network structure, with the network's interstitial spaces filled with unbonded individual water molecules. The core structural element shared by all three models is the idea of liquid water forming hydrogen bonds in transient, distorted tetrahedra. All models also allow individual water molecules to rapidly change their bonding configurations by exchanging one hydrogen bond for a new one, all the while keeping the system's overall degree of hydrogen bonding and structure constant at constant temperature. The more modern computer models exhibit elements of each of the older models in many ways. There is evidence for water molecules to have moved to locations not supported by a conventional hydrogen-bonded network and for hydrogen bonds to have changed orientations. Numerous modeling studies have been able to accurately simulate the behaviors of water that have been observed. While hydrogen bonding is undoubtedly crucial, as the simpler models may imply, well-defined structures do not appear in the computer models, which generate time-averaged images. We can now talk about water's apparent anomalous low viscosity. Since the hydrogen-bonded arrangements of water molecules are highly dynamic and allow individual molecules to change their hydrogen-bonding relationships with neighbouring molecules within a timeframe of nano- to picoseconds, this attribute is easily reconcilable with the types of structures that have been described. This facilitates mobility and fluidity. It is believed that a portion of liquid water's extraordinarily high heat capacity is due to the energy needed to break more hydrogen bonds as temperature rises. Since the majority of the molecules in the vapor are thought to be monomers, the high enthalpy of vaporization indicates that most or all of the remaining hydrogen bonds are broken as the liquid vaporizes.

Naturally, temperature affects the amount of hydrogen bonding that occurs between water molecules. The nearest neighbor distance for ice at 0°C is 2.76 Å, and the coordination number, or number of nearest neighbors, is 4.0. Melting takes place when the enthalpy of fusion is input. The enthalpy of fusion represents the breaking of certain hydrogen bonds (an increase in the distance between nearest neighbors) and the straining of other hydrogen bonds as water molecules transition to a fluid state with, generally speaking, more compact associations. The coordination number rises with temperature, starting at 4.0 in ice at 0°C and going up to 4.4 in water at 1.50°C and 4.9 at 83°C. The closest neighbors are farther apart at the same time, going from 2.76 Å in ice at 0°C to 2.9 Å in water at 1°50°C, and finally to 3.05 Å at 83°C. Thus, it is clear that the transition from ice to water is accompanied by two changes in density: an increase in the average number of nearest neighbors (increased density) and a decrease in the distance between them (decreased density), with the latter occurring more frequently during the phase change and resulting in the well-known net increase in density associated with melting. When the temperature rises above the melting point, the density reaches a maximum at 31.8°C and then starts to progressively decrease. It is evident that above 3.98°C, the effect of increasing the distance between nearest neighbors (thermal expansion) predominates, while the effect of an increase in coordination number is more pronounced at temperatures between 0°C and 3.98°C.

2.2.4. Phase Relationships of Pure Water

Only the molecular and structural characteristics of water—that is, interactions and interrelationships at the microscopic and submicroscopic levels—have been taken into consideration up to this point. It is now appropriate to talk about the observable phase behavior of water because this will help us understand the qualities of pure water as well as later discussions on how aqueous solutions behave under various pressure and temperature conditions. Temperature and pressure have an impact on pure water phase relationships that need to be taken into account. The phase diagram for pure water is displayed in Figure 2.9a. The vapor–liquid equilibrium line and the pressure dependence of stable forms of ice are particularly relevant to the field of food science. As was previously mentioned, various types of ice have been identified, each of which is stable in a specific area of the temperature–pressure diagram. Ice I is the only phase of ice that is relevant under the temperature and pressure conditions used in food processing. Notably, the melting point of ice Ih decreases with increasing pressure for ice I (Figure 2.9b). Specifically, take note that the melting point of ice Ih is below -20°C at a pressure of 270 MPa. This information is useful for methods like pressure shift freezing, which involves cooling food to -20°C under high pressure. Water remains liquid in these circumstances because the material is above the freezing point of ice, despite a decrease in enthalpy (heat content). When pressure is released, freezing occurs very quickly because ice is the stable form of water under the new pressure and temperature conditions, and ice has the same heat content as a sample with a significant fraction of water. Another method of thawing a frozen material has been designed in reverse, where a high enough pressure is applied to the material to cause water to become the stable phase without causing a temperature change. Thawing occurs instantly because pressure can be applied nearly instantaneously. It is possible to remove the applied pressure without ice forming once the temperature rises above 0°C. As a result, it is possible to achieve a thawing process where the phase change is uniform throughout the material as opposed to progressive, which reflects the geometric pattern of temperature change regulated by heat transfer processes. Other forms of ice, such as ice II, ice III, ice IV, and so on, exist at pressures higher than those where ice I is the stable phase. There is no need to continue the discussion of these forms because they are not present in any circumstances related to the handling or processing of food.

When examining equilibria in vapor–liquid systems, keep in mind that pure liquid water's vapor pressure rises from 610 Pa at 0°C to 101,323 Pa at 100°C. When the equilibrium form of HOH is ice, it is also feasible to measure the vapor pressure of water above undercooled liquid water at temperatures below 0°C. These pressures are consistently greater than the water above ice equilibrium vapor pressure at the same temperature. We will talk about the importance of these findings for food science later in this chapter, specifically in relation to the application of relative vapor pressure as a measure of "water availability." Given the significance of water in food, we must create a system or systems to characterize the quantity, kind, and state of the water present. However, in order to comprehend the impact of the molecular environment on the properties displayed by water at both the molecular and bulk levels, we must first take into consideration systems more complex than pure water.

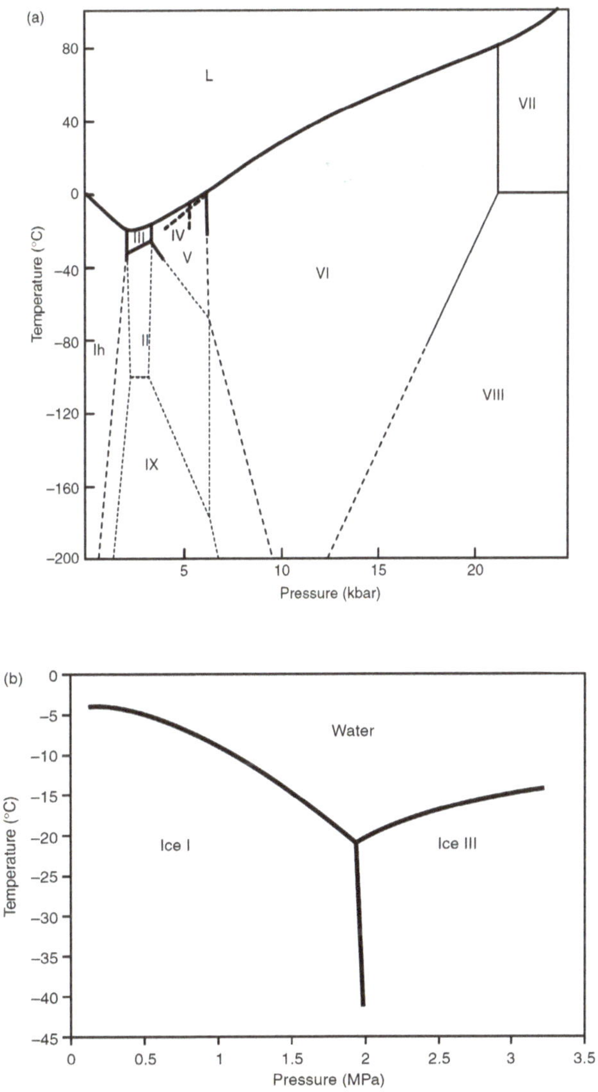

Figure 2.9. *The pressure–temperature phase diagram for (a) pure water; and (b) details of the behavior of liquid water, ice I, and III.*

Source: https://edisciplinas.usp.br/pluginfile.php/7989549/mod_resource/content/1/%C3%81 gua%20 nos%20Alimentos.pdf.

2.3. WATER IN THE PRESENCE OF SOLUTES

There are solutes and water in every food system. Therefore, a discussion of the effects of solutes on the characteristics of a collection of water molecules, including their solvent properties, is required.

2.3.1. Ice in the Presence of Solutes

In an aqueous system, solutes affect the amount of ice present as well as its propagation patterns through kinetic and thermodynamic effects. The amount of ice formed at any given temperature decreases with an increase in the concentration of a specific solute. This is an example of colligative effects and freezing point depression. A straightforward phase diagram for a binary aqueous system, Figure 2.10, illustrates how a binary aqueous solution's freezing point varies with concentration. The kind and quantity of solutes present affect the location, size, structure, orientation, and quantity of ice crystals that form after a specific cooling protocol. Think about how a solute affects the structure of ice, for instance. Luyet and colleagues conducted groundbreaking research on the appearance of ice crystals formed in the presence of different solutes such as albumin, myosin, sucrose, glycerol, and gelatin under a variety of cooling conditions. They came up with a classification scheme based on symmetrical components, morphology, and the cooling velocity needed for the formation of different kinds of visible ice structures. They have classified visible ice structures into four main categories: coarse spherulites, evanescent spherulites, irregular dendrites, and hexagonal forms.

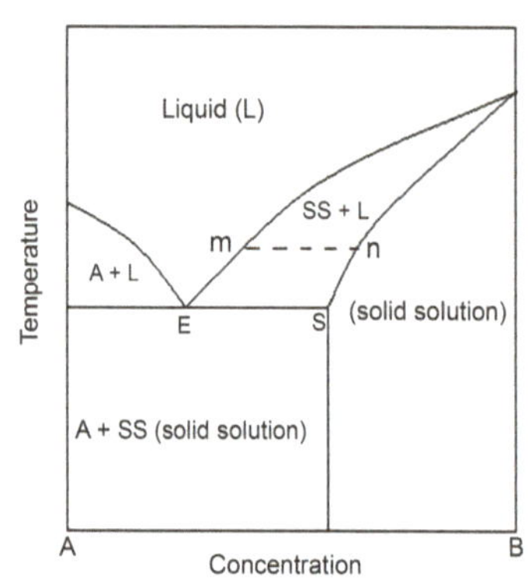

Figure 2.10. *Schematic binary phase diagram for a simple aqueous system.*

Source: https://ars.els-cdn.com/content/image/3-s2.0-B978032395080000008X-f08-06-9780323950800.jpg.

When extremely rapid freezing is avoided and the solute is of a type and concentration that does not unnecessarily interfere with the mobility (ease of spatial reorganization) of water molecules, the hexagonal form—which is the most highly ordered—is found only in foods. For example, high concentrations of gelatin will cause more asymmetrical ice crystal forms. In these preliminary investigations, Luyet and colleagues established unequivocal proof of the presence of a glassy (amorphous) unfrozen phase encircling the ice crystals at low temperatures, and they connected this phase's existence to phenomena like "collapse" in freeze drying. Rey offered comparable insights at about the same time that he was conducting groundbreaking research on the characteristics of the phases involved in the freeze-drying process. These studies were early forerunners in the development of concepts regarding the role of the glassy state in governing the kinetic properties of frozen systems, even though their full significance was not immediately apparent.

2.3.2. Water–Solute Interactions in Aqueous Solutions

2.3.2.1. Macroscopic Level

It is appropriate to introduce some general observations about the behavior of water before delving into a description of molecular interactions between water and solutes. Food qualities are greatly influenced by the presence of water, and these properties vary as the amount of water in the food varies. Water's effects on system properties are described by concepts like water binding, hydration, and water holding capacity. These historical terms, which are frequently based on macroscopic observations, might be regrettable when considered in the context of our knowledge of the underlying molecular processes that they are assumed to reflect. Nevertheless, it is crucial to introduce these descriptive ideas since they have contributed significantly to our growing comprehension of the characteristics and influences of water in food. The terms "water binding" and "hydration" were once frequently used to characterize the water's inclination to bond with hydrophilic materials, such as biological materials. These terms relate to the macroscopic level when used in this context. Even though more technical terms like "water binding potential" have quantitative definitions, they are still only applicable at the macroscopic level. Several factors influence the degree and strength of water binding or hydration, such as the nonaqueous constituent's nature, the composition of salts, pH, and temperature. The ability of a matrix of molecules, typically macromolecules present at low concentration, to physically entrap large amounts of water in a way that inhibits exudation under the application of an external force, often gravitational force, is referred to as "water holding capacity." Common food matrices that do this kind of entrapment of water include starch and pectin gels as well as plant and animal tissue cells. Tissue foods do not easily release physically entrapped water, even when they are chopped or minced. However, during processing, this water exhibits characteristics that are similar to those of pure water. It is easily extracted when drying, quickly transformed into ice when freezing, and easily accessible when used as a solvent. The movement of individual molecules is therefore essentially similar to that of water molecules in a diluted salt solution, despite the fact that its bulk flow is severely restricted. Food quality is significantly impacted when foods' ability to physically trap water, or their capacity to hold water, is compromised. The majority of the water found in tissues and gels falls into this category. Syneresis of gels, thaw exudation from previously frozen foods, and inferior performance of animal tissue in sausage due to the decline in muscle pH that accompanies normal physiological events postmortem are examples of quality defects associated with impairment of water holding capacity. The physical relocation of water molecules in space is the primary cause of the quality defect in all cases; significant changes may not always be reflected.

2.3.2.2. Molecular Level: General

When solutes and water are intimately mixed, the properties of both constituents are changed relative to their unmixed state. These alterations are the consequence of molecular interactions and are thus reliant on the molecular makeup of the solute. Ions, or charged groups, mostly use electrostatic forces to interact with water. These could improve or obstruct the water molecules' typical geometric orientations. Water and predominantly hydrophilic solutes interact intensely, changing the structural relationships and mobility of nearby water while also altering the reactivity and, occasionally, the structure of the hydrophilic solutes. On the other hand, the hydrophobic groups present in the added solutes exhibit weak interactions with nearby water, suggesting a preference for nonaqueous environments. Nonetheless, there may be significant structural repercussions from this weak interaction.

2.3.2.3. Molecular Level: "Bound Water"

Although "bound water" is a term that is frequently used, it is not well defined and does not relate to a single, cohesive thing. It does not always even allude to water in the true sense in a solute-bonding sense. As multiple, frequently contradictory definitions show, it is challenging to come up with a single, coherent descriptive term for bound water, and there is no agreement on which definition — if any — is the best. As a result of the term's controversy, frequent misuse, and lack of understanding, many scientists have recommended that its use be stopped. While it is true that taking this step would help with more accurate communication, the term "bound water" appears so frequently in the literature that it needs to be discussed and its serious limitations acknowledged.

The following partial list of definitions that have been proposed for bound water illustrates why the use of the term has created such confusion:

- Bound water is the equilibrium water content of a sample at some appropriate (and arbitrary) temperature and low humidity.
- Bound water is the water which does not contribute significantly to permittivity at high frequencies and therefore has its rotational mobility restricted by the substance with which it is associated.
- Bound water is that which remains unfrozen at some arbitrary low temperature, usually −40°C or lower.
- Bound water is that which is unavailable as a solvent for additional solutes.
- Bound water is that which produces line broadening in experiments involving proton nuclear magnetic resonance.
- Bound water is that which moves with the macromolecule in experiments involving sedimentation rates, viscosity, or diffusion.

- Bound water is that which exists in the vicinity of solutes and other nonaqueous substances and has apparent properties differing significantly from those of "bulk" water in the same system.

While there is some merit to each of these definitions when applied correctly, few of them will yield the same result when applied independently to the same system. Additionally, there are situations in which the total water content of the system — even when it exceeds the amount of bound water that has been determined — will determine the bound water value that is obtained using a specific technique and definition. The water that exists near solutes and other nonaqueous constituents and that, as a result of its location, exhibits apparent properties that are significantly altered from those of "bulk water" in the same system can be conceptually understood as bound water, albeit imperfectly. Instead of being "immobilized," bound water can be thought of as having "hindered mobility" in comparison to bulk water. This kind of water makes up a very small portion of the water in a typical high-water food; it roughly corresponds to the first layer of water molecules that are next to hydrophilic groups in space. But keep in mind that there are variations in this population of water molecules. The section on molecular mobilities in frozen systems will go into more detail on the topic of bound or hindered water.

2.3.2.4. Interactions of Water with Ions and Ionic Groups

The mobility of water molecules appears to be more hampered or influenced by individual ions as well as the ionic groups of organic molecules than by any other kind of solute. While electrostatic water-ion bonds are stronger than hydrogen bonds between molecules of water, they are still far weaker than covalent bonds. Dissociable solutes can upset the recognized normal structure of pure water, which is based on a hydrogen-bonded, typically tetrahedral arrangement. Dipole–ion interactions occur between water and simple inorganic ions.

The NaCl ion pair's hydration is demonstrated by the example in Figure 2.11. The only molecules that are shown are the first-layer water molecules in the plane of the paper, which are oriented by the ions' radial electrical fields. Because of the competing structural influences of the first-layer water surrounding the charged ions and the more distant, tetrahedrally oriented "bulk-phase" water that is remote from the influence of the radial electrical fields surrounding the ions, it is thought that a second layer of water molecules beyond this first layer exists in a structurally perturbed state in a diluted solution of the ions in water. Bulk-phase water would not exist in concentrated salt solutions, where individual ion electrical fields would be expected to overlap, and the ions would dominate the water structure.

A wealth of data suggests that certain ions in diluted aqueous solutions have the effect of "net structure breaking," making the solution more fluid than pure water, while other ions have the opposite effect, making the solution less fluid than pure water. This isn't meant to imply that the "structure" exists in pure water. The phrase "net structure" describes a wide range of structures, including both novel and conventional water organization. Since "normal" water structure lacks radial symmetry, it is evident that all ions are disruptive from the perspective of "normal" water structure. The ability of a given ion to alter net structure is related closely to its polarizing power (charge divided by radius) or simply the strength of its electric field. Small and/or multivalent ions (such as Li^+, Na^+, H_3O^+, Ca^{2+}, Ba^{2+}, Mg^{2+}, Al^{3+}, F^-, and OH^-) have strong electric fields and are net structure promoters. The structure imposed by these ions more than compensates for any loss in normal water structure. These ions strongly interact with four to six first-layer water molecules, causing them to be less mobile, and to pack more densely than HOH molecules in pure water. Ions that are large and monovalent, such as K^+, Rb^+, Cs^+, NH^+_4, Cl^-, Br^-, I^-, NO^-_3, BrO^-_3, IO^-_3, and ClO^-_4 have rather weak electric fields and are net structure breakers although the effect is very slight with K^+. These ions disrupt the normal structure of water and fail to impose a compensating amount of new structure.

Figure 2.11. *Likely arrangement of water molecules adjacent to sodium chloride ion pair.*

Source: https://encrypted-tbn0.gstatic.com/images?q=tbn:ANd9GcS8KW_Dg74ZKgaRs5euyyJPTnSDSPGljd3Pvw&s.

Naturally, ions have significant impacts beyond their impact on the composition of water. Ions have a significant impact on the "degree of hospitality" extended to other solutes present in the aqueous medium as well as to substances suspended in the medium due to their varied capacities to hydrate (compete for water), govern the thickness of the electrical double layer surrounding colloids, and alter the permittivity of the aqueous medium. The Hofmeister or lyotropic series was the first to recognize this; it arranged ions according to how well they could influence different properties like colloid stability or salt proteins in or out. The fact that there is a strong correlation between the empirically derived Hofmeister series and a series based on the anticipated structural effects of the different ions is encouraging.

2.3.2.5. Interaction of Water with Neutral Groups Capable of Hydrogen Bonding (Hydrophilic Solutes)

Water-ion interactions and water-water hydrogen bond interactions are both somewhat stronger than interactions between water and nonionic, hydrophilic solutes. The strength of the hydrogen bonds between water and solute determines the first-layer water (i.e., water right next to the hydrophilic species), which may or may not have different characteristics from bulk-phase water, such as decreased mobility. At first glance, one might assume that solutes with hydrogen bonding capabilities would improve or at the very least not alter the regular structures of pure water. This oversimplified assumption fails to recognize how crucial spatial orientation and location are to the development of a functional hydrogen-bonded network. In certain cases, it is discovered that the solute's hydrogen-bonding site distribution and orientation are geometrically incompatible with those of pure water. The typical three-dimensional, tetrahedrally oriented structure of pure water is frequently disrupted by such solutes. One small hydrogen-bonding solute that could significantly disrupt the normal structure of water due to geometric factors is urea.

On the other hand, some molecules might contain hydrophilic groups that are capable of forming hydrogen bonds at orientations and spacings that are consistent with the water's hydrogen bonding structures. The simple carbs serve as an illustration. They have been found to have equatorial hydroxyl groups with spatial relationships resembling those of clusters of water molecules (see Figure 2.12). This level of compatibility may even raise the overall hydrogen bonding concentration. It should be noted that while there is a close correspondence and possibly facilitated interaction, it is not claimed that there is an exact correspondence in all circumstances due to the temperature dependence of the oxygen atom spacings in water and individual molecules.

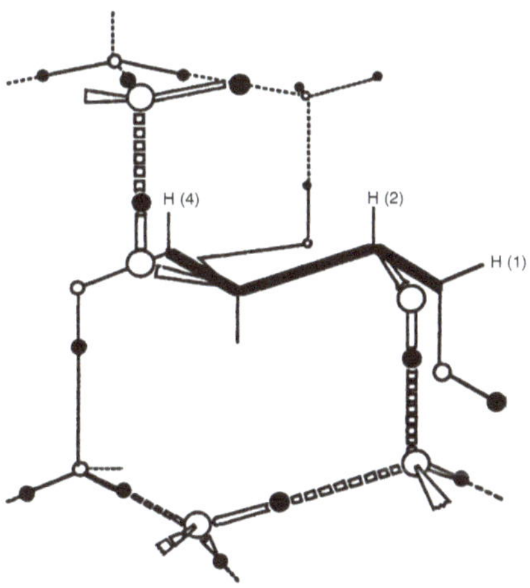

Figure 2.12. *Possible association of d-glucose with tetrahedrally arranged water molecules. The side view of the pyranose ring is represented by a heavy line. Oxygen and hydrogen atoms of water molecules are represented by open and filled circles, respectively.*

Source: https://edisciplinas.usp.br/pluginfile.php/7989549/mod_resource/content/1/%C3%81gua%20nos%20Alimentos.pdf.

It is crucial to realize that adding a hydrogen-bonding solute that upsets the regular structure of water may not have a substantial impact on the total number of hydrogen bonds per mole of water. This is feasible because water-solute hydrogen bonds have the potential to replace broken water-water hydrogen bonds. Such behaving solutes have minimal impact on the "net structure."

Hydrogen bonding between water molecules and different potentially eligible groups (e.g., amide, imino, carbonyl, amino, and hydroxyl groups, among others). Water bridges are occasionally the result of this, in which a single water molecule interacts with two hydrogen-bonding sites on one or more solutes. Diagram showing how water forms hydrogen bonds with two different types of functional groups in proteins (dotted lines). Hydrophilic groups in many crystalline macromolecules are spaced at lengths that are strikingly similar to the nearest neighbor oxygen spacing in pure water, as has been noted for certain sugars. If this spacing is maintained in the hydrated macromolecule, this would increase the stability (lifetime) of the cluster and tend to promote cooperative hydrogen bonding involving both first- and second-layer water.

2.3.2.6. Interaction of Water with Nonpolar Substances

The mixing of water and hydrophobic substances such as hydrocarbons, rare gases, and the apolar groups of fatty acids, amino acids, and proteins is, not surprisingly, a thermodynamically unfavorable event ($\Delta G > 0$). However, the free energy of this process is positive not because ΔH is positive, which is typically the case for solutes with a low solubility, but rather because $T\Delta S$ is negative.

This reduction in entropy, considered an indicator of increased "order," is thought to occur because of special structures in water that form in the vicinity of these incompatible apolar entities. This process has been termed "hydrophobic hydration."

Because hydrophobic hydration is thermodynamically unfavorable, it follows that the system adjusts in an attempt to minimize the association of water with the apolar entities that are present.

Thus, if two separated apolar groups are present, their incompatibility with the aqueous environment serves to encourage their association, thereby lessening the water–apolar interfacial area, a process that is thermodynamically favorable ($\Delta G < 0$). This process, a partial reversal of hydrophobic hydration, is referred to as "hydrophobic interaction" and, in its simplest form, can be represented as:

$$R_{(hydrated)} + R_{(hydrated)} \rightarrow R_{2\,(hydrated)} + H_2O \quad (2)$$

Where R is an apolar group.

Because water and apolar groups exist in an antagonistic relationship, water structure adjusts to minimize contact with apolar groups. The type of water structure that was believed to exist in the layer next to apolar groups.

2.3.2.6.1. Clathrate Hydrates

A clathrate hydrate is an ice-like inclusion compound in which a small apolar molecule known as the guest molecule is physically trapped by water, the host substance, in the form of a hydrogen-bonded cage. These compounds are interesting because they show the most extreme way that water can form a structure in response to an apolar substance, and because natural microstructures of the same kind can be found in biological materials. Crystalline clathrate hydrates are possible. With enough pressure, clathrate crystals can be grown to observable sizes with ease, and some of them can withstand temperatures higher than 0°C. The low molecular weight guest molecules in clathrate hydrates are characterized by sizes and shapes that align with the dimensions of the host water cages, which consist of 20–74 water molecules. Alkyl ammonium, sulfonium, and phosphonium salts; rare gases; short chain primary, secondary, and tertiary amines are typical guests. Low molecular weight hydrocarbons and halogenated hydrocarbons are also common. Water and the guest typically interact directly through weak van der Waals forces; in fact, the guest molecule is frequently free to rotate inside the cavity. Water's remarkable attempt to reduce contact with hydrophobic groups results in clathrate hydrates. Although the water structure of a clathrate hydrate appears to differ greatly from that of ice at first glance, this difference is actually due to a slight alteration in the hydrogen bond's geometry. If one looks along the direction of an oxygen-oxygen link in ice, one will see that the hydrogen bonds between the tetrahedral coordination of the water molecules are in a staggered conformation, whereas in a clathrate hydrate, the geometry of the tetrahedral coordination of the water molecules is in the eclipsed form. Three of a water molecule's four hydrogen bonds can aid in forming the curved surface of a cage (much like a geodesic surface) in structures where the bond orientation rotation of 60° produces, with the fourth hydrogen bond projecting out normally to this surface. As a result, there aren't any hydrogen bonds extending into the interior cavity, which means there aren't any adverse interactions with any apolar groups there. As mentioned, there is free rotation of a small apolar molecule inside the cavity. It's also crucial to remember that, when the free energies of the clathrate cage structure and the ice structure are compared (assuming there are no guest molecules in the cages), the ice structure is marginally more stable than the empty clathrate. Therefore, a crystalline structure with far higher stability than ice can arise from the presence of a suitable guest that stabilizes the cavity through steric interactions.

There is proof that hydrogen-bonded formations that resemble crystalline clathrate hydrates but are less widely distributed in space (i.e., hydrogen-bonded multimers with eclipsed orientation), which could be found naturally in biological matter. If this is the case, localized structures like these have far more significance in food science than crystalline hydrates because they could affect the stability, reactivity, and conformation of molecules like proteins. For instance, it has been proposed that partially exposed hydrophobic protein groups may have clathrate structures. Further evidence for the potential reorientation of the water-water hydrogen bond into a "clathrate-type" orientation in response to the presence of the nonpolar entity comes from molecular dynamic simulations of aqueous systems containing nonpolar species. The more straightforward pictorial model is consistent with the shift in average bond direction, whereas the results of molecular modeling do not depict structures with hydrogen-bonding geometries similar to the clathrate-type orientations. Hydrogen bonds typically lie tangential to molecular surfaces in models involving hydrophobic solutes.

2.3.2.6.2. Interaction of Water with Complex Molecular Structures

Despite the difficulty in determining the arrangement of water molecules around organic molecules through experimentation, this is a field of active research with valuable data obtained. The computer-simulated cross section of myoglobin and the hydrated pyranose sugar ring. With full occupancy of the hydration sites and a separation of 2.8 Å between them, the primary hydration shell of myoglobin would contain about 360 HOH molecules. Inevitable interactions and interferences arise from the coexistence of polar, hydrophilic, and hydrophobic regions within a large molecule. For instance, the inevitable pairing of water with specific hydrophobic groups in proteins affects the functionality of the protein significantly. The presence of apolar side chains on approximately 40% of the amino acids in typical oligomeric food proteins means that the extent of these inevitable contacts could be fairly large. Alanine's methyl group, phenylalanine's benzyl group, valine's isopropyl group, cysteine's mercaptomethyl group, and the leucines' secondary butyl and isobutyl groups are some examples of these nonpolar groups. Hydrophobic interactions can also involve the apolar groups of other compounds, including fatty acids, alcohols, and free amino acids; however, the outcomes of these interactions are not as significant as those involving proteins. It is thermodynamically unfavorable for protein apolar groups to be exposed to water, which facilitates the association of hydrophobic groups or hydrophobic interactions.

Protein folding is thought to be primarily fueled by hydrophobic interaction, which leads many hydrophobic residues to take up positions deep within the protein. However, it is estimated that nonpolar groups in globular proteins still normally occupy about 40–50% of the surface area despite such hydrophobic interactions. Because of these hydrophobic groups that are positioned on the surface, hydrophobic interactions are also thought to be crucial for preserving the tertiary structure (subunit relationships, etc.).) of the majority of proteins. Therefore, it is crucial for understanding the structural complexity of proteins that a drop in temperature results in a strengthening of hydrogen bonds and a weakening of hydrophobic interactions. The importance of water's hydrogen-bonding associations has been confirmed by recent applications of molecular modeling to the solvation effects of solutes. These applications show that solutes primarily affect the hydrogen-bonding associations that occur within the pure solvent, in particular, causing changes that resemble the changes that changes in temperature and pressure induced in the pure solvent and which are reflected in the solvent water's equation of state. The description of aqueous solutions and water given above lays a solid basis for comprehending the various functions that water plays in food chains as well as the impact that water's properties and quantity have on the chemistry and microbiology of food. In the conversation that follows, we'll look at the usefulness of several methods for comprehending the specific function that water plays in defining food stability and qualities.

2.4. WATER ACTIVITY AND RELATIVE VAPOR PRESSURE

Food perishability and its water content have long been known to be correlated, albeit imperfectly—a relationship that dates back to prehistoric times. The main goals of concentration and dehydration processes are to reduce a food's water content while concurrently raising its solute concentration and lowering its perishability. On the other hand, it has also been noted that food varieties with identical water contents vary greatly in terms of perishability. It follows that water content by itself is not a valid predictor of perishability. This condition can be partially attributed to variations in the strength of the associations that water has with nonaqueous constituents. Strongly associated water is expected to be less able to support degradative processes like the growth of microorganisms and hydrolytic chemical reactions than weakly associated water. The term "water activity" (a_w) was developed to reflect the intensity with which water associates with various nonaqueous constituents. Experience shows that food stability, safety, and other properties can be predicted far more reliably from aw than from water content. Even so, a_w is not a totally reliable predictor. The reasons for this will be explained in a later section. Despite this lack of perfection, a_w correlates sufficiently well with rates of microbial growth and rates of many degradative reactions to make it a useful indicator of potential product stability and microbial safety. The fact that aw is specified in some U.S. federal regulations dealing with good manufacturing practices for food attests to its usefulness and credibility, and also dictates that this topic be fully explored.

2.4.1. Definition and Measurement

Lewis and Randall rigorously derived the notion of substance "activity" from the laws of thermodynamics, and Scott pioneered its application to foods. It is sufficient here to state the following Equation:

$$a_w = \left(\frac{f}{f^0}\right)_T \quad (3)$$

Where f is the fugacity of the solvent (fugacity being the escaping tendency of a solvent from solution) and f° is the fugacity of the pure solvent in some defined standard state. The subscript T refers to the measurement being at constant temperature. At low pressures (e.g., ambient) the difference between f / f° and p/p° is less than 1%, so defining aw in terms of p/p° is clearly justifiable. Thus;

$$a_W = (p/p^o)_T \tag{2}$$

It is important to be aware that this equality is based on the assumption of thermodynamic equilibrium. Since with foods this assumption is generally violated, Eqn. (2) must be considered approximate, and the proper expression is:

$$a_W \approx (p/p^o)_T \tag{3}$$

In food science, it is more appropriate to use the term (p/p°)T instead of aw because p/p° is a term that can be measured easily and is not always equivalent to aw. Here, this procedure will be adhered to. (p/p°)T is also known as relative vapor pressure (RVP), and these terms will be used interchangeably. Although it is more scientifically sound to use RVP instead of aw because RVP does not imply equilibrium, readers should be aware that aw is a widely used term.

There are two main reasons why the aw–RVP approach is not a perfect estimator of food stability: solute-specific effects and assumptions that underlie Eqn. (2) are violated. Fortunately, most of the time, breaking one of Eqn. (2)'s presumptions won't significantly reduce RVP's value as a technological tool.

There is an exception when dry products are made through water absorption as opposed to desorption (hysteresis effect). We'll talk about this later. However, in cases where the theoretical models rely on the validity of these assumptions, violating Eqn. (2)'s assumptions renders RVP useless as a tool for mechanistic interpretation. Solute-specific effects can render RVP an unreliable indicator of food stability and safety in a few potentially crucial cases. Even if all of the presumptions underlying Eqn. (2) are true, this can still happen. Foods with the same RVP but different solute compositions will display different stabilities and other properties in these kinds of circumstances. Anyone using RVP as a tool to assess the safety and stability of food must always keep this very important point in mind.

Relative vapor pressure is related to percent equilibrium relative humidity (%ERH) of the product environment as follows:

$$\text{RVP} = (p/p^o)_T = \%\text{ERH}/100 \tag{4}$$

There are two significant aspects of this relationship. First, while percent ERH is a property of the atmosphere established in steady state with the sample, RVP is an intrinsic property of the sample. Keep in mind that the presence of equilibrium is not always guaranteed by the existence of a steady state. Secondly, the equality delineated in Eqn. (4) is contingent upon the establishment of equilibrium between the product and its surroundings. Even with very small samples (<1g), establishing equilibrium takes a long time and is nearly impossible for large samples, particularly at temperatures below 20°C. Since ERH implies equilibrium, it might therefore be more appropriate to use a modified form of this relationship, which more correctly states that the measured RVP is related to the steady-state relative humidity (SSRH) of the product environment:

$$\text{RVP} = (p/p^o)_T \approx \%\text{SSRH}/100 \tag{5}$$

A small sample's relative vacuum pressure (RVP) can be ascertained by subjecting it to a closed chamber for a duration long enough to produce apparent equilibrium, or constant weight, and then monitoring the chamber's relative humidity or pressure. Different kinds of instruments are available to measure relative humidity (electric hygrometers, dew point instruments) and pressure (manometers). RVP can also be calculated using knowledge of the freezing point depression, but this only directly relates to the temperature at the freezing point. Based on numerous joint investigations, the accuracy of aw calculations is approximately ±0.005. A small sample can be adjusted to a particular RVP by putting it in a closed chamber kept at a consistent temperature, keeping the sample atmosphere at a known and constant relative humidity using a suitable saturated salt solution or another comparable technique, and storing it until the sample weight remains constant.

2.4.2. Temperature Dependence

Relative vapor pressure is temperature dependent, and the Clausius–Clapeyron equation in modified form provides a means for estimating this temperature dependence. This equation, though based on a_w, can often be applied to RVP and takes the form:

$$\frac{d \ln a_W}{d(1/T)} = \frac{-\Delta H}{R} \tag{6}$$

Where T is absolute temperature, R is the gas constant, and ΔH is the isosteric heat of sorption at the water content of the sample. By rearrangement, this equation can be made to conform to the generalized equation for a straight line. It is then evident that a plot of ln aw vs. 1/T (at constant water content) should be linear, and if the same holds true for ln(p/p°)T vs. 1/T, then the equation can be used to estimate effective sorption heats for the purposes of comparison. These relationships assume that equilibrium has been achieved, which is often not the case.

Linear plots of $\ln(p/p°)_T$ vs. 1/T for native starch at various moisture contents are shown in Figure 2.13. It is apparent that the degree of temperature dependence is a function of moisture content. At a starting (p/p°)T of 0.5, the temperature coefficient is 0.0034 K^{-1} over the temperature range 275–313 K (2–40°C). On the basis of the work of several investigators, temperature coefficients for (p/p°)T (temperature range 5–50°C at a starting (p/p°)T of 0.5) range from 0.003 to 0.02 K^{-1} for high carbohydrate or high protein foods. Thus, depending on the product, a 10 K change in temperature can cause a 0.03–0.2 change in $(p/p°)_T$. This behavior can be important for 'packaged food because it will undergo a change in RVP with a change in temperature, causing the temperature dependence of its stability to be greater than that of the same product unpackaged. The $(p/p°)_T$ vs. Not all 1/T relationships are linear across large temperature ranges. At the beginning of ice formation, for instance, they usually show sharp breaks. Consider the definition of RVP as it relates to subfreezing temperatures before interpreting data at such temperatures. As was already mentioned, at temperatures below 0°C, undercooled water can exist as metastable. The issue that thus emerges is whether the vapor pressure of ice or undercooled water should be used to represent the denominator term p°, the vapor pressure of the pure solvent.

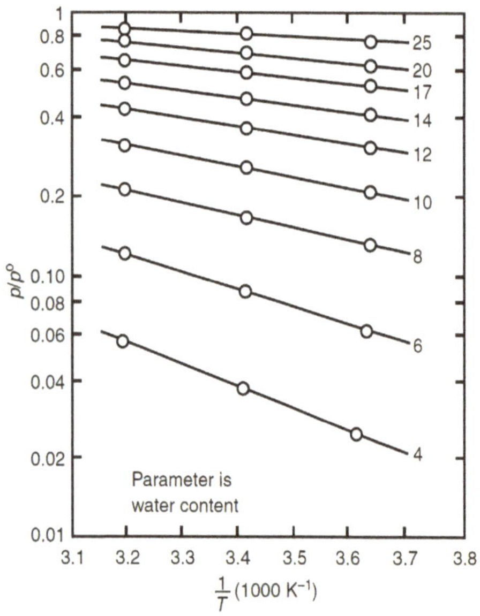

Figure 2.13. *Relationship between RVP and temperature for native starch of different water contents. Water content values displayed after each line are expressed as g HOH/g dry starch.*

Source: https://edisciplinas.usp.br/pluginfile.php/7989549/mod_resource/content/1/%C3%81gua%20nos%20Alimentos.pdf.

A more appropriate choice for the standard state is the vapor pressure of undercooled water, even though pure ice is the equilibrium form of pure water at these temperatures. This is because (1) the RVP values at subfreezing temperatures can then, and only then, be directly compared with those at above-freezing temperatures, and (2) choosing the vapor pressure of ice as the defining p° would result in a situation where the RVP would be unity at all subfreezing temperatures for samples containing ice. Because the partial pressure of water in any frozen food is equal to the vapor pressure of ice at the same temperature, as can be seen from standard thermodynamic relationships, this second point is the outcome.

Because the vapor pressure of undercooled water has been measured down to −15°C, and estimated by extrapolation to −30°C and the vapor pressure of ice has been measured to much lower temperatures, it is simple to calculate RVP values for frozen foods based on the undercooled water standard state:

$$a_w = \left[\frac{p_{ff}}{p^o(\mathrm{UCW})}\right]_T = \left[\frac{p_{ice}}{p^o(\mathrm{UCW})}\right]_T \qquad (7)$$

where p_{ff} is the partial pressure of water in partially frozen food, p°(UCW) is the vapor pressure of pure undercooled water, and p_{ice} is the vapor pressure of pure ice, all at the same temperature T. RVP values calculated from the vapor pressures of ice and undercooled water. These values are identical to those of frozen foods at these same temperatures. A plot of $\log(p/p^o)_T$ vs. 1/T for a typical aqueous solution, illustrating that (1) the relationship is linear at subfreezing temperatures, (2) the influence of temperature on RVP is typically far greater at subfreezing temperatures than at above-freezing temperatures, and (3) a sharp break occurs in the plot at the freezing point of the sample. Similar behavior is to be expected of biological systems.

There are two key differences to be aware of when comparing RVP values at temperatures above and below freezing. Firstly, at temperatures above freezing, RVP depends on both temperature and sample composition, with the former being more important. Since RVP values are unaffected by the type or ratio of solutes present in the presence of an ice phase, RVP values at subfreezing temperatures become independent of sample composition and depend only on temperature. Thus, any subfreezing event that depends on the type of solute present (e.g., diffusion-controlled processes, catalyzed reactions, and reactions impacted by chemicals that change pH and oxidation-reduction potential, cryoprotective agents, or antimicrobial agents) cannot be reliably predicted using the RVP value. Consequently, RVP values at subfreezing temperatures are far less useful than RVP values at above-freezing temperatures as markers of physical and chemical events. As a result, RVP at a temperature above freezing cannot be predicted using information about RVP at a temperature below freezing. Also take note of the fact that estimating aw or RVP through freezing point depression essentially involves figuring out the curve's break point. Second, the definition of RVP in terms of food stability also shifts when the temperature is sufficiently altered to form or melt ice. For instance, microorganisms will not grow and chemical reactions will proceed slowly in a product at –15°C ($(p/p^o)_T = 0.86$). On the other hand, some chemical reactions will happen quickly and some microorganisms will grow at moderate rates at +20°C and $(p/p^o)_T = 0.86$. The molecular mobility approach to understanding food stability relationships was developed in part due to the lack of usefulness of RVP as an indicator of product stability at subfreezing temperatures. Since the ideas developed in the molecular mobility approach can help clarify some of the relationships between moisture sorption isotherms and product stability, it is better to wait to discuss the relationship between moisture content and sample RVP and how this relates to food stability.

CASE STUDY

The word "water" Originated from the old English word "wæter." The formula for water is H_2O. Water is a pure, colorless, odorless and transparent chemical substance which surrounds 71% of the earth's surface like the oceans, rivers, sea, streams and lakes. Also, water is important for life to exist.

- **Properties of Water**

Attraction of Cohesion and Adhesion: takes place to attract the polar molecules. Cohesion tends to attract bond hydrogen to other water molecules together as water's polarity while in adhesion, it attract between molecules of different substances which it can also form hydrogen bond with.

High-Speed IC Heat of Water: Temperature of two properties can be moderated by water. The quantity of energy consumed nor lost by a gram of substance by 1°C can change it's temperature. The more the water molecules bond together the more friction and heat leads to the increase of temperature.

Water Increase the Heat of Evaporation: In order to moderate temperature, water's high heat of evaporation is needed therefore the quantity of energy helps transform liquid to a gaseous form.

Ice's lower Density: Ice crystals are formed from the hydrogen bond of water molecules when the temperature is cold which helps maintain and stabiles the ice crystal. The solid form of water is ICE.

- **Water Resources**

Water is also a natural resources which is essential for use in households, agriculture, recreation, industries and environmental activities. All living organisms needs water to exist. According to science, Salt water on earth is 97% while freshwater contains only 3% also water covers 71% of the earth surface. The water bodies include rivers, oceans, sea lakes and streams. Another source of water is rainwater and underground water.

- **Water Recycle**
  - Agriculture water is highly required for plant growth and production through rainfall, flood or irrigation, the artificial supply of water to the soil. Water also serves as means of survival to animals.
  - Transportation Water on a large surface like rivers, oceans, sea etc., are often used as means of transportation of people, goods and services from one place to another
  - Hydroelectric power water serves as source of power supply to supply generate electricity and heat. For example, the kanji dam.

- Domestic Use Generally water serves a great purpose in our daily activities at home like bathing, drinking, washing, food preparation etc. Water is one of the important necessities of life.
- Industrial Use Water is required for chemical processing and manufacturing of fertilizer and other chemicals substances, also serves as means of production of food and beverage.
- Mineral Deposits Water bodies consists of several mineral resources such as crude oil, salt, titanium, limestone, placer gold, diamond, tin, gypsum etc., which could also serve as an economy resources.
- Recreation and tourism Activities such as golf, swimming, fishing, picnicking etc., can be carried out on the water bodies provided with excellent facilities.
- Employment Opportunities are created in the tourism centers for individual to work. Also, in terms of water transportation like the sailors, marine, the shipbuilders and the fishermen.
- Source of food Most creatures under the water bodies serves as food to man such as seaweed, fishes, crab, oyster, octopus, shrimp and lot more, they are mainly found in salt water.

CLASS ACTIVITY

EXPERIMENTS WITH WATER AND ICE

Experience Science in a playful way through water play. Changing textures provides a great foundation for detailed observations and conversations.

- **Materials**

For this activity you will need:

  - Various plastic containers, pipes and water
  - During wintertime, containers can be placed outdoors and at other times in the freezer

- **Learning Goals**

With this activity, you will learn the following skills:

  - Experience Science in a playful way through water play. Changing textures provides a great foundation for detailed observations and conversations.
  - **Curiosity:** Encourage children to question and wonder about the world and its changes
  - **Fine Motor:** handling slippery ice
  - **Science:** understanding that water can exist as a solid (ice) and also as a liquid (water) or melt
  - Learn about Igloo's and differing lifestyles
  - Process Description

Step 1: Explain how the temperature can affect water and change it's shape or even how it feels

Step 2: Children can experiment with water in different sized containers and pipes

Step 3: During cold days, the water that remains in the different containers will freeze and children can observe the shapes of the ice and discuss the melting process

Step 4: Food coloring can be added before the freezing process. Discuss what happens to the water when food coloring is added. Allow children to experiment and play with the water and ice. What differences do they see?

Step 5: Depending on the age and stage of the child, watch short videos on how igloos are built and how strong ice can be. This can feed into how many Eskimos used to love in igloos and how different people around the world live.

SUMMARY

- Most of the fresh foods contain large amounts of water. It is one of the major components in the composition of many foods. Each food has its characteristic amount of this component. The effect of water on structure, appearance and taste of foods as well as their susceptibility to spoilage depends on its amount, location, and orientation.
- Water has unusually high boiling point, surface tension, permittivity, heat capacity, and heat of phase transition values. Other unusual attributes of water include expansion upon solidification, high thermal conductivity compared to that of other liquids, and moderately large thermal conductivity of ice compared to that of other nonmetallic solids.
- This physically entrapped water does not flow from food even when it is cut or minced. But this water behaves almost like pure water during food processing operations like drying, freezing, etc., it is also available as a solvent.
- Interactions between water and nonionic, hydrophilic solutes are weaker than that of the interactions between water and ions and of the almost same strength as that of the hydrogen bonds between water molecules.
- The molecular formula of water is H_2O. You know that oxygen has six valence electrons and hydrogen has one valence electron. According to Lewis's formulation, to complete the octet, oxygen shares one electron each with the two hydrogen atoms.
- The liquid water has a hexagonal lattice structure. When the temperature of liquid water at room temperature is lowered, its density increases like that of any other substance and reaches a maximum value of 1.00 at around 4°C.
- The water content or the moisture content of a food influences its appearance, texture and flavor. It varies a great deal in different food items. The green leafy vegetables contain more than 90% of water while it may be negligible in oils and fats like ghee, butter, oil etc.
- The water present in foods interacts with major hydrophilic nutrients such as carbohydrates, proteins, water-soluble vitamins, and minerals etc., by means of polar hydrogen-bonding interactions.
- Individual ions and also the ionic groups of organic molecules appear to hinder or influence the mobility of water molecules to a greater degree than any other types of solutes. The strength of electrostatic water–ion bonds is greater than that of water–water hydrogen bonds, but still much less than that of covalent bonds.
- A clathrate hydrate is an ice-like inclusion compound wherein water, the host substance, forms a hydrogen-bonded cage-like structure that physically entraps a small apolar molecule known as the guest molecule.

REVIEW QUESTIONS

1. Describe the arrangement of water molecules in the liquid state.
2. What is the chemical equation for the dissociation of water molecules?
3. How do water molecules arrange themselves in a pure water system at room temperature?
4. What is the significance of water's high heat capacity and surface tension?
5. Explain the phase diagram of water and the significance of the triple point.
6. Discuss the process of hydration and its significance in solution chemistry.
7. Explain how water's structure changes in the presence of nonpolar substances.

REFERENCES

1. Ball, P. (2001). *Life's Matrix, a Biography of Water*. University of California Press: Berkeley, CA.
2. Berendsen, H. J. C. (1971). The molecular dynamics of water in biological systems. *Proceedings of the First European Biophysics Congress, 1*, 483–488.
3. Bertoluzza, A., Fagnano, C., Morelli, M. A., Tinti, A., & Tosi, M. R. (1993). The role of water in biological systems. *Journal of Molecular Structure, 297*, 425–437.
4. Brady, G. W., & Romanov, W. J. (1960). Structure of water. *Journal of Chemical Physics, 32*, 106.
5. Chaplin, M. (2001). Water, its importance to life. *Biochemical and Molecular Education, 29*, 54–59.
6. Cheftel, J. C., Levy, J., & Dumay, E. (2000). Pressure-assisted freezing and thawing: Principles and potential applications. *Food Reviews International, 16*, 453–483.
7. Cheftel, J. C., Thiebaud, M., & Dumay, E. (2002). High pressure–Low temperature processing of foods: A review. In *Advances in High Pressure Bioscience and Biotechnology II*, Proceedings of the International Conference on High Pressure Bioscience and Biotechnology, Dortmund, Germany, September 16–19 (pp. 327–340).
8. Davidson, D. W. (1973). Clathrate hydrates. In F. Franks (Ed.), *Water, A Comprehensive Treatise* (pp. 115–234). Plenum Press: New York.
9. Eagland, D. (1975). Nucleic acids, peptides and proteins. In F. Franks (Ed.), *Water, a Comprehensive Treatise* (pp. 305–518). Plenum: New York.
10. Errington, J. R., & Debenedetti, P. G. (2001). Relationship between structural order and the anomalies of liquid water. *Nature, 409*, 318–321.
11. Fennema, O. (1973). Water and ice. In O. Fennema, W. D. Powrie, & E. H. Marth (Eds.), *Low Temperature Preservation of Foods and Living Matter* (pp. 1–77). Marcel Dekker: New York.
12. Franks, F. (1975). The hydrophobic interaction. In F. Franks (Ed.), *Water, a Comprehensive Treatise* (pp. 1–94). Plenum Press: New York.

13. Franks, F. (1988). Protein hydration. In F. Franks (Ed.), *Characteristics of Proteins* (pp. 127–154). Humana Press: Clifton, NJ.
14. Franks, F. (2000). *Water, A Matrix of Life*. Royal Society of Chemistry: London.
15. Geiger, A. (1981). Molecular dynamics simulation study of the negative hydration effect in aqueous electrolyte solutions. *Berichte der Bunsen-Gesellschaft, 85*, 52–63.
16. Gerstein, M., & Lynden-Bell, R. M. (1993). Simulation of water around a model protein helix. 1. Two-dimensional projections of solvent structure. *Journal of Physical Chemistry, 97*, 2982–2990.
17. Guillot, B. (1991). A molecular-dynamics study of the far infrared spectrum of liquid water. *Journal of Chemical Physics, 95*, 1543–1551.
18. Guillot, B. (2002). A reappraisal of what we have learned during 3 decades of computer simulations on water. *Journal of Molecular Liquids, 101*, 219–260.
19. Hobbs, P. V. (1974). *Ice Physics*. Clarendon Press: Oxford.
20. Koga, K., Tanaka, H., & Nakanishi, K. (1994). On the stability of clathrate hydrates encaging polar guest molecules: contrast in the hydrogen bonds of methylamine and methanol hydrates. *Molecular Simulation, 12*, 241–252.
21. Koga, K., Tanaka, H., & Nakanishi, K. (1994). Stability of polar guest-encaging clathrate hydrates. *Journal of Chemical Physics, 101*, 3127–3134.
22. Kuntz, I. D., & Kaufmann, W. (1974). Hydration of proteins and polypeptides. *Advances in Protein Chemistry, 28*, 239–345.
23. Lewin, S. (1974). *Displacement of Water and its Control of Biochemical Reactions*. Academic Press: London.
24. Luck, W. A. P. (1981). Structures of water in aqueous systems. In L. B. Rockland & G. F. Stewart (Eds.), *Water Activity: Influences on Food Quality* (pp. 407–434). Academic Press: New York.
25. Luyet, B. (1960). On various phase transitions occurring in aqueous solutions at low temperature. *Annals of the New York Academy of Sciences, 85*, 549–569.
26. Luyet, B. J. (1966). Anatomy of the freezing process in physical systems. In H. T. Meryman (Ed.), *Cryobiology* (pp. 115–138). Academic Press: New York.
27. Morgan, J., & Warren, B. E. (1938). X-ray analysis of the structure of water. *Journal of Chemical Physics, 6*, 666–673.
28. Rapatz, G., & Luyet, B. (1972). Patterns of ice formation and rates of ice growth in gelatin solutions. *Biodynamica, 11*, 117–123.
29. Rey, L. R. (1958). *Etude Physiologique et Physico-Chimique de l'Action des Basses Temperatures sur les Tissues Animaux*. Universite de Paris: Paris, France.
30. Rontgen, W. C. (1882). VIII Uber die constitution des flussigen wassers. *Annalen der Physik und Chemie, 281*, 91–97.
31. Sceats, M. G., & Rice, S. A. (1982). Amorphous solid water and its relationship to liquid water: a random network model for water. In F. Franks (Ed.), *Water, A Comprehensive Treatise* (pp. 83–214). Plenum Press: New York.

32. Schmidt, S. J. (2004). Water and solids mobility in foods. *Advances in Food and Nutrition Research, 48*, 1–101.

33. Simatos, D., & Rey, L. (1965). Freeze-drying of biologic materials. *Federation Proceedings, 24*, 213–215.

34. Stillinger, F. H. (1980). Water revisited. *Science, 209*, 451–457.

35. Tait, M. J., Suggett, A., Franks, F., Ablett, S., & Quickenden, P. A. (1972). Hydration of monosaccharides. Study by dielectric and nuclear magnetic relaxation. *Journal of Solution Chemistry, 1*, 131–151.

CHAPTER 3

Carbohydrates and Lipids

LEARNING OBJECTIVES

After studying this chapter, you will be able to:

- Describe function, structure, metabolism of carbohydrates
- Understand categories, biological functions and metabolism of lipids

KEY TERMS FROM THIS CHAPTER

Cholesterol absorption	Dietary fiber	Eukaryotic cells
Galactose	Glucose	Glycemic index (GI)
Glycemic load (GL)	Lipids	Metabolic pathways
Monosaccharides	Organic compounds	Polyketides
Polysaccharides	Postprandial glycemia	
Processed foods	Saccharides	

INTRODUCTION

A carbohydrate is a biomolecule made up of carbon (C), hydrogen (H), and oxygen (O) atoms. It typically has a hydrogen-oxygen atom ratio of 2:1 (like in water), giving it the empirical formula $C_m(H_2O)_n$, where m and n can differ. However, this does not imply that the H has covalent bonds with O; in the case of CH_2O, for instance, the H has a covalent bond with C but not with O. However, not all substances that are classified as carbohydrates automatically do so (such as formaldehyde and acetic acid), nor do all carbohydrates automatically meet this exact stoichiometric definition (such as fucose, uronic acids, and deoxy-sugars).

The word is most frequently used in biochemistry, where it is a synonym for saccharide, a class of molecules that comprises cellulose, starch, and sugars. The four chemical groups that make up saccharides are oligosaccharides, polysaccharides, disaccharides, and monosaccharides. The smallest (lower molecular weight) carbohydrates are called sugars; these are called monosaccharides and disaccharides. The names of the monosaccharides and disaccharides frequently end in the suffix -ose, which was initially derived from the word glucose and is used for practically all sugars, such as fructose (fruit sugar), sucrose (cane or beet sugar), ribose, lactose (milk sugar), and so on, despite the complexity of the scientific nomenclature of carbohydrates.

There are many functions that carbohydrates fulfill in living things. Polysaccharides function as both structural elements (such as chitin in arthropods and cellulose in plants) and energy stores (such as starch and glycogen). The fundamental building block of the genetic molecule known as RNA and an essential part of coenzymes such as ATP, FAD, and NAD is the 5-carbon monosaccharide ribose. A constituent of DNA is the corresponding deoxyribose. Numerous other significant biomolecules that are essential for the immune system, fertilization, blood clotting, development, and prevention of pathogenesis are found in saccharides and their derivatives. (Carroll, Wang, Turro, & Koberstein, 2008). Essential to diet, carbohydrates can be found in many different types of foods, both natural and processed. Polysaccharide starch can be found in large quantities in potatoes, wheat, corn, and rice cereals as well as processed foods like pasta, pizza, and bread that are made with cereal flour. The main forms of sugars found in the human diet are lactose, which is found in large amounts in milk, table sugar (sucrose, which is extracted from sugarcane or sugar beets), and glucose and fructose, which are found naturally in honey, many fruits, and some vegetables. Drinks and a lot of prepared foods like cakes, biscuits, and jam frequently have table sugar, milk, or honey added to them.

One of the primary constituents of insoluble dietary fiber is cellulose, a polysaccharide present in the cell walls of all plants. Cellulose and insoluble dietary fiber, despite not being digestible by humans, generally aid in the maintenance of a healthy digestive system by promoting regular bowel movements. Additional polysaccharides found in dietary fiber include inulin and resistant starch, which nourish certain bacteria in the large intestine's microbiota and allow the bacteria to break down the fiber into short-chain fatty acids.

A wide range of organic substances are referred to as lipids, and they include fats, waxes, sterols, fat-soluble vitamins (like A, D, E, and K), monoglycerides, diglycerides, phospholipids, and more. Lipids have several uses, such as energy storage, signaling, and serving as structural elements of cell membranes. Lipids are used in nanotechnology, the food and cosmetics industries.

Lipids are classified as small molecules that are either hydrophobic or amphiphilic. Their amphiphilic nature enables them to form structures in an aqueous environment, such as membranes, vesicles, and multilamellar/unilamellar liposomes. Ketoacyl and isoprene groups are two different kinds of biochemical subunits or building blocks that can contribute entirely or partially to the formation of biological lipids. This method allows for the classification of lipids into eight groups: polyketides (produced from the condensation of ketoacyl subunits); sterol lipids and prenol lipids (produced from the condensation of isoprene subunits); fatty acyls, glycerolipids, glycerophospholipids, sphingolipids, saccharolipids, and polyketides.

Figure 3.1. *Lipids and carbohydrates.*

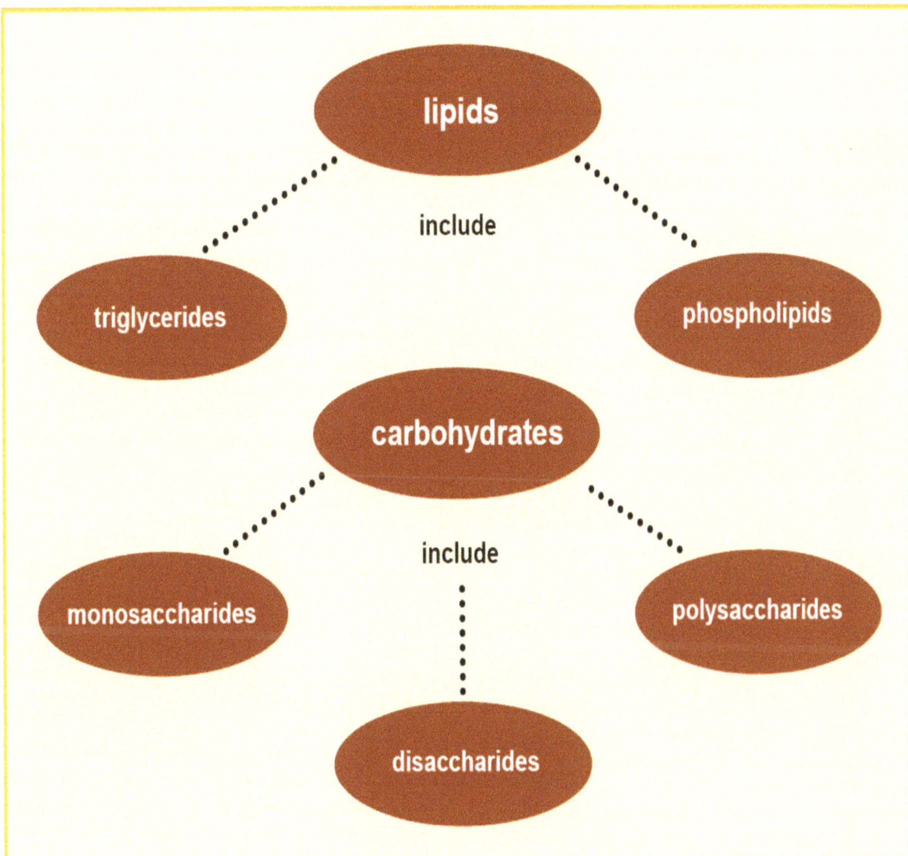

Source: https://www.rcsdk12.org/cms/lib04/NY01001156/Centricity/Domain/5368/Lipids%20and%20Carbohydrates.pdf.

3.1. CARBOHYDRATES

The primary energy source that the human body uses is carbohydrates. The brain uses glucose mostly. Moreover, red blood cells only use glucose. The absence of the enzyme cellulase prevents the human body from breaking down dietary fiber. The body uses glucose as its primary energy source. Glycogen is a form of glucose storage found in the liver and skeletal muscles. When the body uses up more glucose than is consumed, it gets turned into fat. The ribosome is involved in the synthesis of deoxyribonucleic acid. Polyhydroxy alcohols, or carbohydrates, have an aldehyde or keto group that has the potential to be active. Based on the number of carbon atoms they contain, carbohydrates can be divided in monosaccharides, disaccharides, oligosaccharides, and polysaccharides. Monosaccharides are not hydrolyzable into more basic forms. Upon hydrolysis, disaccharides yield two monosaccharides. There are two types of polysaccharides: heteropolysaccharides and homopolysaccharides.

3.1.1. Structure

Chemistry used to refer to any compound with the formula $C_m(H_2O)_n$ as a carbohydrate. According to this definition, formaldehyde (CH_2O) was deemed the simplest carbohydrate by some chemists, while glycolaldehyde was declared the winner. Currently, the term is most commonly understood in the context of biochemistry, which includes many biological carbohydrates that deviate from this formula and excludes compounds with only one or two carbons. For instance, although the aforementioned representative formulas appear to represent the widely recognized carbohydrates, ubiquitous and abundant carbohydrates frequently depart from this. For instance, chemical groups like N-acetyl (found in chitin), sulfate (found in glycosaminoglycans), carboxylic acid, and deoxy modifications (found in fucose and sialic acid) are frequently seen in carbohydrates.

Monosaccharides, or simple carbohydrates with the general formula $(CH_2O)_n$, where n is three or more, are the building blocks of natural saccharides. A typical monosaccharide is an aldehyde or ketone with numerous hydroxyl groups added, usually one on each carbon atom that is not a part of the aldehyde or ketone functional group. Its structure is $H–(CHOH)_x(C=O)–(CHOH)_y–H$. The monosaccharides glucose, fructose, and glyceraldehydes are examples. Some biological substances that are commonly referred to as monosaccharides, such as uronic acids and deoxy-sugars like glucose, do not, however, conform to this formula, and many chemicals that do, such as formaldehyde CH_2O and inositol $(CH_2O)_6$, do not consider as monosaccharides.

Monosaccharides in their open-chain form frequently coexist with their closed-ring form, which is created when a hemiacetal with a new C–O–C bridge is formed by the reaction of the aldehyde/ketone carbonyl group carbon (C=O) and hydroxyl group –OH.

There are numerous ways in which monosaccharides can be joined to form polysaccharides, also known as oligosaccharides. Numerous carbohydrates have one or more modified monosaccharide units, meaning that one or more groups have been added, subtracted, or replaced. For instance, chitin is made up of repeating units of N-acetyl glucosamine, a nitrogen-containing form of glucose; deoxyribose, a component of DNA, is a modified form of ribose.

3.1.2. Monosaccharide

Simple sugars, or monosaccharides, are the most basic sugar molecules and the building blocks (monomers) upon which all other carbohydrates are constructed. This is the structural unit of carbohydrates, to put it simply. Typically, they are crystalline organic solids that are colorless and soluble in water. Only a small percentage of monosaccharides taste sweet, despite their name as sugars. The formula for most monosaccharides is $(CH_2O)^x$, however not all molecules with this formula are monosaccharides. Galactose, fructose (levulose), and glucose (dextrose) are a few types of monosaccharides. The building blocks of polysaccharides (like cellulose and starch) and disaccharides (like sucrose, lactose, and maltose) are called monosaccharides. In everyday speech, table sugar is called sucrose, a disaccharide made up of one molecule of each of the two monosaccharides, d-glucose and d-fructose.

With the exception of those at the end of the chain, every carbon atom that hosts a hydroxyl group is chiral. As a result, several isomeric forms with the same chemical formula are produced. For example, although both galactose and glucose are aldohexoses, their molecular makeup and physical forms differ.

In order to provide chemical energy to living things through the processes of glycolysis and the citric acid cycle, the monosaccharide glucose is essential to metabolism. The dehydrated condensate of two glucose molecules is maltose.

3.1.2.1. Structure and Nomenclature

Monosaccharides, with a few notable exceptions (such as deoxyribose), have the chemical formula $(CH_2O)_x$, where typically $x \geq 3$. The number x of carbon atoms in a monosaccharide determines its classification: triose (3), tetrose (4), pentose (5), hexose (6), heptose (7), and so forth.

Hexose, or glucose, is utilized as a fuel and in the synthesis of starch, glycogen, and cellulose. Pentose sugars are ribose and deoxyribose, which are found in RNA and DNA, respectively. Mannoheptulose, sedoheptulose, and the ketoses are examples of heptoses. Due to their high instability, monosaccharides containing eight or more carbons are not frequently found. If a monosaccharide has more than four carbons, it exists as a ring in aqueous solutions.

3.1.2.1.1. Linear-Chain Monosaccharides

The carbon skeleton of simple monosaccharides is linear and unbranched, with one hydroxyl (OH) group on each of the remaining carbon atoms and one carbonyl (C=O) functional group on each carbon atom. As a result, the elemental formula of a simple monosaccharide is $C_xH_{2x}O_x$, and its molecular structure can be expressed as $H(CHOH)_n(C=O)(CHOH)_mH$, where $n + 1 + m = x$.

Conventionally, the carbon atoms are numbered along the backbone from 1 to x, beginning at the end nearest the C=O group. The most basic type of sugar and the simplest unit of carbohydrates are called monosaccharides.

If the carbonyl is at position 1 (that is, n or m is zero), the molecule begins with a formyl group H(C=O)– and is technically an aldehyde. In that case, the compound is termed an aldose. Otherwise, the molecule has a ketone group, a carbonyl –(C=O)– between two carbons; then it is formally a ketone, and is termed a ketose. Ketoses of biological interest usually have the carbonyl at position 2.

A more general nomenclature for open-chain monosaccharides combines a Greek prefix to indicate the number of carbons (tri-, tetr-, pent-, hex-, etc.) with the suffixes "-ose" for aldoses and "-ulose" for ketoses. In the latter case, if the carbonyl is not at position 2, its position is then indicated by a numeric infix. So, for example, $H(C{=}O)(CHOH)_4H$ is pentose, $H(CHOH)(C{=}O)(CHOH)_3H$ is pentulose, and $H(CHOH)_2(C{=}O)(CHOH)_2H$ is pent-3-ulose.

3.1.2.1.2. Open-Chain Stereoisomers

It is possible for two monosaccharides to be different stereoisomers even though their molecules have different spatial orientations and equivalent molecular graphs (same chain length and carbonyl position). This only occurs in the presence of a stereogenic center in the molecule, namely a chiral carbon atom linked to four different molecular sub-structures. The handedness of those four bonds allows for one of two possible spatial configurations. Every carbon in a basic open-chain monosaccharide is chiral, with the exception of the first and last atoms of the chain and, in the case of ketoses, the carbon containing the keto group.

For instance, there is only one stereoisomer of the triketose H(CHOH)(C=O)(CHOH)H (glycerone, dihydroxyacetone) because it lacks a stereogenic center. The central chiral carbon of the other triose, the aldose $H(C{=}O)(CHOH)_2H$ (glyceraldehyde), is bonded to groups –H, –OH, $-C(OH)H_2$, and –(C=O)H. As a result, it exists as two stereoisomers, or left and right gloves, whose molecules are mirror images of one another. Monosaccharides that have four or more carbons usually have more than two stereoisomers because they can have multiple chiral carbons. Two times the total number of chiral carbons (c), we get a bound on the number of distinct stereoisomers with the same diagram.

An acyclic monosaccharide's skeletal formula can be systematically drawn using the Fischer projection, which precisely specifies the handedness of each chiral carbon. The positions (left or right) of the chiral hydroxyls (the hydroxyls attached to the chiral carbons) in the Fischer diagram can be used to identify each stereoisomer of a simple open-chain monosaccharide.

In contrast to their mirror images, the majority of stereoisomers are chiral in themselves. Two mirror-image isomers are different in the Fischer projection because all of their chiral hydroxyl positions are flipped from right to left. In non-chiral environments, mirror-image isomers are chemically identical, but they typically exhibit very different biochemical characteristics and natural occurrences.

Even though the majority of stereoisomers can be arranged in pairs of mirror-image forms, some non-chiral stereoisomers have chiral centers but are exactly like their mirror images. This occurs when the two halves of a symmetric molecular graph are mirror images of one another, as in the case of the 3-ketopentoses

$H(CHOH)_2(CO)(CHOH)_2H$. Then mirroring is the same as rotating half a turn. For this reason, despite the molecule having two chiral carbons, there are only three unique 3-ketopentose stereoisomers.

Even in non-chiral environments, distinct stereoisomers that are not mirror-images of one another typically have different chemical properties. Therefore, a unique monosaccharide name may be assigned to each mirror pair and each non-chiral stereoisomer.

For instance, the term glucose refers to a particular pair of mirror-image aldohexoses even though there are 16 different aldohexose stereoisomers. One of the two glucose isomers has the hydroxyl at left on C3, and at right on C4 and C5, in the Fischer projection, whereas the other isomer has the opposite pattern. Conventional three-letter abbreviations for these particular monosaccharide names, such as "Glu" for glucose and "Thr" for threose, are used.

A monosaccharide having n asymmetric carbons typically has 2^n stereoisomers. Compared to a ketose monosaccharide of the same length, an aldose monosaccharide has one more open chain stereoisomers. Each ketose has $2^{(n-3)}$ stereoisomers, where n is greater than two.

Each aldose has $2^{(n-2)}$ stereoisomers, where n is greater than or equal to two carbons. When –OH and –H groups are arranged differently at asymmetric or chiral carbon atoms, these are also called epimers (this does not apply to those carbons having the carbonyl functional group).

3.1.2.1.3. Configuration of monosaccharides

The two stereoisomers of glyceraldehyde, even in solution, will gradually rotate the polarization direction of linearly polarized light as it passes through them, just like many other chiral molecules do. The prefixes d- and l- designate the two stereoisomers based on the direction of rotation: d-glyceraldehyde rotates the polarization axis in a clockwise manner, whereas l-glyceraldehyde rotates it in a counterclockwise manner.

Other monosaccharides can also be distinguished between two specific stereoisomers that are mirror images of one another using the d- and l-prefixes. The chiral carbon farthest from the C=O group is taken into consideration for this purpose.

All four of its bonds need to be attached to –H, –OH, –C(OH)H, and the remaining molecules. The isomer obtains the d-prefix if the molecule can be rotated in space such that the directions of those four groups coincide with those of the analog groups in the C2 of d-glyceraldehyde. It is given the l-prefix otherwise.

The configuration at the carbon atom second from bottom in the Fischer projection is indicated by the prefixes d- and l-, which stand for the right and left sides, respectively, of the hydroxyl.

It should be noted that the arrangement at all chiral centers combined produces the combined effect of the d- and l-prefixes, which do not indicate the direction of rotation of polarized light. Nonetheless, the light will always be rotated by the two enantiomers by the same amount and in opposite directions.

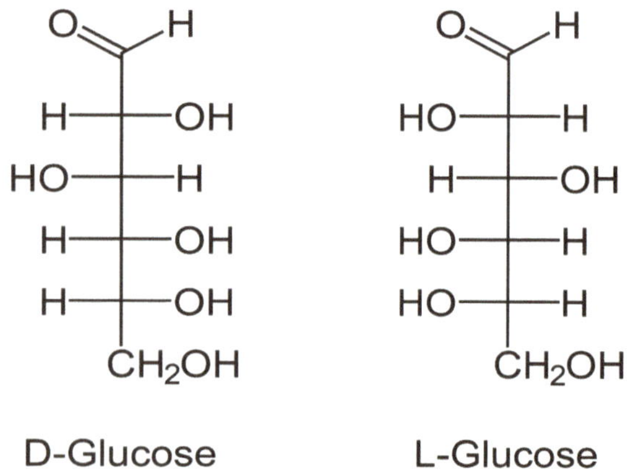

Figure 3.2. *d- and l-glucose.*

Source: https://upload.wikimedia.org/wikipedia/commons/thumb/0/06/DL-Glucose.svg/1024px-DL-Glucose.svg.png.

3.1.2.2. Use in Living Organisms

The primary fuel for metabolism is monosaccharides, which are also utilized in biosynthesis and as an energy source (glucose being the most significant in nature as it is a byproduct of photosynthesis in plants). Monosaccharides are frequently transformed into more space-efficient (i.e., less water-soluble) forms, frequently polysaccharides, when they are not immediately needed. Glycogen is this storage form found in many animals, including humans, particularly in the liver and muscle cells. Starch serves the same purpose in plants. Cellulose, the most prevalent carbohydrate, serves as a structural element in the cell walls of numerous types of algae and plants. RNA is made up of ribose. Deoxyribose is a DNA constituent. One part of lyxoflavin that is present in the human heart is lyxose. Within the pentose phosphate pathway, ribulose and xylulose are produced. Galactose, a constituent of lactose, is present in glycoproteins in numerous tissues and galactolipids in plant cell membranes. Human metabolism involves mannose, particularly when it comes to the glycosylation of specific proteins. Fruit sugar, or fructose, is present in semen and is metabolized in the liver before being directly absorbed into the intestines during digestion. It is present in a wide variety of plants and humans. In order for an insect to sustain continuous flight, its main sugar, trehalose, is quickly hydrolyzed into two glucose molecules.

3.1.3. Disaccharide

Similar to monosaccharides, disaccharides are simple sugars that are soluble in water. They are created when two monosaccharides are linked by a glycosidic bond. Lactose, maltose, and sucrose are three typical examples. One of the four chemical groups that make up carbohydrates is called disaccharides (monosaccharides, disaccharides, oligosaccharides, and polysaccharides). With the general formula $C_{12}H_{22}O_{11}$, the three most prevalent disaccharides are lactose, maltose, and sucrose. All three contain 12 carbon atoms. Atomic configurations within the molecule are the cause of the variations in these disaccharides. A condensation reaction, which involves the removal of a water molecule from the functional groups only, joins monosaccharides to form a double sugar. A kind of enzyme known as a disaccharidase facilitates the hydrolysis process that splits a double sugar into its two monosaccharides. Breaking down the sugar consumes a water molecule, just as building the larger sugar ejects one. In metabolism, these reactions are essential. A matching disaccharidase (lactase, maltase, and sucrase) aids in the breakdown of each disaccharide.

3.1.3.1. Classification

There are two functionally different classes of disaccharides:

- Lactose, maltose, and cellobiose are examples of reducing disaccharides; each has one hemiacetal unit, the other being occupied by the glycosidic bond, which prevents it from acting as a reducing agent. Reducing disaccharides are those in which one monosaccharide, the reducing sugar of the pair, still has a free hemiacetal unit that can function as a reducing aldehyde group. The Woehlk test or Fearon's test on methylamine are excellent methods for identifying them.
- non-reducing disaccharides, in which the acetal linkage between the anomeric centers of the constituent monosaccharides forms a bond. As a consequence, neither monosaccharide has a hemiacetal unit that can function as a reducing agent. Because their respective hemiacetal carbon atoms have a glycosidic bond, sucrose and trehalose are examples of non-reducing disaccharides. When storage stability is crucial, the non-reducing sugars' lower chemical reactivity than that of the reducing sugars might be advantageous.

3.1.3.2. Formation

When two monosaccharide molecules are joined by newly formed vacant bonds on the monosaccharides, a hydroxy group from one molecule and a hydrogen nucleus (a proton) from the other are displaced, forming a disaccharide molecule. The term "dehydration reaction" (also known as condensation reaction or dehydration synthesis) is more convenient for this type of process because it refers to the removal of the water molecule from the product. For instance, one molecule of each of the monosaccharides glucose and galactose condenses to form milk sugar (lactose), whereas glucose and fructose condense to form sucrose, the disaccharide found in sugar cane and sugar beet. Another common disaccharide, maltose, is made up of two glucose molecules that are condensed.

Glycosidic bonds are created by the dehydration process which joins monosaccharides to disaccharides and also to more complex polysaccharides.

3.1.3.3. Properties

On the component monosaccharide, any hydroxy group can form a glycosidic bond. Therefore, different bond combinations (regiochemistry) and stereochemistry (alpha- or beta-) result in disaccharides that are diastereoisomers with different chemical and physical properties, even if both component sugars (e.g., glucose) are the same. Disaccharides can have a sweet taste and sticky feel, be crystalline, or dissolve in water depending on the components of the monosaccharide. Disaccharides can function as functional groups by joining forces with other organic substances to form glycosides through glycosidic bonds.

3.1.4. Nutrition

Simple sugars contain 3.87 kilocalories of energy per gram of carbohydrate, while complex carbohydrates found in most other foods contain 3.57 to 4.12 kilocalories per gram.

Processed foods or refined plant-based foods, such as sweets, candies, and table sugar, honey, soft drinks, breads and crackers, jams and fruit products, pastas, and cereal for breakfast, are linked to relatively high carbohydrate levels. Unrefined foods—such as rice, beans, tubers, and unrefined fruit—generally have higher fiber content and lower digestible carbohydrate content. Foods derived from animals typically contain the fewest carbohydrates, despite the high lactose content of milk.

Most types of carbohydrates are not metabolized by organisms to produce energy. A readily available and almost universal energy source is glucose. Although many organisms can also metabolize other mono- and disaccharides, glucose is frequently metabolized first. For instance, in Escherichia coli, when lactose is present, the lac operon will express the enzymes needed for its digestion; however, when glucose and lactose are present together, the lac operon is repressed, causing the glucose to be used first. Polysaccharides are another typical energy source. While most organisms are able to metabolize glucose from starches with ease, cellulose and other polysaccharides such as chitin and arabinoxylans are difficult for most to do so.

Certain bacteria and protists have the ability to metabolize these types of carbohydrates. Microorganisms, for instance, are used by termites and ruminants to break down cellulose. Despite their poor digestibility, these complex carbohydrates are a vital component of the human diet known as dietary fiber. A few advantages of fiber include improved digestion.

According to the Institute of Medicine, adults in the United States and Canada should obtain 45–65% of their energy from whole-grain carbohydrates. Together, the World Health Organization and the Food and Agriculture Organization advise that 55–75% of total energy come from carbohydrates, with only 10% coming directly from sugars, or simple carbohydrates. This recommendation should be included in national dietary guidelines. The conclusion of the 2017 Cochrane Systematic Review was that the idea that whole grain diets can prevent cardiovascular disease was not well supported by the available data.

3.1.4.1. Classification

Carbohydrates are frequently classified as simple or complex by nutritionists. The precise line that separates these groups, though, is sometimes unclear. In order to differentiate sugars from other carbohydrates—which were thought to be more nutritious—the term "complex carbohydrate" was first used in the US Senate Select Committee on Nutrition and Human Needs publication Dietary Goals for the United States. Nevertheless, the report included fruit, vegetables, and whole grains in the complex carbohydrate category even though they may contain both sugars and polysaccharides. This confusion continues because, in contrast to processed carbohydrates, which only offer energy, some nutritionists now use the term "complex carbohydrate" to describe any type of digestible saccharide found in a whole food that also contains fiber, vitamins, and minerals. Generally speaking, though, carbohydrates are categorized chemically into two groups: simple, which are sugars (monosaccharides and disaccharides), and complex, which are polysaccharides (or oligosaccharides).

Regardless, the distinction between simple and complex chemicals is not very useful in assessing the nutritional value of carbohydrates. Certain complex carbs, like starches, raise blood sugar more slowly than others, such as fructose, which is a type of simple carb. Numerous factors influence the rate of digestion, such as the chemistry of the carbohydrate, individual differences in metabolism, and the other nutrients consumed with the carbohydrate. Occasionally, carbohydrates are separated into two categories: those that are absorbed in the small intestine and those that are unavailable and move to the large intestine, where the gastrointestinal microbiota ferments them.

The 2010 Dietary Guidelines for Americans published by the USDA recommend consuming moderate to high amounts of carbohydrates from a balanced diet that includes six one-ounce servings of grain foods per day, at least half of which should come from whole grain sources and the remaining portion from enriched sources.

The concepts of glycemic load and glycemic index (GI) were created to describe how food behaves during human digestion. They rate foods high in carbohydrates according to how quickly and how much of an impact they have on blood glucose levels. The term "glycemic load" refers to the total amount of absorbable glucose in food, whereas the term "glycemic index" describes how quickly food is absorbed. Similar but more recent is the insulin index, which rates foods according to how they affect blood insulin levels, which are influenced by certain amino acids and glucose (or starch) in food.

3.1.4.2. Health Effects of Dietary Carbohydrate Restriction

Diets low in carbohydrates may miss out on the health benefits of high-quality carbohydrates, which are present in whole grains, legumes and pulses, fruits, and vegetables, as well as an increased intake of dietary fiber. Constipation, headaches, and halitosis were listed as negative diet effects in a meta-analysis of mediocre quality. When total calorie intake is decreased, diets restricting carbohydrates can help achieve weight loss just as effectively as diets low in fat in the short term. Calorie restriction, not the proportion of macronutrients in a diet, is ultimately what determines whether weight loss or maintenance is successful. There is no clinical evidence to support the claims made by diet advocates that low-carb diets have a metabolic advantage and that carbohydrates cause excessive fat accumulation by raising blood insulin levels. Furthermore, while two reviews suggested that carbohydrate restriction may improve lipid markers of cardiovascular disease risk, it is unclear how low-carbohydrate diets affect cardiovascular health. (Schwartz, Seeley, Zeltser, et al., 2017)

Although carbohydrate-restricted diets are not any more effective than a traditional healthy diet in delaying the onset of type 2 diabetes, they can be a good way to help individuals with the disease lose weight or manage their blood sugar levels. The regular application of low-carbohydrate diets to the management of type 1 diabetes is not well supported by research. Rather than focusing on carbohydrates or other macronutrients, the American Diabetes Association advises people with diabetes to adopt a generally healthy diet.

The ketogenic diet, an extremely low-carb diet, was first developed as a medical diet for treating epilepsy. It gained popularity in the early 21st century due to celebrity endorsement, but it also carried a risk of unfavorable side effects, including low energy and increased hunger, sleeplessness, nausea, and gastrointestinal discomfort. (Gjuladin-Hellon, Davies, Penson, & Amiri Baghbadorani, 2019)

3.1.5. Carbohydrate Metabolism

The entire set of biochemical reactions that give rise to break down, and repurpose carbohydrates in living things is known as carbohydrate metabolism.

Many vital metabolic pathways depend on carbohydrates. Through the process of photosynthesis, plants create carbohydrates from carbon dioxide and water, which enables them to internally store solar energy. Cellular respiration is used by animals and fungi to break down the stored carbohydrates in plants so that the cells can use the energy. For use in numerous cellular functions, both plants and animals momentarily store the released energy as high-energy molecules like adenosine triphosphate (ATP).

A wide range of carbohydrates are available to humans; however, digestion reduces complex carbohydrates to their simpler monomers, or monosaccharides, which include glucose, fructose, mannose, and galactose. Following resorption in the stomach, the monosaccharides are transferred to the liver via the portal vein, where they undergo further conversion into glucose along with all other non-glucose monosaccharides, such as fructose and galactose. Blood sugar, or glucose, is transported to the tissues' cells where it is either stored as glycogen or broken down by cellular respiration. Carbon dioxide and water are the byproducts of the metabolism of glucose and oxygen in cellular respiration (aerobic respiration), which releases energy.

3.1.5.1. Metabolic Pathways

3.1.5.1.1. Glycolysis

The process of glycolysis involves dividing a glucose molecule into two pyruvate molecules and storing the energy produced in the process as nicotinamide adenine dinucleotide (NADH) and adenosine triphosphate (ATP). Glycolysis is used by almost all organisms to break down glucose.

These pathways vary between organisms primarily in two categories: glucose regulation and product use. In certain organs and tissues, the only way to produce energy is through glycolysis. Both anaerobic and aerobic respiration use this same pathway.

There are 10 steps in total, divided into two phases of glycolysis. Two ATP molecules must be broken down in the first stage. Chemical energy from the intermediates is converted into ATP and NADH during the second phase. One glucose molecule is broken down into two pyruvate molecules, which can then be further oxidized to obtain additional energy in subsequent processes.

The process of glycolysis can be controlled at various stages by means of feedback regulation. The third step is the one that is most regulated. The purpose of this regulation is to prevent the body from producing too many pyruvate molecules. Furthermore, made possible by the regulation is the conversion of glucose molecules into fatty acids. Numerous enzymes are involved in the process of glycolysis. The process is regulated by the enzymes' upregulation, downregulation, and feedback.

3.1.5.1.2. Gluconeogenesis

A metabolic process called gluconeogenesis (GNG) produces glucose from specific carbon substrates that are not carbohydrates. It is a process that is found in all living things, including bacteria, fungus, plants, and other microbes. Gluconeogenesis mostly takes place in the liver and, to a lesser degree, in the kidney cortex in vertebrates. It is one of the two main ways that humans and many other animals maintain blood sugar levels and prevent low levels (hypoglycemia); the other is the breakdown of glycogen (glycogenolysis). Because rumen organisms typically metabolize dietary carbohydrates, gluconeogenesis happens in ruminants regardless of fasting, low-carb diets, exercise, etc. The process takes place in many other animals during fasting, starvation, low-carb diets, or vigorous exercise.

Substrates for gluconeogenesis in humans can originate from any non-carbohydrate source that can be transformed into pyruvate or glycolysis intermediates. These substrates include glucogenic amino acids (but not ketogenic amino acids) for the breakdown of proteins; glycerol and odd-chain fatty acids (but not even-chain fatty acids; see below) for the breakdown of lipids (such as triglycerides); and lactate from the Cori cycle for other parts of metabolism. Acetone made from ketone bodies can also function as a substrate during extended fasting, opening a channel from fatty acids to glucose. Even though the liver does the majority of the gluconeogenesis, diabetes and extended fasting increase the kidney's relative contribution to gluconeogenesis. Until the gluconeogenesis pathway is connected to the hydrolysis of ATP or guanosine triphosphate (GTP), it is largely endergonic; at that point, the process becomes exergonic. For instance, four molecules of ATP and two molecules of GTP are needed for the spontaneous progression of the pathway that leads from pyruvate to glucose-6-phosphate. Through beta-oxidation, fatty acid catabolism provides these ATPs.

3.1.5.1.3. Glycogenolysis

The breakdown of glycogen is referred to as glycogenolysis. This process happens in the kidneys, muscles, and liver to produce glucose when needed. Glycogen branches off into individual glucose molecules, which are then broken down into glucose-1-phosphate. After that, this molecule can be changed into glucose-6-phosphate, a glycolysis pathway intermediate.

Glycolysis can then proceed with the production of glucose-6-phosphate. When glucose comes from glycogen, only one ATP molecule is needed for glycolysis to occur. Alternatively, if needed, glucose-6-phosphate can be changed back into glucose in the kidneys and liver, raising blood sugar levels.

When blood glucose levels are lowered, or hypoglycemia, the liver's glucagon stimulates the breakdown of glycogen. In between meals, the liver's glycogen can serve as a backup supply of glucose. primarily supports the central nervous system, liver glycogen. During exercise, adrenaline promotes the skeletal muscle's breakdown of glycogen. Glycogen stores in the muscles guarantee a quick source of energy for movement.

3.1.5.1.4. Glycogenesis

The process of creating glycogen is known as glycogenesis. This process is capable of converting glucose in humans into glycogen. Glycogen is a highly branching structure made up of the core protein Glycogenin and branches made of linked glucose units. Because of its branching, glycogen becomes more soluble and more glucose molecules are available for simultaneous breakdown. The kidney, skeletal muscles, and liver are the main sites of glycogenesis. Like most synthetic pathways, the glycogenesis pathway uses energy because it needs to use both an ATP and a UTP for every glucose molecule that is introduced.

3.1.5.1.5. Pentose Phosphate Pathway

An additional pathway for oxidizing glucose is the pentose phosphate pathway. The liver, adipose tissue, testis, adrenal cortex, mammary glands, phagocytes, and red blood cells are among the organs where it manifests. While converting NADP to NADPH, it also yields products that are utilized in other cell processes. Variations in glucose-6-phosphate dehydrogenase activity control this pathway.

3.1.5.1.6. Fructose Metabolism

Before fructose can enter the glycolysis pathway, it needs to go through a few additional steps. Certain tissues contain enzymes that can give fructose a phosphate group. Fructose-6-phosphate, an intermediate in the glycolysis pathway that can be directly broken down in those tissues, is produced by this phosphorylation. The kidney, adipose tissue, and muscles all contain this pathway. Fructose-1-phosphate is created by liver enzymes and then enters the glycolysis pathway where it is cleaved into dihydroxyacetone phosphate and glyceraldehyde.

3.1.5.1.7. Galactose Metabolism

One glucose molecule and one galactose molecule make up lactose, also known as milk sugar. Galactose is transported to the liver where it is converted to glucose after being separated from glucose. Galactokinase phosphorylates galactose with a single ATP molecule. The phosphorylated galactose is subsequently transformed into glucose-1-phosphate, which can subsequently be broken down through glycolysis, and finally glucose-6-phosphate.

3.1.5.2. Energy Production

The cells can access energy through a number of steps in the metabolism of carbohydrates and store it more temporarily in ATP. During this process, the cofactors NAD+ and FAD are occasionally reduced to form NADH and FADH2, which are what power the synthesis of ATP in other processes. While a molecule of FADH2 only produces one point five molecules of ATP, a molecule of NADH can produce one to two point five molecules.

About 30–32 molecules of ATP are typically needed for the full breakdown of one glucose molecule by aerobic respiration, which includes glycolysis, the citric-acid cycle, and oxidative phosphorylation, the last of which provides the most energy. One gram of carbohydrate can be oxidized to produce about 4 kcal of energy.

3.1.5.3. Hormonal Regulation

Glucoregulation is the maintenance of steady levels of glucose in the body.

The pancreas secretes hormones that control how glucose is metabolized overall. The main hormones responsible for preserving a constant blood glucose level are insulin and glucagon, whose release is regulated by the quantity of nutrients available at the time. The quantity of glucose that cells break down is determined by the amount of insulin released into the blood and the sensitivity of the cells to the insulin. Elevated glucagon secretion stimulates the enzymes responsible for glycogenolysis while suppressing the enzymes involved in glycogenesis. On the other hand, when blood insulin levels are high, glycogenesis is increased and glycogenolysis is inhibited.

The level of circulatory glucose, or blood sugar as it is colloquially known, along with the duodenum's ability to detect nutrients, are the primary factors that determine how much insulin or glucagon is produced. Low blood glucose levels cause the release of glucagon, while high blood glucose levels encourage the production of insulin by cells. Dietary carbs control most aspects of metabolism through insulin, since dietary carbohydrate intake largely determines the level of circulatory glucose. In humans, fat is stored in adipose tissue cells, insulin is produced by beta cells in the pancreas, and glycogen is stored in the liver and released when needed. No glucose is released into the blood from the internal glycogen stores of muscle cells, independent of insulin levels.

3.1.6. Relation Between Carbohydrate Types and Health Outcomes

Formally speaking, carbohydrates are defined as having a 1:2:1 ratio of carbon, hydrogen, and oxygen. In actuality, dietary carbohydrates are made up of substances such as sugar alcohols (sorbitol) that can either be oxidized into pyruvate or directly digested or metabolically converted into glucose. Numerous categorization schemes for carbohydrates have been implemented, with differing implications for health consequences.

3.1.6.1. Chain Length

Depending on the extent of polymerization, carbohydrates can be divided into four groups: monomers, disaccharides, oligomers, and polymers (starch). The length of the carbohydrate polymer is thought to control the rate of absorption and digestion, and consequently the rise in blood glucose following a meal. Thus, diabetics were advised to steer clear of sugar and to prioritize starchy foods. However, studies conducted starting half a century ago did not find a significant correlation between postprandial glycemia or insulinemia and the length of the carbohydrate chain. Compared to some high-sugar foods, such as whole fruits, modern starchy foods like bread, potatoes, and rice significantly increase blood glucose and insulin levels. However, some commonly consumed starches (such as legumes, whole kernel grains, pasta, and long fermentation sourdough bread) release glucose more slowly due to the starch's protection from digestion (gelatinization) or the slowing down of gastric emptying caused by the presence of organic acids.

3.1.6.2. Glycemic Index and Glycemic Load

Despite the fact that only carbohydrates directly raise blood glucose, which is the primary factor influencing insulin secretion, population studies indicate that the type of carbohydrates is more significant in predicting the risk of chronic disease than the total amount of carbohydrates as a percentage of dietary energy. Whole fruits, legumes, and minimally processed grains are linked to lower risk, while refined grains, potatoes, and beverages with added sugar are linked to higher risk. This discrepancy could be partially explained by variations in the effects of particular carbohydrates on postprandial hyperglycemia and hyperinsulinemia, which are linked to the development of type 2 diabetes, heart disease, and possibly obesity. The glycemic index (GI) and glycemic load (GL) are two empirical metrics that have been introduced to rank foods based on their effects on blood glucose. The GI uses a standard measurement of the amount of available carbohydrates to compare foods. The glycemic impact of foods, meals, and entire diets can be compared as they would be consumed in real life thanks to glycemic load (GI multiplied by the amount of carbs in a typical serving). It has been demonstrated that GI is a more accurate indicator of glycemic response than carb, protein, and fat content.

Table 3.1. Carbohydrate Content and Glycemic Index of Representative Foods

Food	Serving Size (g)	Available Carbohydrate (g) Per Serving	Glycemic Index	Glycemic Load
Rice, jasmine, boiled	120	32	86	28
Instant oat porridge/oatmeal	250	26	79	21
Rice, basmati, boiled	120	30	57	17
Potato, boiled	150	20	78	16
Breakfast cereal, flaked	30	22	72	16
Pasta, white or brown, boiled	120	31	49	15
Bread, white or brown	40	19	75	14
Traditional oat porridge	250	24	55	13
Fruit juice	250 mL	24	50	12
Fruit, tropical	120	16	58	9
Barley, boiled	120	34	28	9
Bread, wholemeal	40	13	54	7
Legumes, boiled	150	22	31	7
Fruit, temperate	120	14	42	6
Pumpkin, boiled	75	8	64	5
Milk	250 mL	12	32	4
Nuts	30	7	25	2

Source: https://www.ncbi.nlm.nih.gov/pmc/articles/PMC5996878/table/tbl1/?report=objectonly.

Higher energy adjusted GI or total GL has been linked to an independent risk for type 2 diabetes in both men and women, cardiovascular morbidity and mortality, including stroke in women, and specific types of cancer in both sexes, according to prospective observational studies. However, some have questioned the reliability and consistency of these findings.

3.1.6.3. Fiber and Resistant Starch

Plant carbohydrates known as fiber or non-starch polysaccharides are indigestible by human enzymes. To varying degrees, fiber and resistant starch serve as a substrate for the fermentation of colonic microbes, which produces short chain fatty acids that directly fuel the colonic epithelium and affect the sensitivity of the liver to insulin. There are two types of fiber: soluble (viscous or not) and non-soluble. These types of fiber have different effects on metabolism and gastrointestinal absorption. In contrast to insoluble fibers, which have restricted metabolic effects, viscous fibers, like those found in vegetable gums, fruits, legumes, and psyllium, slow down digestion, lower postprandial glycemia, and reduce cholesterol absorption.

3.1.6.4. Added and Free Sugar

Sugars that are added to food during manufacturing, processing, or preparation are referred to as added sugars. The natural sugars found in unsweetened fruit juices are also included in the more recent term, "free sugars"; otherwise, the two terms are interchangeable. Only lactose that occurs naturally in milk products and sugars found in food's cellular structure (such as whole fruits) would be prohibited under this definition.

The majority of health authorities concur that the obesity epidemic has been exacerbated by excessive consumption of added sugars, especially in the form of sugar-sweetened beverages. Changes in the consumption of sugary drinks are directly linked to changes in energy intake and body weight in the higher quality prospective observational studies. Additionally, it was discovered in two sizable randomized controlled trials that removing sugar-filled beverages from adolescent diets at one year of age and younger children at 18 months of age decreased body weight. Reduced intake of added sugars is linked to a modest decrease in body weight, whereas higher intake is linked to a gain that is comparable, according to meta-analyzes of trials conducted on adults following unrestricted diets. However, body weight remained unchanged when sugars were substituted for other carbohydrates on an isocaloric basis.

There is ongoing discussion regarding the possible mechanisms linking sugar to weight gain. Numerous researchers have emphasized fructose's possible function. Fructose is mainly metabolized in the liver and gut, where it can also promote insulin resistance, inflammation, and de novo lipogenesis in specific experimental scenarios. It has been questioned, nevertheless, whether these findings apply to normal consumption patterns. Furthermore, eating a lot of fruits that contain a lot of fructose is linked to having a healthy metabolism, indicating that fructose's food source matters as well. It is unknown how much added sugar and other carbohydrates contribute to the obesity epidemic in relation to one another. Indeed, added sugar significantly reduces the number of calories in typical Western diets compared to high GL starchy foods (without fructose). Since the 1990s, Australia has seen a steady decrease in the consumption of added sugars and drinks with added sugar, despite a sharp increase in the mean body mass index among adults and children.

Higher intakes of added sugars are linked to higher triglycerides, total cholesterol, blood pressure, and other cardiovascular disease risk factors, in addition to body weight, according to meta-analyzes of randomized trials. Non-alcoholic fatty liver disease (NAFLD), an obesity-related ailment that has become a significant public health concern, is especially concerning. In multiple clinical trials, lowering the intake of fructose or sugar led to a decrease in intrahepatic fat. All of these studies do, however, have design flaws, such as the absence of a control group and confounding by inadvertent weight loss. During a half-year study, individuals who drank beverages with added sugar had greater amounts of hepatic and ectopic fat than those who did not, despite no differences in body weight between the diet groups.

3.1.6.5. Composite Quality Indices

In addition to the mechanisms suggested by these broad categorization schemes, foods high in carbohydrates may have diverse other effects on health. Whole plant foods contain a wide range of compounds that may have negative effects (lectins, phytates) or clearly positive effects (vitamins, minerals, antioxidant and anti-inflammatory phytochemicals). In the end, diet needs to be thought of holistically, with adjustments to one food group's consumption influencing others. In population studies, the apparent healthfulness of particular foods will depend on the nature of these exchanges. A number of indices for the quality of the entire diet and carbohydrates (based on GI, fiber, whole-grain consumption, and other factors) have been proposed in recognition of this challenge.

3.2. LIPIDS

Hydrogen, carbon, and oxygen atoms make up lipids, which are organic substances that support the structure and operation of living cells. Since water is a polar molecule, these organic compounds are nonpolar molecules that are only soluble in nonpolar solvents. These molecules, which are produced in the liver of the human body, are present in fried foods, oil, butter, whole milk, cheese, and some types of red meat.

3.2.1. Categories

Lipids have been classified into eight categories by the Lipid MAPS consortium as follows:

3.2.1.1. Fatty Acyls

In a process known as fatty acid synthesis, an acetyl-CoA primer is chain-elongated with malonyl-CoA or methylmalonyl-CoA groups to create fatty acyls, a broad term used to describe fatty acids, their conjugates, and derivatives. They are composed of a hydrocarbon chain that ends in a carboxylic acid group. This configuration gives the molecule a nonpolar, water-insoluble hydrophobic end and a polar, hydrophilic end. One of the most basic types of biological lipids is the fatty acid structure, which is frequently utilized as a building block for lipids with more intricate structural makeup. The carbon chain, which is normally four to 24 carbons long, can be linked to functional groups that contain oxygen, halogens, nitrogen, and sulfur. It can also be saturated or unsaturated. A cis or trans geometric isomerism, which profoundly alters the molecule's configuration, may occur in fatty acids that have double bonds.

More double bonds in the chain exacerbate the bending of the fatty acid chain caused by cis-double bonds. The most prevalent fatty-acyl chain in plant thylakoid membranes, 18-carbon linolenic acid, has three double bonds that make it extremely flexible even in cold environments. These double bonds also cause linolenic acid to exhibit prominent, sharp peaks in high resolution 13-C NMR spectra of chloroplasts. Consequently, this has a significant impact on the composition and operation of cell membranes. While some naturally occurring and partially hydrogenated fats and oils do contain the trans isomeric form, the majority of fatty acids are found in their cis configuration (Mashaghi, Jadidi, Koenderink, & Mashaghi, 2013).

The eicosanoids, which are mainly derived from arachidonic acid and eicosapentaenoic acid and include prostaglandins, leukotrienes, and thromboxanes, are examples of biologically significant fatty acids. In biological systems, docosahexaenoic acid is also crucial, especially about vision. Fatty esters and fatty amides are two more important lipid classes that fall under the fatty acid category. Important biochemical intermediates, such as wax esters, fatty acid thioester derivatives coenzyme A, fatty acid thioester ACP derivatives, and fatty acid carnitines, are examples of fatty esters. N-acyl ethanolamines, which include the cannabinoid neurotransmitter anandamide, are among the fatty amides (Hunter, 2006).

3.2.1.2. Glycerolipids

Triglycerides, which are glycerol's fatty acid triesters, are the most well-known type of glycerolipid. Glycerolipids are made up of mono-, di-, and tri-substituted glycerols. Triacylglycerol and triglyceride are sometimes used interchangeably. Each of the three hydroxyl groups of glycerol is esterified in these compounds, usually by a different fatty acid. These fats make up the majority of the storage fat found in animal tissues because they serve as an energy store.

The first steps in the metabolism of fat are the hydrolysis of triglyceride ester bonds and the release of glycerol and fatty acids from adipose tissue. Glycosylglycerols, which are defined by the presence of one or more sugar residues linked to glycerol through a glycosidic bond, are representative of additional subclasses of glycerolipids. The seminolipid from mammalian sperm cells and the digalactosyldiacylglycerols present in plant membranes are two examples of structures in this category. (Honke, Zhang, et al., 2004).

3.2.1.3. Glycerophospholipids

Glycerophospholipids, often known as phospholipids (although sphingomyelins are also included in this category), are widely distributed in nature and play a crucial role in cell signaling and metabolism in addition to being essential parts of the lipid bilayer of cells. Glycerophospholipids are found in relatively high concentrations in neural tissue, including the brain, and changes in their makeup have been linked to a number of neurological conditions. Based on the makeup of the polar headgroup at the sn-3 position of the glycerol backbone in eukaryotes and eubacteria, or the sn-1 position in archaebacteria, glycerophospholipids can be classified into several classes.

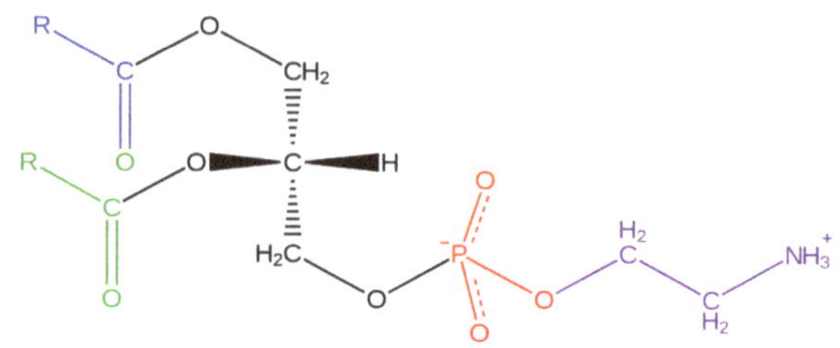

Figure 3.3. *Phosphatidylethanolamine.*

Source: https://upload.wikimedia.org/wikipedia/commons/thumb/c/c7/Phosphatidyl-ethanolamine.svg/1920px-Phosphatidyl-ethanolamine.svg.png.

Examples of glycerophospholipids found in biological membranes are phosphatidylcholine (also known as PC, GPCho or lecithin), phosphatidylethanolamine (PE or GPEtn) and phosphatidylserine (PS or GPSer). Certain glycerophospholipids in eukaryotic cells, such as phosphatidylinositols and phosphatidic acids, are not only essential for cellular membranes and act as binding sites for intra- and intercellular proteins, but they also either precede or are membrane-derived second messengers in and of themselves.

Long-chain fatty acids usually acylate one or both of these hydroxyl groups; however, archaebacteria also produce dialkylether variants of glycerophospholipids as well as alkyl-linked and 1Z-alkenyl-linked (plasmalogen) glycerophospholipids.

3.2.1.4. Sphingolipids

A complex group of molecules known as sphingolipids are united by a shared structural characteristic: a sphingoid base backbone that is created from scratch from the amino acid serine and a long-chain fatty acyl CoA. Ceramides, phosphosphingolipids, glycosphingolipids, and other compounds can then be synthesized from these building blocks. A common term for the main sphingoid base found in mammals is sphingosine. A significant subclass of sphingoid base derivatives with an amide-linked fatty acid is called ceramides (N-acyl-sphingoid bases). Typically, the fatty acids have chains of 16–26 carbon atoms and are either mono-unsaturated or saturated.

Sphingomyelins, or ceramide phosphocholines, are the main phosphosphingolipids found in mammals, while ceramide phosphoethanolamines are primarily found in insects, and phytoceramide phosphoinositols and mannose-containing headgroups are found in fungi.

A glycosidic bond connects one or more sugar residues to the sphingoid base to form the diverse family of molecules known as glycosphingolipids. Simple and complex glycosphingolipids, such as gangliosides and cerebrosides, are examples of these.

3.2.1.5. Sterols

Sterols, including cholesterol and its derivatives, glycerophospholipids, and sphingomyelins, are constituents of membrane lipids. Bile acids and their conjugates, which are produced in the liver of mammals and are oxidized derivatives of cholesterol, are additional instances of sterols. β-sitosterol, stigmasterol, and brassicasterol are examples of plant-equivalent phytosterols; the latter is also employed as a biomarker for algal growth. Ergosterol is the main sterol found in the membranes of fungal cells.

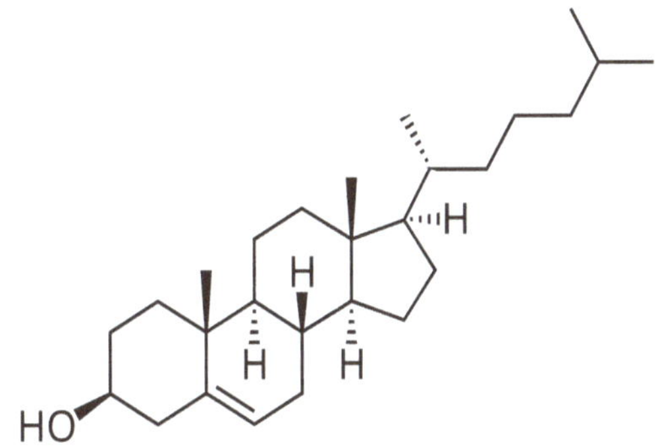

Figure 3.4. *Chemical structure of cholesterol.*

Source: https://pubchem.ncbi.nlm.nih.gov/compound/Cholesterol#section=2D-Structure.

Sterols are steroids with a hydroxyl group substituted for one of the hydrogen atoms at position 3 in the carbon chain. They share the same fused four-ring core structure as steroids. As hormones and signaling molecules, steroids play distinct biological roles. The estrogen family is included in the 18-carbon (C18) steroids, while the androgens, such as testosterone and androsterone, are included in the 19-carbon (C19) steroids. The progestogens, along with glucocorticoids and mineralocorticoids, are members of the C21 subclass. The cleavage of the B ring of the core structure characterizes the secosteroids, which are made up of different forms of vitamin D.

3.2.1.6. Prenol

Prenol lipids are manufactured using five-carbon-unit precursors isopentenyl diphosphate and dimethylallyl diphosphate, primarily through the mevalonic acid (MVA) pathway. These basic isoprenoids (linear alcohols, diphosphates, etc.) are created by the consecutive addition of C5 units and are categorized based on the number of terpene units. Structures with more than 40 carbons are recognized as polyterpenes. Carotenoids are significant simple isoprenoids that serve as antioxidants and as precursors of vitamin A. Another biologically crucial group of molecules includes quinones and hydroquinones, which have an isoprenoid tail connected to a quinonoid core of non-isoprenoid origin. Vitamin E and vitamin K, as well as the ubiquinones, are illustrations of this class. Prokaryotes synthesize polyprenols (known as bactoprenols) in which the terminal isoprenoid unit attached to oxygen remains unsaturated, while in animal polyprenols (dolichols) the terminal isoprenoid is reduced.

3.2.1.7. Saccharolipids

Compounds that have fatty acids bonded to a sugar backbone to create structures that work with membrane bilayers are known as saccharolipids. The glycerol backbone found in glycerolipids and glycerophospholipids is replaced in saccharolipids by a monosaccharide. The acylated glucosamine precursors of the Lipid A component of the lipopolysaccharides in Gram-negative bacteria are the most well-known saccharolipids. Disaccharides of glucosamine that have undergone up to seven fatty-acyl chain derivatizations are the typical lipid A molecules. Kdo2-Lipid A, a hexa-acylated disaccharide of glucosamine glycosylated with two 3-deoxy-D-manno-octulosonic acid (Kdo) residues, is the minimal lipopolysaccharide needed for growth in *E. coli*.

3.2.1.8. Polyketides

The synthesis of polyketides involves the polymerization of acetyl and propionyl subunits by both traditional enzymes and multimodular, iterative enzymes that have mechanistic similarities with fatty acid synthases. They have a wide range of structural variations and are made up of numerous secondary metabolites and natural products derived from bacterial, fungal, plant, and marine sources. The backbones of many polyketides are cyclic molecules that are frequently altered further by oxidation, glycosylation, methylation, hydroxylation, and other processes. Polyketides, or polyketide derivatives, are widely used antimicrobial, antiparasitic, and anticancer agents. Examples of these include erythromycins, tetracyclines, avermectins, and antitumor epothilones.

3.2.2. Biological Functions

3.2.2.1. Component of Biological Membranes

Organelles that are membrane-bound and divided into compartments are a characteristic of eukaryotic cells and perform various biological tasks. In animal cells, the plasma membrane physically divides the intracellular components from the extracellular environment. Glycerophospholipids are the primary structural element of biological membranes, including the cellular plasma membrane and the intracellular membranes of organelles. Amphipathic molecules with both hydrophobic and hydrophilic regions, glycerophospholipids are composed of a glycerol core connected by ester linkages to two "tails" derived from fatty acids and by phosphate ester linkages to one "head" group. Although glycerophospholipids make up the majority of the lipid content in biological membranes, sphingomyelin and sterols—primarily cholesterol in animal cell membranes—as well as other non-glyceride lipid components are also present. The most common lipids in photosynthetic tissues, such as those of higher plants, algae, and some bacteria, are galactosyldiacylglycerols and sulfoquinovosyldiacylglycerol, which are phosphate group-free and play a crucial role in the membranes of chloroplasts and related organelles in plants and algae.

Despite having a distinct lipid composition, plant thylakoid membranes contain a large amount of non-bilayer forming monogalactosyl diglyceride (MGDG) and little phospholipids. However, magnetic resonance and electron microscope studies have demonstrated that chloroplast thylakoid membranes contain a dynamic lipid-bilayer matrix.

A type of lipid bilayer in the lamellar phase is a biological membrane. The hydrophobic effect occurs when the glycerophospholipids mentioned above are in an aqueous environment and the formation of lipid bilayers is an energetically preferred process. The polar heads of lipids in an aqueous system align with the polar, aqueous environment, while the hydrophobic tails minimize their contact with water and tend to group together, forming a vesicle.

This biophysical interaction can lead to the formation of lipid bilayers, liposomes, or micelles, depending on the concentration of the lipid. Additional groupings are also noted and contribute to the variability of amphiphile (lipid) behavior. In biophysics, phase behavior is a field of study. The hydrophobic effect is the mechanism by which micelles and bilayers form in the polar medium. Since the polar molecules (water in an aqueous solution, for example) are unable to form hydrogen bonds with the lipophilic regions of the amphiphile, they become more ordered around the dissolved lipophilic substance when it dissolves in a polar environment. Therefore, the dissolved lipophilic molecule is surrounded by an ordered clathrate cage made of water molecules in an aqueous environment.

In models of abiogenesis, the beginning of life, the formation of lipids into protocell membranes is a crucial step.

3.2.2.2. Energy Storage

Adipose tissue contains triglycerides, which are a major source of energy storage for both plants and animals. They contribute significantly to aerobic respiration's energy supply. About 38 kJ/g (9 kcal/g) are released during the full oxidation of fatty acids, whereas only 17 kJ/g (4 kcal/g) are released during the oxidative breakdown of proteins and carbohydrates. Animal adipocytes, or fat cells, are made to continuously synthesize and break down triglycerides. The hormone-sensitive enzyme lipase is primarily responsible for this process. Triglycerides are used as fuel by migratory birds that must travel great distances without eating.

3.2.2.3. Signaling

There is evidence now available demonstrating the critical role that lipid signaling plays in cell signaling. Members of various lipid categories have been identified as signaling molecules and cellular messengers. Lipid signaling can be triggered by activation of nuclear or G protein-coupled receptors. These include prostaglandins, a type of fatty-acid derived eicosanoid involved in inflammation and immunity; steroid hormones like estrogen, testosterone, and cortisol, which modulate a host of functions like reproduction, metabolism, and blood pressure; and oxysterols like 25-hydroxy-cholesterol, which are liver X receptor agonists. Sphingosine-1-phosphate, a potent messenger molecule derived from ceramide, is a potent messenger molecule involved in regulating calcium mobilization, cell growth, and apoptosis. These also include diacylglycerol and the phosphatidylinositol phosphates (PIPs) and diacylglycerol. It is well known that phosphatidylserine lipids play a role in signaling cells or fragments of cells to be phagocytosed. They achieve this by being exposed to the extracellular face of the cell membrane subsequent to the activation of scramblases, which causes the orientation of the phospholipids to be scrambled, and the inactivation of flippases, which only places them on the cytosolic side. Following this, other cells phagocytose the exposed cells or cell fragments after identifying the phosphatidylserines.

3.2.2.4. Other Functions

The isoprene-based lipids known as fat-soluble vitamins A, D, E, and K are vital nutrients that are stored in the liver and fatty tissues and have a variety of uses. Fatty acids are transported into and out of mitochondria, where they undergo beta oxidation, and are metabolized by acyl-carnitines. Important transport functions are also played by polyprenols and their phosphorylated derivatives, in this case, the passage of oligosaccharides across membranes. Polyprenol phosphate and polyprenol diphosphate sugars are involved in extracellular polysaccharide biosynthesis (e.g., bacterial peptidoglycan polymerization), extracytoplasmic glycosylation reactions, and eukaryotic protein N-glycosylation. Cardiolipins are a subclass of glycerophospholipids that are especially prevalent in the inner mitochondrial membrane. They have three glycerol groups and four acyl chains. They are thought to activate oxidative phosphorylation-related enzymes. Steroid hormones are also derived from lipids.

3.2.3. Metabolism

Sterols, membrane phospholipids, and animal and plant triglycerides are the main dietary lipids for humans and other animals. Lipid metabolism creates structural and functional lipids that are unique to each tissue by synthesizing and breaking down lipid stores.

3.2.3.1. Biosynthesis

When an animal's diet contains too many carbohydrates, the extra carbohydrates are transformed into triglycerides. This entails the esterification of fatty acids in the process known as lipogenesis, which produces triglycerides, and the synthesis of fatty acids from acetyl-CoA. Acetyl-CoA units are polymerized and subsequently reduced by fatty acid synthases to produce fatty acids. A series of reactions that add the acetyl group, reduce it to an alcohol, dehydrate it to an alkene group, and then reduce it again to an alkane group lengthens the acyl chains in the fatty acids. The enzymes involved in the biosynthesis of fatty acids are classified into two groups: in bacteria and plant plastids, individual enzymes carry out each step of the pathway, whereas in animals and fungi, a single multifunctional protein catalyzes all of the fatty acid synthase reactions. After that, the fatty acids might be changed into triglycerides, which are then contained in lipoproteins and released from the liver.

A double bond is added to the fatty acyl chain during the desaturation reaction that creates unsaturated fatty acids. For instance, oleic acid is produced in humans when stearoyl-CoA desaturase-1 desaturates stearic acid. Because mammalian tissues are unable to synthesize the doubly unsaturated fatty acid linoleic acid and the triply unsaturated α-linolenic acid, these fatty acids are essential and must be obtained through diet.

Through metabolic pathways, acyl groups in fatty acyl-CoAs are transferred to the hydroxyl groups of glycerol-3-phosphate and diacylglycerol, resulting in the synthesis of triglycerides in the endoplasmic reticulum.

Reactive precursors isopentenyl pyrophosphate and dimethylallyl pyrophosphate provide isoprene units that are assembled and modified to create terpenes and isoprenoids, including carotenoids. There are various methods for producing these precursors. The non-mevalonate pathway in plants and bacteria uses pyruvate and glyceraldehyde 3-phosphate as substrates, whereas the mevalonate pathway in animals and archaea uses acetyl-CoA to produce these compounds. Steroid biosynthesis is a significant reaction that makes use of these activated isoprene donors. In this case, the isoprene units are combined to form squalene, which is subsequently folded into a set of rings to form lanosterol. Then, lanosterol can be transformed into other steroids, including ergosterol and cholesterol.

3.2.3.2. Degradation

The metabolic process known as beta oxidation produces acetyl-CoA by breaking down fatty acids in peroxisomes or mitochondria. Fatty acids are oxidized primarily by a mechanism that is comparable to, but distinct from, the reversal of the fatty acid synthesis process. That is, following phases of dehydrogenation, hydration, and oxidation, two-carbon fragments are gradually removed from the carboxyl end of the acid to form a beta-keto acid, which is then split by thiolysis. Adenosine triphosphate (ATP), CO_2, and H_2O are eventually produced from acetyl-CoA via the electron transport chain and citric acid cycle. Therefore, if there is little or no glucose available, the citric acid cycle can begin at acetyl-CoA when fat is being broken down for energy. The full oxidation of the fatty acid palmitate yields 106 ATP as energy. For their breakdown, unsaturated and odd-chain fatty acids need extra enzymatic processes.

3.2.4. Nutrition and Health

The majority of fat in food is present as cholesterol, phospholipids, and triglycerides. To aid in the absorption of carotenoids and fat-soluble vitamins (A, D, E, and K), some dietary fat is required. Certain essential fatty acids, such as omega-3 alpha-linolenic acid and omega-6 linoleic acid, are required for human and other mammal nutrition because they cannot be produced in the body from simple precursors. These two fatty acids have the same number and location of double bonds but are both 18-carbon polyunsaturated fatty acids. Linoleic acid is abundant in most vegetable oils, including sunflower, corn, and safflower oils. Green plant leaves, as well as some seeds, nuts, and legumes (particularly flax, rapeseed, walnut, and soy), contain alpha-linolenic acid. The longer-chain omega-3 fatty acids docosahexaenoic acid and eicosapentaenoic acid are especially abundant in fish oils. Numerous studies have demonstrated the beneficial effects of omega-3 fatty acid consumption on cardiovascular disease, cancer, and a range of mental illnesses, including dementia, attention-deficit hyperactivity disorder, and depression. On the other hand, it is now widely known that consuming trans fats, like those found in partially hydrogenated vegetable oils, increases the risk of developing cardiovascular disease. When lipids are overcooked due to incorrect cooking techniques, healthy fats can become trans fats.

CASE STUDY

ASSOCIATION BETWEEN ULTRA-PROCESSED FOODS INTAKE WITH LIPID PROFILE: A CROSS-SECTIONAL STUDY

Dyslipidemia is defined as any lipid abnormality including elevated total cholesterol (TC), low-density lipoprotein cholesterol (LDL-C) and triglycerides (TG) and declined high-density lipoprotein cholesterol (HDL-C)1. Prevalence of hypertriglyceridemia, hypercholesterolemia, high non-HDL, and low HDL in Iranian adults are 28%, 26.7%, 39.5%, and 69.2% respectively.

Abnormalities of lipid profile are associated with many clinical outcomes such as type 2 diabetes and cardiovascular diseases (CVDs). It is also the main cause of more than half of the cases of congenital heart disease and more than four million deaths annually. There is a complex interaction between genetic factors and several environmental factors such as smoking, sedentary lifestyle, and socioeconomic situation as determinants of lipid profile. Particularly taking foods rich in calories, carbohydrates, sodium, cholesterol, trans and saturated fatty acids are related to the high concentration of TC and LDL-C. On the other hand, consumption of polyunsaturated fatty acids, vegetables, dietary fibers, milk, and dairy products could decrease the risk of dyslipidemia.

Nova classification system by considering physical, biological, chemical properties of the foods, also the additives for food manufacturing, classifies foods into different groups, including unprocessed and minimally processed food, processed culinary ingredients, processed foods and ultra-processed foods. UPFs are intrinsically fatty, sugary or salty, high calorie, and poor in protein, dietary fiber, micronutrients, and other bioactive compounds, and usually contain no or small number of whole foods. UPFs contain soft drinks; sweets like chocolate, candies, ice cream, biscuits and cakes; packaged leaves of bread; nuggets and sticks, margarine, pastries; pre-prepared food dishes, and many other products.

Evidence also showed that high consumption of UPFs is potentially related to the high rate of obesity and related comorbidities. Assessment of UPFs' effects on lipid profiles and upcoming diseases is an ongoing debate. Most studies reported a positive association of UPFs intake with lipid abnormalities, altered blood lipid factors varies in the studies. While some reported an association between UPFs intake with elevated TG and lower HDL-C others reported positive association of UPFs consumption with high LDL and total cholesterol. As UPFs production and consumption have increased extremely during the last decades, comprehending their potential effects on human health has become a major concern in health care systems. The purpose of this cross-sectional study was to examine the association between UPFs intake and lipid profile in Iranian adults. In addition, as secondary outcomes, we examined the association of UPFs consumption with the intake of food subgroups, macronutrients, and some of the micronutrients. To the best of our knowledge, no previous study has investigated the association between UPFs intake and lipid profile in Iranian adults.

1. **Methods**

- Study Population

This cross-sectional study was performed on 236 individuals with the age range of 20–50 years who were selected by cluster random sampling in health care centers of Shiraz-Iran. For this purpose, Shiraz was divided in four clusters and one health care center was randomly selected in each cluster. Study sample size was calculated by the following formula, in which $\alpha=0.01$, $\beta=0.10$ and $r=\pm 0.25$ (Figure 3.5).

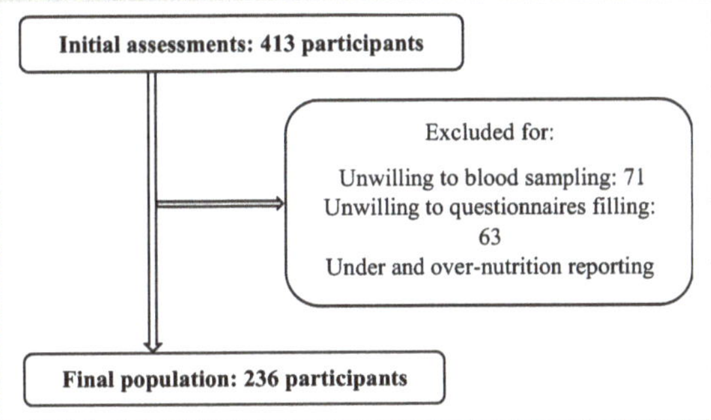

Figure 3.5. *Study flow diagram.*

Source: https://www.nature.com/articles/s41598-023-34451-x.

$$C = 0.5 \times \ln\left[\frac{(1+r)}{(1-r)}\right]$$

$$N = 3 + \left[\left(Z_{1-\frac{\alpha}{2}} + Z_{1-\beta}\right) \div C\right]^2$$

Prerequisites for participating in this study were no history of chronic diseases and no adherence to any special diet. Written consent form was signed by all participants.

- **Dietary assessment**

Food intakes of the participants were evaluated by a 168-item food frequency questionnaire (FFQ) which was previously validated in Iranian populations. The FFQ was filled out based on family food intakes and data was changed to gram. For computing energy, and nutrients intakes we used NUTRITIONIST IV.

To estimate the ultra-processed foods intake, classification of NOVA food group was used. Accordingly, total daily consumption of some foods and beverages items were considered as ultra-processed food (including packaged breads, buns, confectionaries, pastries and sweets, ice cream, biscuits, cakes, soft drinks, industrial fruit drinks, sweetened milk-based beverages, margarine, sauces, dressings, processed meats, fries, and salty snacks). To define the contribution of every food group to overall consumption of ultra-processed foods, the mean daily intake of every 9 subgroups of ultra-processed foods (dairy products, non-dairy beverages, margarine and sauces, cakes and cookies, chips and snacks, breads, fast foods and meats, sweets and others) was divided by daily consumption of ultra-processed foods, then multiplied by 100.

- **Biochemical Assessments**

Serum lipids including TC, TG, HDL-C and LDL-C were measured in blood samples (5 cm^3) taken from the participants commercially accessible enzyme kits (Pars Azmoon, Tehran, Iran).

Socio-demographic and anthropometric assessments

Sex, age, smoking habit and alcohol intake were evaluated by using a questionnaire. Also, we used the International Physical Activity Questionnaire (IPAQ) to assess the physical activity level of the participants. Anthropometric indices [weight (kg), waist circumference (cm) and height (cm)] were measured. Then, according to weight and height, BMI was calculated.

- **Statistical Analysis**

All analysis was done using SPSS for Windows. The level of significance was P-value < 0.05. Normal distribution of the variables was checked by the Kolmogorov–Smirnov test. The relationship between quantitative variables and qualitative variables was evaluated by analysis of variance test and Chi-square test respectively. Crude and adjusted models of logistic regression were used to evaluate the relation between UPFs score with anthropometric index and lipid profile. In adjusted models, the effects of age, energy intake, physical activity, BMI, sex and smoking history were controlled. We dichotomized lipid profile and anthropometric indices then LDL-C more than 130 mg/dL, HDL-C less than 40 mg/dL for male and 50 mg/dL for female, TC more than 200 mg/dL, TG more than 150 mg/dL, non-HDL more than 130 and WC more than 88 cm for female and 102 for male were considered as abnormal.

- **Results**

As shown in Table 3.2, mean age and BMI of study population were 45.97 years and 28.28, respectively kg/m2. Also, 58.90% of the study participants were female. According to Table 3.3 there was no significant difference between age, weight, height, BMI, waist circumference, WHR, TC, LDL, HDL, Non-HDL, physical activity, sex, smoking habits, alcohol intake history and education of the participants in the first and last tertile of UPFs, intake, but the difference was significant for mean of serum TG (P = 0.015).

Table 3.2. Baseline Characteristics of the Study Participants

Variables	N = 236
Age (year)	45.97 ± 11.74
Sex, female (%)	139 (58.90)
Education level, lower than high School (%)	65 (27.54)
Smoking history, yes (%)	28 (11.90)
BMI (kg/m²)	28.28 ± 4.69
Waist circumference (cm)	94.21 ± 11.35
Hip circumference (cm)	101.47 ± 9.64
WHR	0.90 ± 0.07
TG (mg/dL)	121.40 ± 64.44
TC (mg/dL)	179.82 ± 43.10
LDL-c (mg/dL)	109.33 ± 33.74
HDL-c (mg/dL)	38.01 ± 11.08
Non HDL-c	141.81 ± 40.53
Energy (kcal/day)	2,772.84 ± 1,054.18
Protein (g/day)	90.00 ± 37.12
Carbohydrate (g/day)	432.03 ± 168.19
Total fat (g/day)	82.70 ± 38.95

Source: https://www.nature.com/articles/s41598-023-34451-x/tables/1.

Table 3.3. Baseline Characteristics According to Tertile of UPFs

Variables	**UPFs**			
	T_1 (n = 95)	**T_2 (n = 79)**	**T_3 (n = 62)**	**P-value**
Age (year)	46.47 ± 10.92	46.68 ± 12.91	44.30 ± 11.41	0.427
Weight (kg)	73.37 ± 14.84	75.12 ± 13.98	76.34 ± 13.86	0.433
Height (cm)	162.23.18 ± 10.08	162.29 ± 9.00	164.53 ± 9.60	0.114
BMI (kg/m²)	28.14 ± 4.85	28.51 ± 4.78	28.18 ± 4.36	0.865
Waist circumference (cm)	93.81 ± 11.68	94.11 ± 11.11	94.95 ± 11.29	0.830
WHR	0.90 ± 0.08	0.89 ± 0.07	0.91 ± 0.07	0.333
TG (mg/dL)	109.92 ± 52.55	120.39 ± 66.75	140.27 ± 74.00	**0.015**
TC (mg/dL)	179.12 ± 47.21	178.60 ± 40.90	182.45 ± 39.65	0.854
LDL-C (mg/dL)	111.42 ± 36.74	106.55 ± 32.82	109.69 ± 30.19	0.638
HDL-C (mg/dL)	38.06 ± 11.59	39.24 ± 11.11	36.37 ± 10.18	0.313

Non-HDL	141.06±44.42	139.36±38.13	146.08±37.43	0.606
Physical activity (Met.h/day)	21.56±4.75	23.60±4.20	17.23±4.46	0.646
Sex (%)				0.735
Male	35.1	29.9	35.1	
Female	35.3	33.1	31.7	
Smoking history (%)				0.956
Yes	32.1	32.1	35.7	
No	35.6	31.7	32.7	
Education				0.093
Less than high school (%)	39.4	34.3	26.3	
High school and higher (%)	31.6	30.1	38.3	

Source: https://www.nature.com/articles/s41598-023-34451-x/tables/2.

Table 3.4. The Study Participants' Nutrients Intakes Among the Tertiles of UPFs

Variable	**UPFs**			
	T_1 (n=95)	**T_2 (n=79)**	**T_3 (n=62)**	**P-value**
Energy (kcal/day)	2,756.65±985.99	2,542.11±876.50	3,091.63±1,275.98	**0.008**
Carbohydrate (g/day)	434.77±164.46	396.90±144.51	472.59±193.16	**0.028**
Protein (g/day)	89.29±33.10	83.85±98.94	98.95±59.02	0.055
Fat (g/day)	80.12±35.40	74.57±30.00	97.02±49.59	**0.002**
Cholesterol (mg/day)	212.82±118.46	203.34±81.51	258.33±167.20	**0.023**
SFA (mg/day)	23.28±11.86	19.87±7.28	26.10±13.72	**0.004**
MUFA (mg/day)	25.76±12.32	23.50±9.57	31.54±17.23	**0.001**
PUFA (mg/day)	18.63±9.28	17.63±8.20	22.98±13.02	**0.005**
Vitamin B_6 (mg/day)	2.22±0.86	2.01±0.74	2.26±0.97	0.159
Vitamin B_9 (μg/day)	717.53±253.30	692.82±223.75	784.71±296.37	0.099
Vitamin B_{12} (μg/day)	3.53±1.84	3.34±1.59	4.35±2.93	**0.013**
Vitamin C (mg/day)	213.13±159.96	168.96±87.76	222.39±122.48	**0.027**
Vitamin E (mg/day)	14.81±7.72	13.42±4.47	16.15±7.29	0.055
Calcium (mg/day)	1,170.07±478.76	1,068.75±423.82	1,191.43±437.45	0.203
Magnesium (mg/day)	481.70±224.14	441.32±207.59	493.62±224.30	0.312
Zinc (mg/day)	13.29±5.83	12.52±5.64	14.13±7.18	0.306
Selenium (mg/day)	130.70±74.55	119.30±61.64	133.08±71.68	0.431

Source: https://www.nature.com/articles/s41598-023-34451-x/tables/3.

As presented in Table 3.5, participants in the last tertile of UPFs intake had significantly higher intakes of processed and red meat (P=0.007) and breads (P=0.022), also lower intake of margarine and sauces (P<0.001). There were no differences in non-dairy beverages, cookies and cakes, dairy products, potato chips and salty snacks, sweets and other food items intakes among the UPFs tertiles.

Table 3.5. Daily Intakes of UPFs Subgroups Across the Tertiles of UPFs

Variables	UPFs			
	T_1 (n=95)	T_2 (n=79)	T_3 (n=62)	P-value
Non-dairy beverage (%)	11.94±1.69	9.75±1.58	11.61±1.53	0.593
Cookies and cakes (%)	19.91±2.16	19.84±2.29	23.48±2.21	0.452
Dairy products (%)	11.34±1.66	10.27±1.18	8.68±1.09	0.376
Potato chips and salty snacks (%)	7.52±1.06	5.42±1.27	6.14±1.17	0.438
Processed meat and fast food (%)	7.01±1.06	10.40±1.49	13.42±1.77	**0.007**
Margarine and sauces (%)	16.55±1.51	15.70±1.37	7.99±0.70	**<0.001**
Sweets (%)	5.42±0.82	5.72±1.31	3.10±0.46	0.091
Bread (%)	5.99±0.96	5.17±0.86	10.57±2.20	**0.022**
Others (%)	14.33±1.93	17.29±2.12	15.01±1.97	0.556

Source: https://www.nature.com/articles/s41598-023-34451-x/tables/4.

We found that, higher UPFs intake was associated with increased OR of serum TG and HDL abnormality in both crude (OR 3.41; 95% CI 1.58, 7.34; P-trend=0.001 and OR 2.99; 95% CI 1.31, 6.82; P-trend=0.010) and adjusted model (OR 3.69; 95% CI 1.67, 8.16; P-trend=0.001 and OR 3.38 95% CI 1.42, 8.07; P-trend=0.009).

- **Discussion:**

We found that higher intake of UPFs was associated with elevated lipid profile abnormality including TG and HDL. In terms of other blood lipids such as TC, LDL and non-HDL, while they were not significantly associated with UPFs intake, all were higher in last tertile of UPFs intake.

In agreement with our study, Lima et al. reported that higher UPFs consumption was associated with higher TG and lower HDL-C levels. Furthermore, a large cohort of older adults in Spain found the same results, and a systematic review and meta-analysis resulted in a negative association between UPFs consumption and HDL-cholesterol levels. On the other hand, a cross-sectional study on Ecuadorian adolescents showed that dietary patterns consisted of processed foods were associated with an increased level of LDL and cholesterol. Also, a longitudinal study in Brazil on preschool children found that intake of UPFs was a predictor of higher total cholesterol and LDL cholesterol but not HDL and TG probably because effects of food habits on serum LDL and total cholesterol in children are stronger than its effects on other dyslipidemia markers.

We found a strong association between the dietary contribution of UPFs and the dietary content of energy, carbohydrates, fat, cholesterol, and SFAs. Moreover, intake of MUFAs and PUFAs increased parallel to UPFs consumption, probably due to high fat content of the UPFs.

With regard to anthropometric indices, although WC and WHR were not significantly associated with UPFs intake, participants in the last tertile of UPFs had higher mean of WC and WHR. Findings of some cohort studies indicated that higher UPFs intake was associated with greater adiposity accumulation, higher BMI, weight gain and incidence of obesity. Besides, another cohort study on overweight and obese participants aged 55–75 years reported that higher UPFs intake was associated with a higher age-related increase in visceral and overall adiposity. Another similar study also showed a positive association between UPFs intake and the incidence of abdominal obesity in adults. A systematic review and meta-analysis and a multi-national cohort study also reported positive associations between higher consumption of UPFs and general and abdominal obesity. We assume that our opposite findings is due to the high mean of our participants' weight, WHR, and WC.

UPFs are often calorie-dense, contain large amounts of fats, saturated fats, trans fats, sodium, and simple sugars with high glycemic index and contain no or small amount of fibers, vitamins, minerals, or other bioactive compounds which naturally exist in fresh foods, so they are nutritionally unbalanced.

Several mechanisms have been suggested to explain the adverse effects of UFPs on lipid profile and health. First of all, due to their intrinsic palatability, overconsumption of UPFs may result in physiological disruption of hunger and satiety patterns. Besides, energy-dense foods usually contain high amounts of free sugar and trans fats which enhance lipogenesis and decrease fatty acids oxidation which lead to their aggregation in tissues and blood circulation and elevated LDL level. Furthermore, partially hydrogenated vegetable oils content of UPFs contain trans fatty acids which have adverse impacts on lipid profile. In addition, scientific evidence support the hypothesis that the interacting effects of substances produced through the high-heat processing of oils determine their health effects. Heat processing also causes degradation of food substances and formation of furans in UPFs. Higher intake of UPFs increase exposure to phthalates which is used in the packing process and could transfer to food. The accumulation of phthalates, bisphenols, furans, and their metabolites may ultimately lead to lower HDL-c and elevated TG levels through disruption of endocrine functions. Evidence has shown that urinary concentration of phthalates and their metabolites was positively associated with TG and negatively associated with LDL levels. On the other hand, the impact of carbohydrates on lipid profile also depends on their sources and processing methods. While consumption of free sugar elevates serum TG, whole grains intake decreases TC,

LDL and TG levels. Consumption of minimally-processed whole grains such as oatmeal, instead of highly processed refined grains may improve lipid profile. Our study has some limitations as follows: First, because of the nature of the cross-sectional study, we were not able to assess causal correlations between UPFs intake and lipid profile. In the second place, the study was done in Shiraz city, so it should be generalized to other Iranian adults with caution. Ultimately, although we have removed the effects of some confounders in our analysis, there may be some others which have not been recognized in this study. In conclusion, our results showed significant associations between ultra-processed foods intake and dietary nutrient profiles which resulted in dyslipidemia as a risk factor for chronic diseases. Findings of the present study highlight a necessity for more evidence, particularly longitudinal, to define the effect of UPFs on lipid profiles.

CLASS ACTIVITY

Fill out the table(lipid table). Try your best to complete all parts of the table. Examine your corrected table and correct any mistakes you may still find.

Molecule	Type of Biomolecule (be Specific)	Functional Groups Found in Molecule	Elements Found in Molecule	C:H:O Ratio (Approx.)	Molecular Arrangement (Description of Main Components)	Type of Linkage (be Specific)	Type of Organisms Where Molecule is Found	Functions in Living Organisms
Cholesterol								
Oleic acid								
Phosphatidylet-hanolamine (cephalin)								
Stearin								

SUMMARY

- A carbohydrate is a biomolecule consisting of carbon (C), hydrogen (H) and oxygen (O) atoms, usually with a hydrogen-oxygen atom ratio of 2:1 (as in water) and thus with the empirical formula $Cm(H_2O)n$ (where m may or may not be different from n), which does not mean the H has covalent bonds with O (e.g., with CH_2O, H has a covalent bond with C but not with O).
- Carbohydrates are central to nutrition and are found in a wide variety of natural and processed foods.
- Lipids are a broad group of organic compounds which include fats, waxes, sterols, fat-soluble vitamins (such as vitamins A, D, E and K), monoglycerides, diglycerides, phospholipids, and others.
- Monosaccharides (from Greek monos: single, sacchar: sugar), also called simple sugars, are the simplest forms of sugar and the most basic units (monomers) from which all carbohydrates are built. Simply, this is the structural unit of carbohydrates.
- A disaccharide is the sugar formed when two monosaccharides are joined by glycosidic linkage Like monosaccharides, disaccharides are simple sugars soluble in water. Three common examples are sucrose, lactose, and maltose.
- Carbohydrate consumed in food yields 3.87 kilocalories of energy per gram for simple sugars, and 3.57 to 4.12 kilocalories per gram for complex carbohydrate in most other foods.
- Carbohydrate metabolism is the whole of the biochemical processes responsible for the metabolic formation, breakdown, and interconversion of carbohydrates in living organisms.
- Carbohydrates are formally defined as containing carbon, hydrogen, and oxygen in the ratio of 1:2:1. In practice, dietary carbohydrates comprise compounds that can be digested or metabolically transformed directly into glucose, or that undergo oxidation into pyruvate, including some sugar alcohols (e.g., sorbitol).
- Fatty acyls, a generic term for describing fatty acids, their conjugates and derivatives, are a diverse group of molecules synthesized by chain-elongation of an acetyl-CoA primer with malonyl-CoA or methylmalonyl-CoA groups in a process called fatty acid synthesis.
- Glycerolipids are composed of mono-, di-, and tri-substituted glycerols, the best-known being the fatty acid triesters of glycerol, called triglycerides.
- Glycerophospholipids, usually referred to as phospholipids (though sphingomyelins are also classified as phospholipids), are ubiquitous in nature and are key components of the lipid bilayer of cells, as well as being involved in metabolism and cell signaling.
- Sphingolipids are a complicated family of compounds that share a common structural feature, a sphingoid base backbone that is synthesized de novo from the amino acid serine and a long-chain fatty acyl CoA, then converted into ceramides, phosphosphingolipids, glycosphingolipids and other compounds.

- Sterols, such as cholesterol and its derivatives, are an important component of membrane lipids, along with the glycerophospholipids and sphingomyelins.
- Prenol lipids are synthesized from the five-carbon-unit precursors isopentenyl diphosphate and dimethylallyl diphosphate, which are produced mainly via the mevalonic acid (MVA) pathway.
- The major dietary lipids for humans and other animals are animal and plant triglycerides, sterols, and membrane phospholipids.

REVIEW QUESTIONS

1. What is the structure of monosaccharides.
2. Why are carbohydrates generally called saccharides? How are they classified according to this term?
3. How many cycles are there in carbohydrate metabolism?
4. What type of bond is involved between two units of a disaccharide and how is it formed?
5. Which class of complex lipids is a component of cell membranes? What is their general structure?
6. What is the relation between carbohydrate types and health outcomes?
7. What are the classification of lipids?
8. Discuss the role of lipids in our diet and their critical functions in the body.
9. Explain the importance of fats to the bioavailability of other nutrients.
10. Discuss the role of fats as an energy source for the body.

REFERENCES

1. Aberg, H., Lithell, H., Selinus, I., & Hedstrand, H. (1985). Serum triglycerides are a risk factor for myocardial infarction but not for angina pectoris. Results from a 10-year follow-up of Uppsala Primary Preventive Study. *Atherosclerosis, 54*, 89–97.
2. Admirand, W. H., & Small, D. M. (1968). The physicochemical basis of cholesterol gallstone formation in man. *Journal of Clinical Investigation, 47*, 1043–1052.
3. Ahrens, E. H., Jr., Blankenhorn, D. H., & Tsaltas, T. T. (1954). Effect on human serum lipids of substituting plant for animal fat in diet. *Proceedings of the Society for Experimental Biology and Medicine, 86*, 872–878.
4. Alderson, L. M., Hayes, K. C., & Nicholosi, R. J. (1986). Peanut oil reduces diet-induced atherosclerosis in cynomolgus monkeys. *Arteriosclerosis, 6*, 465–474.
5. Alhazmi, A., Stojanovski, E., McEvoy, M., & Garg, M. L. (2012). Macronutrient intakes and development of type 2 diabetes: A systematic review and meta-analysis of cohort studies. *Journal of the American College of Nutrition, 31*, 243–258.
6. American Health Foundation. (1979). Conference on the health effects of blood lipids: Optimal distributions for populations. *Preventive Medicine, 8*, 580–759.

7. Armstrong, M. J., Stephan, Z., & Hayes, K. C. (1982). Biliary lipids in New World monkeys: Dietary cholesterol, fat, and species interactions. *American Journal of Clinical Nutrition, 36*, 592–601.

8. Arntzenius, A. C., Kromhout, D., Barth, J. D., Reiber, J. H., Bruschke, A. V., Buis, B., van Gent, C. M., Kempen-Voogd, N., Strikwerda, S., & van der Bruschke, E. A. (1985). Diet, lipoproteins, and the progression of coronary atherosclerosis: The Leiden Intervention Trial. *New England Journal of Medicine, 312*, 805–811.

9. Baudet, M. F., Dachet, C., Lasserre, M., Esteva, O., & Jacotot, B. (1984). Modification in the composition and metabolic properties of human low density and high-density lipoproteins by different dietary fats. *Journal of Lipid Research, 25*, 456–468.

10. Beaton, G. H., Milner, J., Corey, P., McGuire, V., Cousins, M., Stewart, E., de Ramos, M., Hewitt, D., Grambsch, P. V., Kassim, N., & Little, J. A. (1979). Sources of variance in 24-hour dietary recall data: Implications for nutrition study design and interpretation. *American Journal of Clinical Nutrition, 32*, 2546–2549.

11. Beveridge, J. M. R., Connell, W. F., Mayer, G. A., Firstbrook, J. B., & DeWolfe, M. S. (1955). The effects of certain vegetable and animal fats on the plasma lipids of humans. *Journal of Nutrition, 56*, 311–320.

12. Carroll, G. T., Wang, D., Turro, N. J., & Koberstein, J. T. (2008). Photons to illuminate the universe of sugar diversity through bio arrays. *Glycoconjugate Journal, 25*(1), 5–10.

13. Gjuladin-Hellon, T., Davies, I. G., Penson, P., & Amiri Baghbadorani, R. (2019). Effects of carbohydrate-restricted diets on low-density lipoprotein cholesterol levels in overweight and obese adults: A systematic review and meta-analysis (PDF). *Nutrition Reviews (Systematic review), 77*(3), 161–180.

14. Hardy, D. S., Garvin, J. T., & Xu, H. (2020). Carbohydrate quality, glycemic index, glycemic load and cardiometabolic risks in the US, Europe and Asia: A dose–response meta-analysis. *Nutrition, Metabolism, and Cardiovascular Diseases, 30*, 853–871.

15. Honke, K., Zhang, Y., Cheng, X., Kotani, N., & Taniguchi, N. (2004). Biological roles of sulfoglycolipids and pathophysiology of their deficiency. *Glycoconjugate Journal, 21*(1–2), 59–62.

16. Hunter, J. E. (2006). Dietary trans fatty acids: Review of recent human studies and food industry responses. *Lipids, 41*(11), 967–992.

17. Laughlin, M. R., Bantle, J. P., Havel, P. J., Parks, E., Klurfeld, D. M., Teff, K., & Marovada, P. (2014). Clinical research strategies for fructose metabolism. *Advances in Nutrition, 5*, 248S–5S.

18. Livesey, G., & Livesey, H. (2019). Coronary heart disease and dietary carbohydrate, glycemic index, and glycemic load: Dose-response meta-analyses of prospective cohort studies. *Mayo Clinic Proceedings: Innovations, Quality & Outcomes, 3*, 52–69.

19. Mashaghi, S., Jadidi, T., Koenderink, G., & Mashaghi, A. (2013). Lipid nanotechnology. *International Journal of Molecular Sciences, 14*(2), 4242–4282.

20. Rebello, S. A., Koh, H., Chen, C., Naidoo, N., Odegaard, A. O., Koh, W., Butler, L. M., Yuan, J., & van Dam, R. M. (2014). Amount, type, and sources of carbohydrates in

relation to ischemic heart disease mortality in a Chinese population: A prospective cohort study. *American Journal of Clinical Nutrition, 100*, 53–64.

21. Schwartz, M. W., Seeley, R. J., Zeltser, L. M., Drewnowski, A., Ravussin, E., Redman, L. M., & Leibel, R. L. (2017). Obesity pathogenesis: An endocrine society scientific statement. *Endocrine Reviews, 38*(4), 267–296.

22. Tieri, M., Ghelfi, F., Vitale, M., Vetrani, C., Marventano, S., Lafranconi, A., et al. (2020). Whole grain consumption and human health: An umbrella review of observational studies. *International Journal of Food Sciences and Nutrition, 71*, 668–677.

23. van Buul, V. J., Tappy, L., & Brouns, F. J. (2014). Misconceptions about fructose-containing sugars and their role in the obesity epidemic. *Nutrition Research Reviews, 27*, 119–130.

CHAPTER 4

Amino Acids, Peptides, and Proteins

LEARNING OBJECTIVES

After studying this chapter, you will be able to:

- Understand the physicochemical properties of amino acids
- Explain protein structure
- Know protein denaturation
- Discuss functional properties of proteins
- Identify the most common sources of plant and animal proteins
- Focus on unconventional and alternative sources of proteins

KEY TERMS FROM THIS CHAPTER

Absorption	Cells	Enzymes
Glycine	Glycoproteins	Heteroproteins
Homoproteins	Metalloproteins	Nucleic acids
Nucleoproteins	Phosphoproteins	Polymers
Polysaccharides	Proteins	Stereoisomers

INTRODUCTION

A crucial part of biological systems is played by proteins. Enzymes are the only biological agents that carry out the chemical and biochemical processes necessary for a cell or organism to survive, despite DNA holding the information necessary for cell evolution and biological organization. There are thousands of enzymes known to exist. They are all responsible for catalyzing very distinct biological reactions in cells. Apart from their role as enzymes, proteins (like keratin, elastin, collagen, etc.) serve as structural elements in both complex organisms and cells. The chemical composition of proteins is primarily responsible for their diverse range of functions. 20 distinct amino acids combine to form proteins, which are extremely complex polymers. The components are joined by amide bonds that have been substituted. The fact that the amide linkage in proteins is a partial double bond rather than the ether and phosphodiester bonds found in polysaccharides and nucleic acids emphasizes the structural complexity of protein polymers even more. If proteins' composition weren't so complex, they wouldn't be able to carry out the vast array of biological tasks that they do. This complexity results in a wide variety of three-dimensional structural forms that have distinct biological purposes. These macromolecules were given the name proteins—derived from the Greek word proteois, which means of the first kind—to denote their biological significance. Proteins are composed of 50–55% carbon, 6–7% hydrogen, 20–23% oxygen, 12–19% nitrogen, and 0–2–0.2–3.0% sulfur at the elemental level. In ribosomes, protein synthesis takes place. Certain amino acid constituents undergo modification by cytoplasmic enzymes subsequent to synthesis. Certain proteins' elemental composition is altered as a result. Conjugated proteins, also known as heteroproteins, are proteins that have been modified or complexed with nonprotein constituents; homoproteins are proteins that are not enzymatically modified in cells. The term prosthetic groups are frequently used to describe the nonprotein constituents. Conjugated proteins include glycoproteins (ovalbumin, k-casein), lipoproteins (egg yolk proteins, several plasma proteins), phosphoproteins (a- and b-caseins, kinases, phosphorylases), and nucleoproteins (ribosomes) (hemoglobin, myoglobin, and several enzymes). While the other conjugated proteins are noncovalent complexes containing metal ions, lipids, or nucleic acids, glyco- and phosphoproteins have covalently linked carbohydrate and phosphate groups, respectively. The right circumstances allow for the dissociation of these complexes.

Another way to categorize proteins is based on their overall structural arrangement. Hence, polypeptide chains fold on themselves to form globular proteins, which are spherical or ellipsoidal in shape. Conversely, fibrous proteins are rod-shaped molecules made up of linear polypeptide chains that have been twisted (e.g., keratin, elastin, collagen, and tropomyosin). Actin and fibrin are examples of small globular proteins that can aggregate linearly to form fibrous proteins. Globular proteins make up the bulk of enzymes, and fibrous proteins are almost always involved in structural functions.

Serum albumin, transferrin, hemoglobin, enzyme catalysts, structural proteins, contractile proteins (myosin, actin, tubulin), hormones (insulin, growth hormone), antibodies (immuno-globulins), storage proteins (egg albumen, seed proteins), and protective proteins (toxins and allergens) are some of the biological roles that proteins perform. Plant seeds and eggs are the main sources of storage proteins. These proteins provide nitrogen and amino acids needed for seeds and embryos to germinate. The defense mechanism that ensures the survival of specific microorganisms and animals includes the protective proteins.

Although the primary 20 amino acids are the building blocks of all proteins, some proteins may lack one or more of these amino acids. The sequence in which the amino acids are joined by amide bonds accounts for the variations in the structure and function of these thousands of proteins. It is literally possible to create billions of different proteins with distinct characteristics by varying the polypeptide chain length, amino acid type and ratio, and amino acid sequence. Any protein produced by the body can be used as food. However, in practical terms, food proteins can be described as readily absorbed, safe, sufficiently nourishing, functionally useful in food products, and readily available. The main food sources of protein in the past have been milk, meats (including fish and poultry), eggs, cereals, legumes, and oilseeds. But in order to meet future demand, nontraditional sources of proteins for human nutrition must be developed due to the growing global population. However, the cost and capacity of these novel protein sources to replace traditional protein ingredients in processed and homemade foods will determine how suitable they are for use in food. Food proteins' structural and other physicochemical features are linked to their functional qualities. To enhance the functionality of proteins in food and to create new, affordable protein sources that can rival traditional food proteins, a basic understanding of the physical, chemical, nutritional, and functional properties of proteins as well as the changes these properties undergo during processing is necessary.

4.1. PHYSICOCHEMICAL PROPERTIES OF AMINO ACIDS

It speaks of the molecular characteristics or the compound's physical and chemical characteristics that determine its intrinsic chemical reactivity. Stereochemistry, electrostatic qualities, optical qualities, titration patterns, absorption spectra, and isoelectric properties are all included.

4.1.1. Structural Features

All standard amino acids are alpha-amino acid (⊠) with amino group (NH_2), carboxyl group (COO^-) and functional group (R) are all attached to the alpha carbon. Diverse functional groups give amino acids their diversity. Aliphatic, aromatic, carboxylic, sulfur-containing, amino, amide, and hydroxyl groups are examples of functional groups. The R group is just another hydrogen atom found in glycine. Amino acids can have two possible stereoisomers due to the tetrahedral arrangement of bonding orbitals around the ⊠-carbon atom, which allows the four different groups to occupy two distinct spatial arrangements.

4.1.2. Absolute Configuration

The stereochemistry of the molecule provides a description of the amino acid's absolute configuration. The two forms represent an enantiomer class of stereoisomers since they are non-superimposable mirror images of one another. Two widely used DL systems—the DL system and the RS system—can be used to describe the absolute configuration of a molecule. DL system- the absolute configuration of amino acids are simple sugar are determined using this system, it is based on the absolute configuration of glyceraldehyde, a three-carbon sugar. The arrangement of amino acids is defined by their amino groups; an amino acid is considered to be in L configuration if the ⊠-amino group is on the left side, and in D configuration if the amino group is on the right. The L- configuration is present in all common amino acids.

4.1.3. Optical Characteristics

The term "optical characteristic" refers to an amino acid's optical activity, which is defined as a compound's capacity to rotate plane-polarized light. An amino acid's asymmetric

⊠ carbon functions as a chiral molecule. An optically active amino acid has a chiral molecule. The only exception is glycine, which is a symmetric achiral molecule because its functional group is a hydrogen atom, as was previously mentioned. Since the achiral molecule cannot rotate a plane-polarized light, it is optically inactive.

- **Plane Polarized Light:** It is a light wave made up of unidirectional vibrations, meaning that each wave's vibration is in the same plane as its direction. While the direction of wave propagation remains constant, the electric vectors of vibration in an unpolarized light are randomly positioned about the propagation axis.
- **Plane of Polarization:** It refers to the direction of polarization of a polarized wave.

4.1.4. Electrostatic Characteristics

Protein-protein and amino acid-nucleic acid interactions, as well as the location of an amino acid within the tertiary structure of proteins, are examples of properties that are defined by electrostatic characteristics. The side chain of an amino acid is responsible for its electrostatic properties.

The net charge of an amino acid and its relative stability in proteins are determined by the charge on the side chain. Based on their polarity, the side chains of standard amino acids can be divided into polar and non-polar groups. There are two additional categories for polar side chains: charged and uncharged. It is possible to further differentiate side-chain amino acids into positively and negatively charged forms.

4.1.5. Absorption in Amino Acid

The process of obtaining energy in the form of radiation—typically UV radiation in the case of amino acids—is known as absorption. The electrons move into an excited state as a result of this gain. Each biomolecule absorbs light at a particular wavelength. The maximum absorption that corresponds to a wavelength on a graph is known as the absorption peak. Three amino acids exhibit notable ultraviolet absorption because of their aromatic side chains. They are tyrosine, tryptophan, and phenylalanine.

4.1.6. Titration of Amino Acid

Using the titration method, the physicochemical properties of the amino acid can be greatly optimized. Titration curves, which show the gradual addition or removal of protons to form a protonated and deprotonated state, are characteristic of amino acids.

At pH neutrality, amino acids exist in the zwitterionic form. Amino acids move toward a fully deprotonated stage when the pH rises and subsequently increases the concentration of OH-, and vice versa as the pH falls and the H+ concentration rises. A group's propensity to donate protons is expressed in terms of protons (pKa); the carboxylic group's pK value is represented by pK1, and the amino group's by pK_2. The ionic strength, temperature, and microenvironment of the ionizable group all affect the pKa value. Titration curve analysis is a crucial step in determining the isoelectric pH. The pH at which an amino acid has no net charge is known as isoelectric pH (pI);

$$pI = pK_1 + pK_2/2$$

4.2. PROTEIN STRUCTURE

Two or more amino acid residues are joined in various configurations and orientations to form proteins. Peptide bonds, which are created when two amino acids condense and release water molecules in the process, bind the amino acids in proteins together. Although the descriptions of peptides and proteins' structures may sound similar, there is a subtle distinction between the two because of their different sizes, structures, and functions, as will be discussed below.

- Peptides can be further divided into oligopeptides and polypeptides. Peptides are made up of two to 50 amino acids. An oligopeptide is made up of 20–30 amino acid residues, whereas a polypeptide is a chain of 20 amino acids or more connected by peptide bonds. On the other hand, proteins consist of an extended sequence of 50 or more amino acids joined by peptide bonds.
- Proteins have four structural levels: primary (linear), secondary, tertiary, and quaternary, while peptides primarily have a linear structure.
- Proteins perform a broader range of tasks, such as determining structure, facilitating enzymatic reactions, and regulating hormonal processes, whereas peptides are involved in controlling the actions of other molecules.

Different chemical interactions within the structure of proteins allow for the different levels of protein structure.

For instance, a number of non-covalent interactions, including hydrogen bonding, ionic interactions, Van der Waals forces, and hydrophobic interactions, are responsible for the folding of the linear sequence of amino acids into three-dimensional structures. A brief explanation of the four levels of protein structure is provided below.

4.2.1. Primary Structure

The arrangement of amino acids into a linear polypeptide chain is the main structural component.

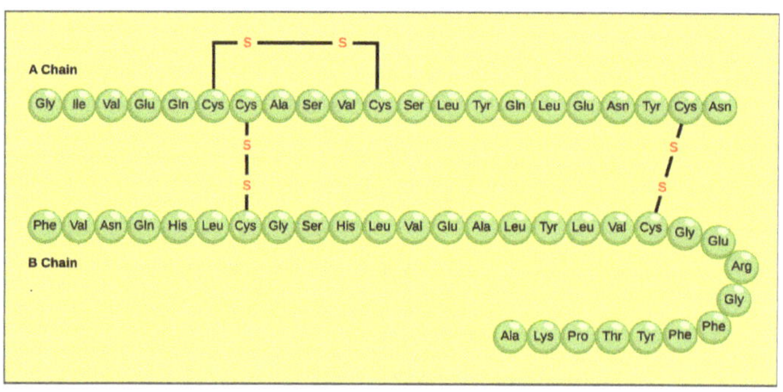

Figure 4.1. *The representation of the primary structure of the protein.*

Source: https://conductscience.com/structure-and-functions-of-proteins/.

Only peptide bonds and disulfide bonds bind the amino acids together in this instance. Deciphering the genetic codes or genes corresponding to the proteins determines the sequence of amino acids in a polypeptide chain, which determines the functions of the proteins.

4.2.2. Secondary Structure

The regular, repeating arrangements of neighboring amino acids within a polypeptide chain are known as secondary structure. This is the first stage of polypeptides folding into three-dimensional structural form, which happens before tertiary and quaternary structures, and it is caused by the formation of hydrogen bonds between molecules.

The most common secondary structures are alpha-helix and beta-sheet.

- **Alpha Helix:** The amino hydrogen (N-H) of the fourth amino acid in the sequence is hydrogen-bonded to the carbonyl group (C=O) of one amino acid in this structural form. As a result, the linear chain is drawn into the shape of a helical ribbon, with three to six amino acids in each turn.
- **Beta Sheet or Strand:** In this instance, two or more polypeptide chain segments are arranged side by side. Either parallel or antiparallel hydrogen bonds hold the strands together.

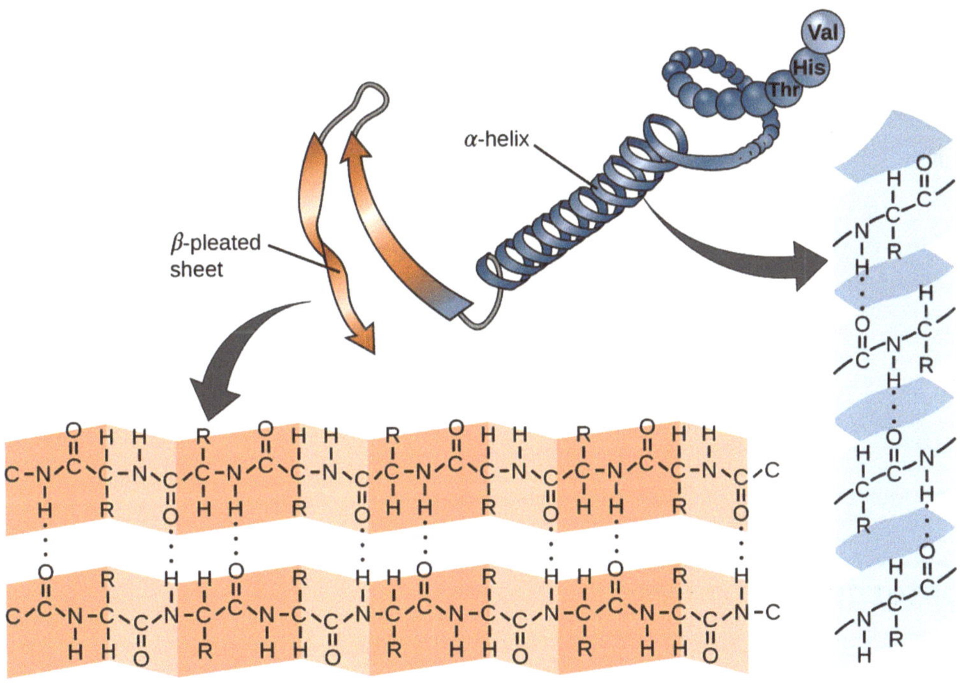

Figure 4.2. *Structural forms of alpha-helix and beta-pleated sheets.*

Source: https://conductscience.com/structure-and-functions-of-proteins/.

4.2.3. Tertiary Structure

Here, one or more domains are involved in the polypeptide chain's further folding into a three-dimensional space. Non-specific hydrophobic interactions cause the beta sheets and alpha helices to fold into a compact globular structure. Within the polypeptide chain, interactions take place between polar, nonpolar, acidic, and basic R groups. Proteins are incapable of carrying out any of their functions when they lose their tertiary structural form. The hydrophilic R groups are found on the surface or outside the structure of proteins in an aqueous environment, whereas the hydrophobic or nonpolar groups travel into the interior of the proteins.

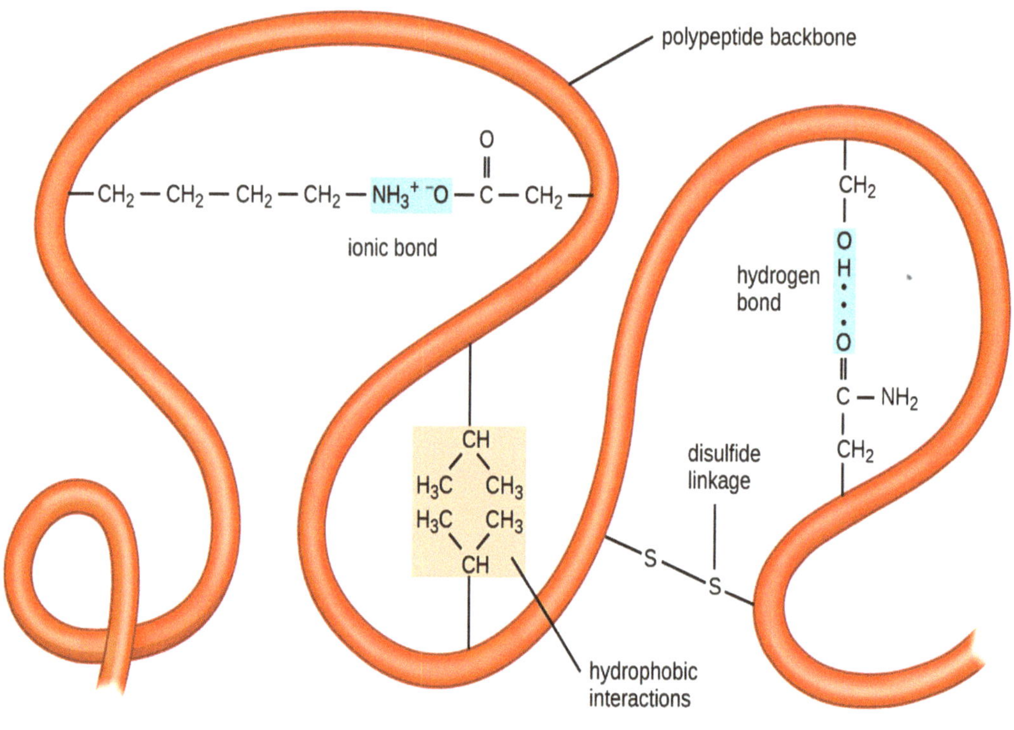

Figure 4.3. *The representation of the tertiary structure of the protein.*

Source: https://conductscience.com/structure-and-functions-of-proteins/.

4.2.4. Quaternary Structure

A quaternary structure is formed by the interaction of multiple polypeptide chains. The polypeptide chains may differ in their amino acid sequences or be identical in multiple copies. The weak interactions that hold several polypeptides together to form a quaternary structure are hydrogen bonding and London dispersion forces. Hemoglobin, which carries oxygen in the blood and is made up of four peptide chains — two alpha and two beta - to form a tetramer, is a well-known example of a quaternary structure.

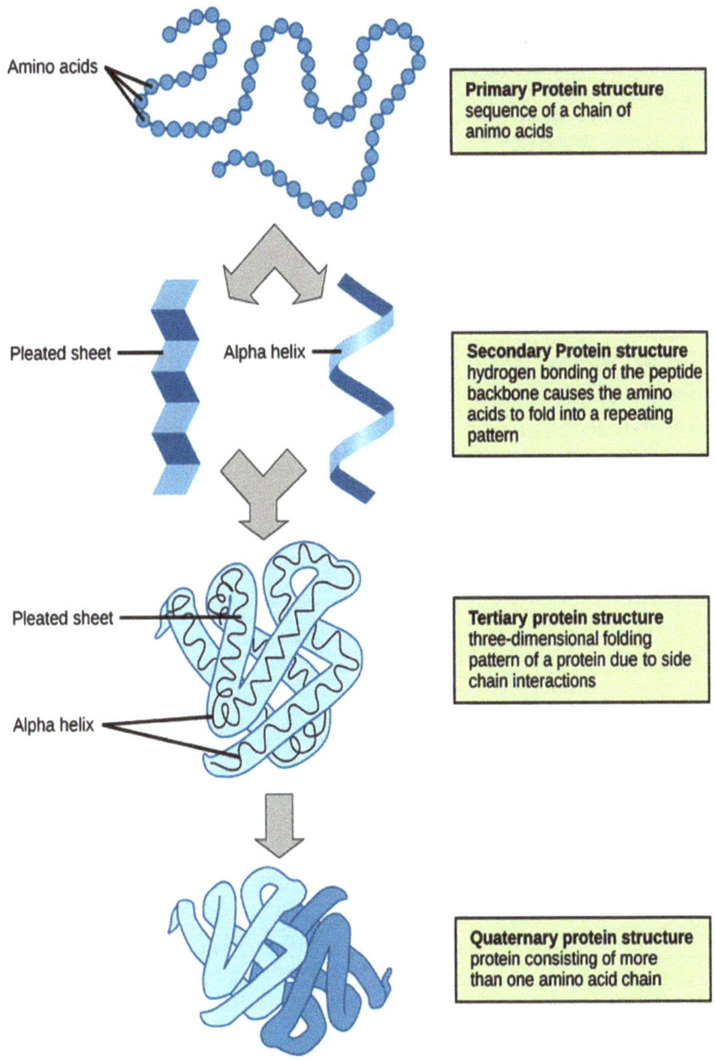

Figure 4.4. *An illustration of the four levels of protein structure.*

Source: https://conductscience.com/structure-and-functions-of-proteins/.

4.3. PROTEIN DENATURATION

Given the complexity of many food systems, it is not surprising that denaturation is a difficult process with a difficult to grasp explanation. Proteins undergo denaturation, which is a structural change that results in the loss of shape and function. When proteins are exposed to heat, acid, or other environmental stimulants, this can happen. When a protein becomes denaturated, it loses its ability to carry out its original function, which can alter the food's flavor and texture. For instance, when enzymes, a class of protein, are denatured, they lose their capacity to catalyze and turn inactive. This is the precise process that occurs when a protein is denaturated.

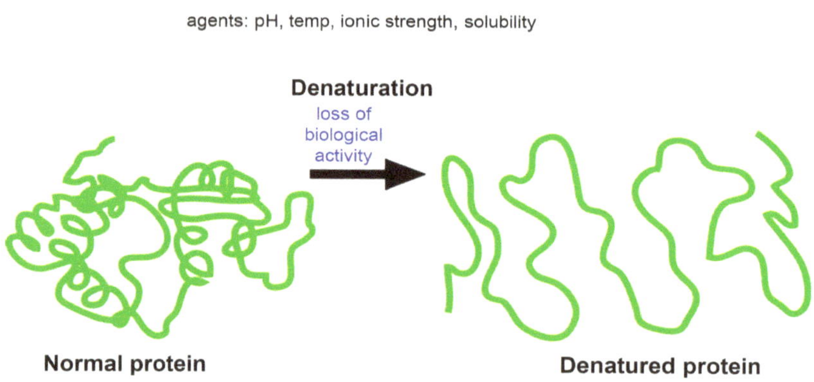

Figure 4.5. *When a protein is exposed to a different environment, such as increased temperature, it unfolds into a single strand of amino acids.*

Source: https://pressbooks-dev.oer.hawaii.edu/humannutrition/chapter/the-role-of-proteins-in-foods-cooking-and-denaturation/.

A protein's secondary, tertiary, and/or quaternary structure alters when it becomes denatured. Its fundamental structure remains unchanged. Therefore, the peptide bonds are not broken. The chemical covalent bonds known as peptide bonds hold the amino acids in a protein chain together. Because of their remarkable strength and stability, peptide bonds are crucial to the stability and functionality of proteins. A repeating series of peptide bonds makes up the backbone of a protein chain, giving the protein its unique stability and shape. Although the protein unfolds when denatured, the amino acid sequence doesn't change.

Denaturation is the result of a treatment that breaks hydrophobic interactions, ionic bonds, or hydrogen bonds within a protein. If the denaturing agent keeps the protein in its unfolded state, denaturation can frequently be undone, enabling the protein to refold into its original conformation after the agent is eliminated. However, if an unfolded protein is stabilized by interaction with other protein chains—for example, with egg proteins during boiling—denaturation may also become irreversible.

4.3.1. Factors That Affect Protein Denaturation in Food

Protein denaturation can be caused by a number of things. Among these are variations in temperature, freezing, ionic strength, pH, and surface conditions.

4.3.1.1. Heat

Mild alterations usually lead to denaturation. Therefore, denaturation of proteins can be induced by mild heat treatments like pasteurization or blanching. Temperatures between 135°F (57°C) and 167°F (75°C) denature meat proteins, which has a major effect on texture, water-holding capacity, and shrinkage. When cutting or grinding, the natural proteins in meat maintain the tissue's moisture content. The proteins lose their natural state and interact with water less when heated. Since the myosin proteins are no longer retaining the water, the juices that leak from a cooked steak are primarily water. When a steak is rested after cooking, some of the water can bind to proteins in new ways, preventing juice from oozing out when the steak is sliced.

Heat typically modifies the polypeptide chains' tertiary structure, which leads to a disorganized configuration. But this is not always the case. Gelatin and milk casein are two examples of proteins that withstand heat well. Furthermore, they are heat treatable without seeming to lose any of their stability. Because of casein's remarkable stability, milk can be sterilized, boiled, and concentrated without coagulating. The heat stability of these proteins is explained by the inhibitions placed on the formation of a folded tertiary structure. The relatively high proline and hydroxyproline content in the heat stable proteins is what causes these restrictions.

Peptide chains without proline are more likely to form intra- and intermolecular hydrogen bonds than chains with multiple proline residues. This illustrates the direct relationship between the amino acid composition of proteins and their secondary and tertiary protein structures, which control the functionality of the protein.

4.3.1.2. Freezing

Protein denaturation can happen at low temperatures as well as during the thermal processing of food. Protein denaturation is also brought on by exposing food to freezing temperatures, which range from 32°F (0°C) to 5°F (–15°C). Compared to proteins derived from plants, muscle proteins are more prone to denaturation. Ice crystals form at low temperatures and have the ability to rupture cell membranes. This is followed by denaturation, which may be caused by surface phenomena at the interface between the two phases (protein solution and ice crystals) or by the freezing of water, which is necessary to keep the protein in its native conformation. One method of reducing denaturation is to freeze the food quickly, a process that is typically accomplished by industrial freezers.

This causes the cell's interior moisture to rapidly cool, which causes tiny ice crystals to form inside the cell. These crystals are often spear-shaped and are separated from one another by proteins. Less dehydration and protein damage occur as a result. On the other hand, slow freezing frequently causes the cell to become dehydrated and its ionic strength to rise. The extracellular ice crystals will grow, distort, and harm the membrane and proteins when water taken from the inside of the cell freezes onto them. Cryoprotectants like polyols or antioxidants are applied before freezing in the fish industry. By doing this, the functionality and stability of proteins are preserved during freezing and storage.

4.3.1.3. pH Change

One significant factor that may affect the denaturation of proteins in food is pH. A food product's pH can have an impact on the charge of the amino acid residues inside a protein molecule, which can change the protein's structure, stability, and functionality. Protein denaturation is the result of a food's protein molecules changing shape in response to pH changes. Food proteins may become denatured if they come into contact with acidic environments, like those found in citrus fruits or vinegar. The protein molecules unfold as a result of the acidic environment, losing their functional structure and being unable to carry out their intended function. But sometimes, like when making cheese, this is advantageous.

Lactic acid bacteria are used in cheese manufacturing to convert lactose, a milk sugar, to lactic acid, thus acidifying milk. As lactic acid concentration rises, milk's pH falls, making casein proteins less soluble and more likely to coagulate. In order to create cheese, the liquid whey must be separated from the solid curds through a process called coagulation. The unique flavor and texture of different cheeses are also enhanced by this acid-induced denaturation of proteins. For instance, to make fresh cheese, milk must be boiled and then acidified with vinegar or citric acid. The proteins become denatured due to the acidity, resulting in the formation of tiny, delicate curds that separate from the liquid whey. On the other hand, hard, aged cheese experiences a prolonged acidification process that leads to a greater degree of protein denaturation, a firmer texture, and a sharper flavor.

4.3.1.4. Ionic Strength Change

The concentration of ions in a solution is referred to as its ionic strength, and it can be raised by adding salt. Protein denaturation is influenced by a solution's ionic strength, which modifies the strength of electrostatic interactions between protein molecules. In general, a food's ionic strength can be increased to lessen the degree of protein denaturation. The reason for this is that solutions with high ionic strength tend to weaken the electrostatic bonds that hold together protein molecules, which in turn makes them less prone to denature and aggregate. Salts can be used in a variety of food processing applications, like meat curing, to cause protein denaturation and gel formation. In some situations, salts raise the ionic strength of the solution, which causes more widespread protein denaturation and aggregation. In certain processed meat products, this can lead to the formation of a gel-like structure, which is desired.

4.3.1.5. Surface Changes

Protein denaturation can be impacted by surface modification because it alters how proteins interact with their environment.

Proteins can interact with other molecules in their environment, other proteins, and water molecules through a variety of interfaces. The stability and structure of the protein may be impacted by these surface interactions. Modifying the protein surface's hydrophobicity—the capacity of a material to repel water—is one way to cause surface alterations. The hydrophilic and hydrophobic surfaces of proteins affect how they interact with other proteins and water molecules. Certain substances or alterations in the environment's polarity may alter the protein surface's hydrophobicity, decreasing its stability and causing denaturation. Protein denaturation may also be impacted by modifications to the protein's polarity. The charge distribution on the surface of proteins can be modified by specific solvents or pH shifts, which can impact the potency and selectivity of electrostatic interactions between proteins and other molecules. The act of beating egg whites causes the most surface changes in food that are frequently noticed. Albumin is a globular, water-soluble protein found in egg whites. The albumin molecules unfold during egg beating, revealing the hydrophobic amino acid residues to the atmosphere. This facilitates the incorporation of air into the mixture by the albumin molecules. The protein consequently becomes more hydrophobic, generating a new surface area where it can interact with adjacent protein molecules' hydrophobic amino acids. A stable foam structure is produced as a result of these protein networks stabilizing the air bubbles and preventing them from escaping. This foam structure's high volume and light texture make it ideal for cakes, meringues, and soufflés.

4.3.2. Applications of Protein Denaturation in Food

Since the food business depends on modifying proteins to create appetizing food products, a food manufacturing perspective provides a better understanding of protein denaturation. But since it happens practically always when cooking and preparing food, it is also crucial. Cooking denaturates protein, which improves the flavor and digestibility of meat. Protein denaturation happens in a number of food manufacturing processes. A few examples are as follows: Cheese production is one of the most popular uses of protein denaturation. Curds are created when milk proteins are denatured by the addition of acid during the cheese-making process. The curds are used to make a range of cheese products once they are extracted from the liquid whey. Denaturation of milk proteins is closely monitored to ensure consistent quality and product characteristics, as it is essential for the formation of cheese.

The production of foams like meringue or whipped cream also requires protein denaturation. To construct a stable foam structure, proteins must denature and unite into a network that can maintain air bubbles. To achieve the ideal foam texture and stability during foam generation, controlling the amount of protein denaturation is essential. Temperature, pH, and mixing duration can all affect how much protein denaturation occurs and, consequently, how good the final product turns out.

The production of emulsions like mayonnaise and salad dressings also uses protein denaturation. Emulsions are frequently made by mixing water and oil, two substances that do not naturally combine. Conversely, denatured proteins can stabilize the emulsion by encasing and holding down oil droplets in a solution based on water. The kind and quantity of proteins used, as well as pH and temperature, all have an impact on the stability of the emulsion.

4.4. FUNCTIONAL PROPERTIES OF PROTEINS

The ability to form or stabilize emulsions (the oil/water interface), foams (the air/water interface), or solubility (combining the connections between water and proteins) are examples of surface properties that proteins can have. Furthermore, gelation, texture, and molding sensory properties (taste and smell) are hydrodynamic properties of proteins that are dependent on intermolecular interactions (Foegeding & Davis, 2011). The specific characteristics of proteins' molecules—such as their size, shape, susceptibility to denaturation, flexibility, amino acid composition, hydrophilicity, and hydrophobicity—as well as the charge distribution within the molecule, the type and quantity of microdomain structures, the molecule's ability to adapt as a whole or as individual domains to changing environmental conditions, and the nature of the interactions between various proteins and other food ingredients—all directly affect the functional properties of proteins (Figure 4.6).

Significant environmental parameters in the protein's environment, such as pH, temperature, pressure, and ionic strength, have an impact on the functional characteristics of proteins (Wijaya, Patel, Setiowati, & Van Der Meeren, 2017). Furthermore, technological processes significantly influence how proteins' functional properties are formed. Proteins form complex systems with other food ingredients that influence the formation of those properties (Lam, & Nickerson, 2013). The majority of hydrophilic functional groups in proteins are found on the molecules' surfaces. The hydrophobic groups are partially contained inside them, though. Within globular proteins, hydrophobic amino acid residues may occupy 40–50% of the molecule's surface. Their precise location within the polypeptide chains influences how protein molecules form their surfaces, their capacity to form oligomers and micellar structures, and their functional characteristics (Udgaonkar, 2015).

Protein structure and function are largely shaped by sulfuric amino acids, and thiol groups can be oxidized to create intra- and intermolecular disulfide bridges. The structure and functions of the proteins are altered by these interactions. Disulfide bridges that bend in both directions stabilize the tertiary structure of proteins and impact their functional characteristics.

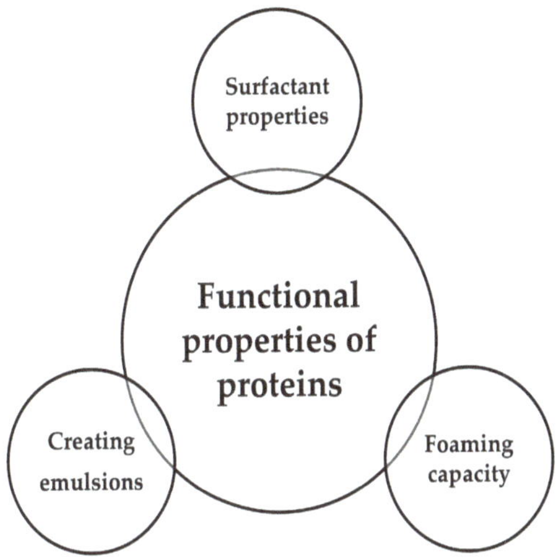

Figure 4.6. *Classification of the main functional properties of proteins.*

Source: Małecki J, Muszyński S, & Sołowiej BG, (2021). Proteins in food systems-bionanomaterials, conventional and unconventional sources, functional properties, and development opportunities. Polymers (Basel), 13(15):2506. doi: 10.3390/polym13152506. PMID: 34372109; PMCID: PMC8347159.

4.4.1. Functional Properties: Solubility

Solubility is one of the functional attributes of proteins that needs special consideration. One of the fundamental functional characteristics of food proteins is their solubility. The application of protein preparations in food technology is primarily determined by this characteristic. Good functional qualities of the protein are frequently linked to its high solubility, which enables manufacturers to produce food products with the desired, repeatable, and predictable qualities. Loss of solubility brought on by harsh food processing conditions is frequently a sign of denaturation and the subsequent cross-linking of proteins. The structure and characteristics of the solvent, temperature, the environment's pH, the concentration and charge of ions, and the type of interactions with other molecules all affect how soluble proteins are.

What matters most about a protein molecule is its surface hydrophobicity and the resulting electric charge that dictates how it behaves in the solvent. The characteristic of the varied electrostatic potential of different protein surface fragments, which determines the protein's spatial shape and behavior toward polar and apolar solvents, is indicated by surface hydrophobicity. Enzymatic modification (density cargo) such as dephosphorylation and chemical modification such as glycosylation are two ways to alter hydrophobic/hydrophilic surface protein molecules. Using proteins contributes to an environment's continued high solubility. This characteristic is especially important in environments that are acidic. For instance, coagulation is avoided when proteins are added to juices and other drinks.

4.4.2. Surface Properties of Proteins

Proteins' capacity to adsorb at the interface, lower surface tension, and form a coherent layer around oil droplets or air bubbles is what causes the phenomenon of them stabilizing emulsions or foam. Phase interface adsorption might not happen if the particle surface is completely hydrophilic. Adsorption, however, might take place if it interacts with the phase's surface and has few hydrophobic residues. Put another way, the statistical likelihood of hydrophobic groups on the surfaces of protein molecules with phase boundaries colliding determines adsorption at the interface between oil and water and air. Protein structures' amphiphilicity and membrane adsorption are influenced by their conformational stability, reorganization potential at the interface, and polar and apolar functional group distribution symmetry or asymmetry. Owing to these characteristics and their superior satiety-promoting potential over that of carbs and fats, proteins can function as a gastrointestinal hormone regulator to heighten feelings of fullness and decrease the amount of calories taken in during meals. Food items like chocolate, cheese puffs, and gelatin foams can be made more innovative by adding foams with bubbles, which also lower the calorific value of the products by making them bulkier than their regular counterparts.

Moreover, aerated gels can be used to create gastronomic structures, release taste, selectively supply bioactive particles, control satiety (just like foams), and form capsular products. As a result, proteins can enhance water absorption, inhibit syneresis, and give food products the correct texture. Immobilized water molecules and additional food ingredients like fibers, carbohydrates, and polyhydric alcohols aid in the creation of a stable gel structure for food products while preserving the gel matrix.

4.4.3. Foaming Capacity

Dispersing air bubbles in the liquid phase produces foams. Protein addition raises the viscosity of the aqueous phase, strengthening the interfacial film and causing foam. By interacting with the water molecules and air, proteins lower the surface tension, which permits the formation of foam bubbles. A coherent, flexible film surrounds the interphase of the air bubble after polar amino acid residues on the surfaces of the protein molecules react with the liquid and non-polar amino acids react with the air following adsorption on the air bubbles. Because the protein molecules' unaltered fragments interact with them, the bubbles stay apart from one another. Up to 99% of the total volume of the foam can be made up of the 0.1–1 mm diameter air bubbles. Surface hydrophobicity, the location of hydrophobic amino acid residues on the surface of the protein, thiol groups, cations and anions, carbohydrates, and lipids are the variables that affect foam formation. The ability of the protein to shield the foam from the effects of gravity and mechanical interactions determines how stable the foam is once it has formed. When electrostatic interaction forces are at their lowest, at a pH that is near to the protein's isoelectric point, stable foam is typically produced. The properties of foaming are enhanced by processes that increase hydrophobicity. Short heating times can enhance the foaming characteristics of the protein. For instance, whey proteins' ability to foam is enhanced by thermal denaturation for 30 minutes at temperatures between 40–60°C. The kind and quantity of protein determine the ideal heating parameters.

4.4.4. Creating Emulsions

Emulsions are dispersion systems made up of one or more immiscible liquids that exist as small droplets that are dispersed and continuous at the same time. Limiting contact and phase separation is preferable when combining oils and solutions with aqueous proteins. The formation of spherical droplets with external energy input results in minimal contact at first. To aid in the emulsion's formation and increase stability, a stabilizing agent is added. Proteins can be used as a stabilizer because they are an emulsifying agent. Protein-stabilized emulsions are energy-efficient because they don't require a lot of energy to form, and they guarantee that hydrophobic groups and water make minimal contact. Depending on the type of protein, different times are needed for a coherent layer to form around the oil droplets and for thermodynamic equilibrium to be reached. These phenomena occur fast in the case of loose, elastic-structured proteins, medium speed in the case of globular proteins, and slowly in the case of compact-structured proteins. The fundamental size of the emulsion is determined by the droplet size of the dispersed phase. In food product emulsions, these drops range in diameter from 0.2 to 10 mm. The size is determined by the energy input during emulsion formation, the emulsifier used, the viscosity difference between the two phases, and the emulsion production method. Products of low quality have drops larger than 10 mm in diameter. The drops in premium products, like mayonnaise, are 2–4 mm in size. The proteins ovalbumin, β-lactoglobulin, β-casein, α-lactalbumin, and lysozyme are frequently used in industry for emulsification.

Unlike low-molecular-weight emulsifiers, adsorption may have an impact on the structure of proteins. As a result, a good emulsifier ought to establish and maintain the newly created interface. Protein-stabilized emulsions exhibit greater stability at pH values other than the proteins' isoelectric point values. The viscosity of the continuous phase, gravity forces, the resulting charge, and the structure of the protein all affect how stable the emulsion is. The energy used during emulsification in the emulsion preparation device primarily controls the degree and kind of emulsion changes that occur over time. The stability of the emulsion is further determined by environmental factors like protein concentration, active acidity, oil/water phase ratio, and ionic strength. The ability of a protein to form and stabilize emulsions can be impacted by conformational changes caused by structural modifications. The emulsifying properties may be positively shaped by an increase in hydrophilicity.

4.4.5. Protein-Based Bionanomaterials

Many examples of solid and functional bionanomaterials, which are defined as a blend of small organic molecules or materials and bio-macromolecules (primarily proteins) can be found in nature. These hybrids, which can be used to create advanced and highly effective photosystems, metalloenzymes, antenna systems, and bionanocomposites, are based on this combination. Nanomaterials are important in the rapidly expanding fields of economics and science. Numerous new physicochemical properties and applications could arise from size reduction. Using methods and techniques to create materials, elements, and devices with at least one controlled dimension in the nanoscale range of 1–100 nm, nanotechnology is an innovative technology. To produce nanomaterials, two methods are employed: top-down and bottom-up. The top-down approach involves a decrease in particle size. The bottom-up approach considers building new structures from pre-existing nanoparticles. Depending on the characteristics of the finished product, the building blocks of the nanostructures created using this method may be atoms, molecules, or nanoparticles.

By altering the building material's size, regulating its exterior and interior characteristics, and enforcing particular guidelines for combining particles into a nanomaterial, it is possible to produce a material with the required properties (Tadros, Izquierdo, Esquena, & Solans, 2004). Applications of nanotechnology are found in all branches of food science, including agriculture, food processing, packaging, nutrition, safety, and nutraceuticals (Figure 4.7).

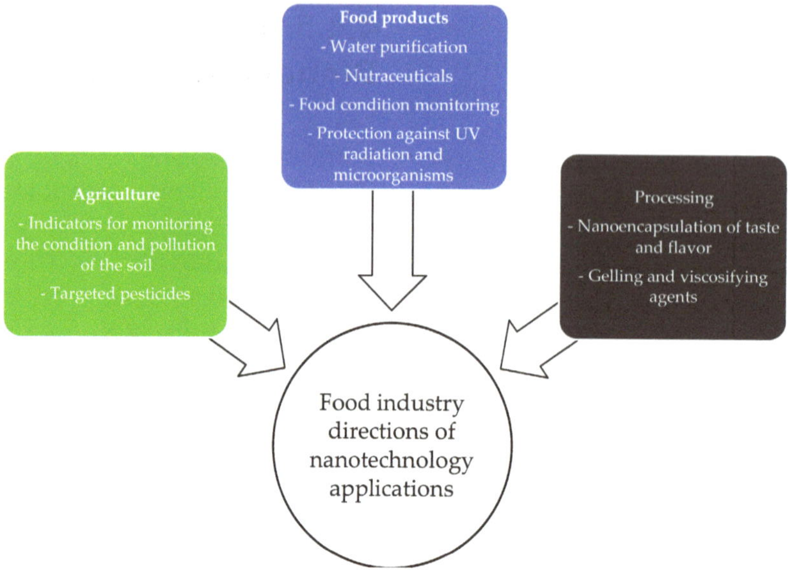

Figure 4.7. *Directions of nanotechnology applications in the food industry (Duncan T.V., 2011).*

Source: Małecki J, Muszyński S, & Sołowiej BG. (2021). Proteins in food systems-bionanomaterials, conventional and unconventional sources, functional properties, and development opportunities. Polymers (Basel), 13(15), 2506. doi: 10.3390/polym13152506. PMID: 34372109; PMCID: PMC8347159.

Novel methods are being utilized in the creation and blueprinting of fresh protein-based bionanomaterials. Beyond their inherent functionality, proteins can be used as highly tunable platforms to accommodate and bind synthetic materials, suggesting new functions for the hybrid system. This is why using proteins in the synthesis of bionanomaterials is significant. Not only can structural proteins be functional and versatile building blocks for design, but protein-based materials are also eco-friendly, long-lasting, biodegradable, and biocompatible when compared to other platforms.

4.5. THE MOST COMMON SOURCES OF PLANT AND ANIMAL PROTEINS

Food proteins are vital nutrients needed to sustain human health and a variety of physiological processes. Proteins are necessary components of food, particularly some conventional plant- and animal-derived protein sources. This section contains a list and description of them.

4.5.1. Soy

The main source of protein is soybeans, which are harvested globally each year at a rate of about 300 million metric tons. For many years, soybeans have been used as a food source. It is the most widely used genetically modified crop in the world, accounting for 80% of GM crop acreage globally, or about 100 million metric tons annually. Because of the uncontrolled modification of transgenic organisms released into the environment, the primary threat associated with genetically modified organisms is disturbance of biodiversity. There may be fewer of a given species as a result of modified plant varieties replacing traditional ones. Plant species and insects become more resistant to the chemical agents used to control them as a result of the establishment of agricultural monocultures, which also leads to a notable increase in their population. "Superweeds" can arise when transgenic plants are crossed with wild plants that grow quickly. Thus, in order to ensure the safety of genetically modified crops and transgenic food products, a thorough examination and elucidation of matters pertaining to GM plant toxicity and health safety, GM plant effects on other organisms, allergenicity of food products derived from GM plants, biological safety, and antibiotic resistance are required.

The largest producer of soy and soy-related products is the United States. Because of their high protein content (35–40%), soybeans are frequently used to produce oil, which results in the production of soy flour. Animal feed is made up primarily of defatted soy flour. The leftover flour is utilized to make a variety of high-protein goods that are meant for human consumption. In the daily diet of humans, soy protein can serve as a suitable replacement for meat and milk proteins due to its relatively high protein content and preferentially balanced amino acid composition. Soy flour, soy protein isolate, soy protein concentrate, textured, and hydrolyzed soy protein are the most popular soy products.

Soy protein concentrate (SPC), which is made from defatted soy flakes devoid of soluble cell wall components and contains about 65% protein, is produced. The most highly refined soy product is soy protein isolate (SPI), which contains about 90% protein. Isoelectric precipitation and alkalic extraction are the processes that form it. The extrusion method yields textured soy protein that has a meat-like texture. The main purpose of this product is to provide a complete or partial substitute for animal protein in a variety of food products, especially for vegans and vegetarians who follow a plant-based diet. The functional properties of soy flour, concentrate, isolates, and soy protein—such as binding to fat and water, emulsifying, foaming, and gelling—make them popular ingredients in food industry preparations.

4.5.2. Wheat

The most important and widely grown crop in the world is wheat. Over a billion people worldwide consume it in different ways. A total of 772 million metric tons of wheat were produced worldwide in 2017, of which 150 million metric tons were used as animal feed. At 150.2 million metric tons, the European Union is the world's largest producer of wheat, according to data from 2017. Russia, China, and India produce about 41% of the wheat produced worldwide. Because wheat and wheat products are consumed worldwide, wheat is a more common source of calories and protein than any other food. Depending on the variety, wheat can have anywhere from 8% to 15% protein. Wheat's amino acid composition is highly unbalanced and depletes essential amino acids like methionine, lysine, and threonine when it is processed into different products. Gluten protein is the primary source of protein in wheat grains. Unlike other plant proteins, gluten has certain useful qualities. Because gluten forms a cohesive, slightly elastic, cross-linked protein structure, wheat can be used to make goods like sourdough. Rinsing the flour with water yields commonly used gluten, which has an approximate protein content of 80%. This method was discovered in the second half of the 19th century. Apart from the traditional technique, there exist various other approaches for obtaining and modifying gluten. For instance, eluted gluten undergoes chemical and enzymatic modification to fortify its structure. With the help of current technologies, wheat protein isolates can have up to 90% of their protein content. Gluten is mostly used in pasta, bread goods, and cereals for breakfast. Gluten is used in the meat and fish industries because of its effective ability to abstract water and fat. Vegetarian and vegan cuisine frequently uses gluten in place of meat proteins. Extruded wheat protein that has been textured is being used more and more to mimic the look and texture of meat products.

Less than 1% of people have gluten-related disorders like celiac disease and gluten ataxia, which are uncommon ailments. Even though these conditions are uncommon, over the past three decades, there has been a noticeable rise in the adoption of a gluten-free lifestyle and the consumption of gluten-free foods. People may eliminate gluten from their diets for a variety of reasons, including the relief of gastrointestinal and non-gastrointestinal symptoms and the belief that removing gluten from the diet could be unhealthy. There is growing evidence that some patients with gastrointestinal symptoms, like those frequently associated with irritable bowel syndrome, may benefit from avoiding gluten.

4.5.3. Milk Proteins (Casein)

Milk's protein content is crucial because it influences the milk's nutritional value and processing suitability. The two main fractions of cow's milk, casein and whey proteins, which make up about 80% and 20% of protein and nitrogen compounds, respectively, and have different physicochemical properties, add up to the milk's average protein content of 3:4%. The foundation for creating different milk-protein preparations is an understanding of these substances and how to use them in practice. Skim milk powder and pasteurized skim milk are made from casein as the raw material. Casein is differentiated by rennet or acid coagulation techniques. The average composition of acid casein is approximately 88% protein, 0.55% fat, 0.33% lactose, and 2.11% ash. Approximately 82.5% protein, 1% fat, 0.5% lactose, and up to 8.5% ash make up rennet casein. All essential amino acids are present in higher concentrations in cow's milk casein than the FAO/WHO standard. The amino acid contents of casein are clearly lower than those of whole hen egg protein in terms of isoleucine, lysine, threonine, tryptophan, valine, and sulfur. Their functional properties, which describe how proteins interact with food ingredients to determine their potential practical application, are a significant quality-distinguishing feature of caseinate preparations. Caseinates' solubility, water absorption, viscosity, gelling, fat binding, emulsification, and foaming are crucial functional characteristics in food processing. A variety of food processing industries, including meat processing, delicatessen production, cereal product production, baking, confectionery, dairy, beverage production, food concentrates, and the preparation of products for specific nutritional needs, use caseinates because of their unique properties. A portion of the population may experience an allergic reaction to milk proteins. An immunologically mediated reaction to the proteins in cow's milk, cow's milk allergy (CMA) can affect the respiratory tract, gastrointestinal tract, skin, or occasionally multiple systems (systemic anaphylaxis). It probably affects 1–3% of the general population, with infants having the highest prevalence and adults having the lowest.

4.5.4. Whey Proteins

Whey proteins are now viewed as nutrients in the production of dietetic food, physiologically active in the production of functional foods, and structure-forming in both traditional and new generation food due to their advantageous physicochemical and biological properties. A 70 kg person's daily requirement for essential amino acids is met by consuming 14 g of whey protein, which is equal to 23 g of casein or 17 g of egg white. Because of their biological activity and potential for use in the production of "functional food," whey proteins are highly sought-after. Whey's dry matter composition supports its use as a raw material for additional processing.

The use of ultrafiltration densification to lower concentrated whey operations' expenses is growing. Whey proteins have several functional qualities in addition to their nutritional ones. Whey proteins are used by the meat industry to enhance flavor, texture, emulsification, gelling, water binding, and nutritional qualities. Whey proteins allow for the partial substitution of meat proteins, as well as the replacement of soy products and other additives that aren't meat, like modified starches. When yogurt is made, adding milk powder and the right amount of functional whey proteins (about 35%) enhances the yogurt's rheological and flavor characteristics and permits the resignation or limitation of non-dairy thickeners like pectin, modified starches, and gelatin. Whey proteins can be used in place of skim milk powder in confections. Whey proteins are also used to make sauces, soups, and nonfat mayonnaise. Their benefits include the capacity to form gels, bind water, and mimic the flavor characteristics of fats in addition to being soluble in a wide range of pH values.

4.5.5. Egg White Proteins

From a nutritional perspective, eggs are especially interesting because they provide important lipids, proteins, vitamins, minerals, and trace elements. They are also a moderate source of calories (about 140 kcal/100 g), have a lot of culinary applications, and are reasonably priced. As a matter of fact, it has been determined that eggs are the most affordable animal source of proteins, iron, vitamin B_{12}, riboflavin, and choline, and the second-cheapest source of zinc and calcium. Whereas lipids, vitamins, and minerals are primarily concentrated in the egg yolk, egg proteins are split evenly between the egg white and egg yolk. When major constituents like water, proteins, lipids, and carbohydrates are taken into account, the relative content of minerals, vitamins, and particular fatty acids in eggs varies between national references but stays globally comparable. In contrast to minor components that are influenced by a variety of factors, including hen nutrition, the major nutrients in eggs are highly stable and depend on the ratio of egg white to yolk. Water, protein, fat, carbs, and ash combined make up about 76.1%, 12.6%, 9.5%, 0.7%, and 1.1% of raw, freshly laid eggs. Around 10% of egg white is made up of protein, with many functionally significant proteins included. These include lysozyme (35%), ovalbumin (54%), ovotransferrin (12%), ovomucoid (11%) and ovomucin (35%). If separated, these major proteins have a high potential for industrial applications. Important requirements for the commercial production and use of egg proteins include their high degree of reproducibility, their non-toxic chemical composition, their sequential separation of multiple proteins, and their simplicity. The separated proteins can be altered with enzymes to suit the requirements of a particular industry. They are utilized in the food and pharmaceutical industries. Lysozyme is primarily used as a food preservative, whereas ovotransferrin serves as a metal transporter, an antibacterial, or an anti-cancer agent. Ovomucin is used as a cancer-inhibiting drug, while ovalbumin is commonly used as a dietary supplement. The main egg allergen is ovomucoid, which is also used as an anti-tumor agent due to its ability to stop tumor growth. These proteins' hydrolyzed peptides exhibit strong antioxidant, metal-binding, angiotensin I converting enzyme inhibitory, and antitumor properties. Egg proteins are therefore novel compounds with a wide range of potential uses, as is the synthesis of bioactive peptides from these proteins. Moreover, egg proteins are extensively utilized in many areas of the food industry for stabilization, foam formation, gelling, coagulating, and water binding because of their functional qualities. Furthermore, they find application in the manufacturing of cheese (ripened varieties), the meat sector (raw meat, fish, pates, baked goods, sausages, and canned meat), confectionery (foam and meringues), the brewing of wine, mead, and beer, and the fat sector.

4.5.6. Gelatin

Made from collagen, gelatin is a highly pure form of animal protein that makes up about 30% of a protein substance present in high-collagen parts of the animal and human body, such as the skin, bones, cartilage, and connective tissue. Collagen is converted during the production process into water-soluble gelatin. The physicochemical characteristics of gelatin, specifically the gel strength of the resulting proteins, are determined by collagen. One protein with a peculiar amino acid composition is collagen. It has notable concentrations of proline and glycine as well as two amino acids, hydroxylysine and hydroxyproline, which are not obtained directly from ribosome translation. The most important characteristic of gelatin is its ability to form gel reversibly, regardless of the method used to produce it. The ability of different species of gelatin to gel is variable. An essential auxiliary material in pharmacy, gelatin functions as a gelling agent, stabilizer, protective colloid, emulsifier, foaming agent, carrier, and binder. It can be used to make granulates, dragées, liquid medications, and capsules. Gelatin is an excellent way to produce low-calorie semi-finished goods in the food industry. It works well as a stabilizing and binding additive in meat, fish, yogurt, and jelly products. When confectionery products are frozen and thawed, the use of gelatin enhances their structural integrity.

4.5.7. Protein Hydrolysates—Food and Feed Additive and Use in Other Industrial Products

All substances exhibiting biological activity can be utilized as natural additives to simulate the functional characteristics of food products. Unsaturated fatty acid oxidative changes can occur in a variety of groceries. These alterations could degrade the quality of the product; even slight oxidation can have a substantial negative impact on sensory value. Lipid oxidation products, or free radicals, have the potential to be harmful or carcinogenic as well. The raw ingredients of food are refined through enzymatic protein hydrolysis. Enzymatic protein hydrolysis is used in food preparation technology to enhance product consistency, in addition to its function in high-fat products. Compared to parental proteins, the constituents of hydrolysates have superior solubility, emulsifying, foaming, and dispersing abilities. Hydrolysates also enhance the taste and ability of products to absorb water. Protein chains that have been shortened typically have higher nutritional value and better flavor. However, a bitter taste may result from severe protein degradation. Peptide size, quality, and share of free amino acids can be used to create mixtures of well-known, well-established, sophisticated, and designed hydrolysate compositions. Additionally, since treated products have a comparatively longer shelf life, using protein hydrolysates makes financial sense. This technique is frequently applied in the processing of meat to improve the products' consumer appeal and availability. Protein hydrolysates, on the other hand, have a far broader application and serve particular purposes in the feed sector. Because of their effects on feed digestibility and gastrointestinal flora, they have a positive influence on animal welfare and performance traits. Additionally, the paper, pharmaceutical, and cosmetics industries (especially those with higher fat content) use protein hydrolysates. Paint, biodegradable materials, adhesives, binders, coatings, special mechanical qualities and barriers, nanomaterials, and biopolymers are among the other applications for protein hydrolysates that are becoming more and more common.

4.6. UNCONVENTIONAL AND ALTERNATIVE SOURCES OF PROTEINS

Proteins derived from non-traditional sources, like plants and insects, have garnered significant attention in the development of novel food products that have a reduced environmental impact and provide a means of feeding the world's expanding population. Regretfully, in comparison to the substantial amount of knowledge that has been gathered over the years regarding many established proteins and protein fractions, there is little information available on these emerging sources of protein.

4.6.1. Rice

After wheat, rice has the second-largest harvest and plant consumption worldwide. Harvests occur every year in a range of 480 million metric tons. Asia is known for having the highest rice consumption rates. Because of its amino acid composition, rice protein is preferred over other cereal proteins and has the second-highest concentration of lysine, the limiting amino acid in cereal. At about 8%, rice has a comparatively low protein content. Nonetheless, because of its high consumption, it is one of the main sources of protein in Southeast Asia and the southern countries. Because high-protein rice ingredients are hypoallergenic, they are being used more and more in Europe and the US to make gluten-free food. Rice proteins are more soluble at pH values between 4 and 5, and their solubility increases with pH changes. The primary factor influencing rice proteins' capacity to dissolve at particular pH levels is their high glutelin content. Glutelin makes up a sizable portion of all the proteins in rice bran and is the predominant protein fraction in endosperm. Antioxidant, antihypertensive, antineoplastic, and anti-obesity qualities can also be found in preparations made from rice proteins and certain fractions of them. The use of rice proteins in dietary supplements and sports nutrition products is growing in popularity. This protein can be added to bread, biscuits, high-protein bars, or edible films to enhance their nutritional and functional value. It can also be used as a substitute for the widely used casein, whey, or soy.

4.6.2. Corn

One of the most important crops used in industry, especially in the US, is corn. Maize varies in its protein content from 9% and 12%. Roughly half of the harvest in the

US is utilized to make animal feed. The remainder is utilized in the manufacturing of spirits, syrup (glucose syrup), starch, oil, maize protein, and many other products. Furthermore, maize is used in the food industry as an intermediary in the production of foods like tortillas and chips. A wet, milled maize product known as corn gluten flour is created and utilized in the production of corn protein. One of the few vegetable proteins made commercially is corn gluten, which is mostly utilized to make food bags and polymers.

Zein is not frequently utilized in cornmeal and other human food products as a source of protein. Its use in the food industry is restricted by its low water solubility and lack of some essential amino acids, including tryptophan and lysine. As a result, its primary application is in animal feed. Zein's low solubility in water is caused by a high percentage of non-polar and hydrophobic amino acid residues that are hidden inside the protein structure. Zein makes up roughly ≈45–50% of corn protein and is the main storage protein in corn. The zein isolate's negative nitrogen balance and low water solubility prevent it from being used directly for human consumption. Commercial interest has been piqued by zein and its resins' capacity to form hard, glossy, hydrophobic, fat-proof coatings as well as their resistance to microbiological growth. Zein has the potential to be used in chewing gum, fiber, adhesives, coatings, ceramics, ink, textiles, and biodegradable plastics. Zein has great potential in the biodegradable plastic, specialty food, and pharmaceutical sectors — but only if production costs can be brought down. By altering the amino acid composition of this protein, its market value and range of applications could be expanded to allow for its use in the food industry.

The enhanced solubility of corn peptides is due to their structural characteristics. This property is primarily manifested in the flow phenomenon, the formation of homogenous solutions without precipitation, and the high solubility over a broad pH range. Corn peptides are also widely used in the acidic beverage industry due to their good solubility even in harsh environments like low pH. Corn peptide drinks are also popular because they have a high glutamate content, a pleasant taste, and low viscosity. They can also enhance brain function.

4.6.3. Quinoa

Quinoa (*Chenopodium quinoa* Willd) is an annual plant species. Quinoa seeds have a higher bran fraction content than grains like wheat or corn, which contributes to their high protein and fat content. Compared to most cereals, quinoa proteins have a higher proportion of exogenous amino acids and a well-balanced amino acid composition. Quinoa is a great source of vitamins, minerals, and dietary fiber. It also contains flavonoids, phytosterols, and polyphenols. Quinoa is a versatile ingredient in food and other industries because of its emulsifying, foaming, gelling, and solubility. Quinoa oil has a high vitamin E and omega-6 content. Quinoa starch's functional qualities, such as its ability to modify solution viscosities and maintain structural stability during freezing, make it useful in a wide range of cutting-edge industrial applications.

4.6.4. Beans

Polysaccharides, oligosaccharides, polyphenols, and bioactive peptides are abundant in beans. Because they are legumes, beans naturally contain a high percentage of protein—up to 30%. Because it contains bioactive peptides that can be used to make metal-chelating, anti-inflammatory, hypotensive, and anti-diabetic medications, bean protein is becoming more and more popular. Furthermore, due to their high dietary fiber content, beans can reduce intestinal mucosal irritation by aiding in the removal of harmful substances from the intestines, particularly the large intestine. Additionally, the high fiber content of beans can facilitate easier defecation and lessen the symptoms of hemorrhoids and constipation. Legumes are a great source of dietary fiber and protein. On the other hand, they contain a wide range of antinutritional factors, including lectins, goitrogenic factors, proteinaceous compounds, neurotoxins, phytic acid, and amylase inhibitors. These elements affect the nutrients' bioavailability and absorption by humans and other animals that eat beans. Reducing the antinutritional factors of beans can be accomplished through the breeding of crop varieties with lower concentrations of these factors, the use of enzymes to lower the concentration of these factors, and the use of indigenous techniques like soaking, cooking, and germination.

4.6.5. Lupine

Scientific research indicates that lupine is virtually starch-free (up to 2%) and has a slightly higher protein content than other legumes that are commonly consumed by humans. The structure of lupines is usually dicotyledonous. Compared to most domesticated species of cereals and legumes, its thick skin makes up about 30% of the seed weight. The lupine seed fibers are made up of 40:60% soluble and insoluble fractions, while the shell is composed of cellulose and hemicellulose. Depending on the type, variety rounds, growth conditions, and type of soil, lupine's crude protein content can range from 28% to 42%. Albumin and globulins (38% and 35%, respectively), 4.3% glutelin, and 0.6% prolamine make up the majority of the protein composition in fractional lupine. Because of the easy digestion of this protein combination, lupine has a major advantage over its rivals (beans, soy, and peas) [Villarino, Jayasena, Coorey, Chakrabarti-Bell, & Johnson, 2016]. Furthermore, because lupine's fractional protein contains fewer inhibitors of the proteolytic enzyme than other legumes, it is easier to digest than products that contain soy or peas in their composition. Along with having a complete protein composition, lupine is also a good source of vitamins. It contains water-soluble vitamins like pyridoxine, folic acid, riboflavin, and thiamine as well as fat-soluble oils and provitamins like tocopherol, sterols, and carotenoids, and much lower inhibitor content. Protein preparations from wolves have been successfully applied to dairy or meat production technology, bakery and confectionery recipes, and sports nutrition formulas. In recent years, a wide range of lupine products have been developed, such as low-fat lupine flour and malt, lupine oil, lupine protein isolates and concentrates, and lupine flour extrudate.

4.6.6. Sunflower

Sunflower, along with soybean, rapeseed, cottonseed, and peanuts, is one of the most produced oilseed crops. The FAO estimates that 56.07 Mt of sunflower seeds were produced worldwide in 2019. In conclusion, 10–27% of a sunflower seed's mass is protein. However, the percentage rises to 40% for mechanically extracted seeds and 50% for using an organic solvent to remove the oil when making sunflower meal. The protein content of dehulled seeds can range from 53–66%. Protein bodies and protein storage vacuoles of embryonic and endosperm cells are the main locations for sunflower proteins. The intact proteins that make up 87–99% of the nitrogen in sunflower seeds. The remaining 1–13% comes from nitrogenous materials such as amino acids, peptides, or other substances. Total carbohydrate content in sunflowers varies from 4% to 18%, with sunflower seed carbohydrates having a low starch content (approximately 0.42%).

With the exception of Lys, sunflower protein composition is in line with FAO recommendations. Additionally, sunflower protein has a lower concentration of sulfur-containing amino acids—Met and Cys in particular—than rape protein. There is a reasonable balance between the amounts of basic (18%) and acidic (20%) amino acids. Unfortunately, because of its distinct aftertaste and dark color, this protein can only be used in small amounts in food products. But you can easily hide the boring color by covering the product with glaze or chocolate. Similar to this, the amount of sunflower protein present determines how strong the aftertaste is. Thus, without reducing the protein content, combining with other, better-tasting proteins may help to lessen the unpleasant aftertaste.

4.6.7. Insects

Almost two billion people on the planet still eat insects as part of their traditional diet. Insects are a topic of great interest because they have been a part of the human diet in some regions for centuries, specifically as an alternative source of protein. In many regions of Asia, Latin America, and Africa, there are communities that consume insects for food. With their vast biodiversity, insects (also known as invertebrates) account for 95% of all animal species. They have been used as food by humans since prehistoric times and can be eaten at various stages of their lives, including eggs, larvae, pupae, and adults. Coleoptera (31%), Lepidoptera (18%), Hymenoptera (14%), Orthoptera (13%), and Hemiptera (10%) are the primary orders of insects consumed. The EU member states that make up the Committee on Plants, Animals, Food and Feed (Novel Food and Toxicological Safety section) approved a draft law in May 2021 that would have allowed the sale of yellow mealworms (*Tenebrio molitor*) as novel foods. The primary nutritional component of insects is their high protein content, but they also contain notable concentrations of other vital nutrients like lipids, vitamins, minerals, and beneficial fatty acids (Huang, Feng, Xiong, Wang, Wang, Wang, & Yang, 2019). Insect proteins have higher amounts of high-quality nutritional protein, higher total protein levels, and a 50–80% essential amino acid profile when compared to plant and meat proteins. Generally speaking, the proportion of unsaturated fatty acids in insect lipids is higher than that of saturated fatty acids.

Insects also contain a variety of minerals, including calcium, phosphorus, magnesium, manganese, copper, iron, zinc, potassium, and sodium. They also include riboflavin, pantothenic acid, biotin, and occasionally folic acid, along with a wide range of lipophilic vitamins. In light of this, insect proteins present a promising raw material for future study and practical application. Unfortunately, there are currently a lot of questions surrounding the use of insects as food. When it comes to safety specifically, the common risks associated with eating insects are allergic, parasitic, and microbiological. Therefore, more investigation and testing are needed regarding the production technology and the safety of their use (De Castro, Ohara, dos Santos Aguilar, & Domingues, 2018).

4.6.8. Algae

Except for lichens and embryophyte terrestrial plants, algae are a diverse group of species that are classified as photosynthetic, unicellular, or multicellular organisms that produce oxygen. Marine algae can have a protein content of 20–60% of dry matter, depending on the type and location of harvest. Macrophages can be categorized into three primary groups based on their pigmentation: *Rhodophyta* (red algae), *Phaeophyta* (brown algae), and *Chlorophyta* (green algae). Microalgae are a very diverse group as a result. Many species are used for a variety of biotechnological applications, including animal feed and biofuel.

Moreover, because of their high vitamin and mineral content, *Arthrospira platensis* (spirulina) and *Chlorella vulgaris* (chlorella) are marketed as functional foods. Algae are frequently contrasted with other protein sources, like soybean or egg, and are widely recognized as a rich source of protein due to their composition meeting FAO requirements. Tryptophan and lysine are the limiting amino acids in the majority of algae species, while aspartic acid and glutamic acid make up a sizable portion of the total amino acid content in many seaweed species, which is largely responsible for the unique "umami" flavor that is associated with seaweed. Microalgae are now commonly taken in the form of tablets, powder, and capsules as dietary supplements. But they're also added to a number of functional foods, such as beer, high-protein bars, pasta, bread, biscuits, drinks, and sweets. For instance, AlgaVia® is a business that sells algae products and uses *Chlorella protothecoides* to produce an algal powder that is high in protein and fat. High concentrations of algae protein in food produce a distinct aftertaste and cause the product—especially high-protein bars—to harden significantly, so they might not be useful in similar goods. But this kind of protein works well as an addition to feed, particularly for chickens. Several species, including *Chlorella* sp., have shown that adding microalgae as a protein source to poultry feed can enhance health, productivity, and value. *Arthrospira* sp. *Porphyridium* sp. as well as *Haematococcus* sp. Increased viability, better general health, and lower plasma concentrations of fatty acids, cholesterol, and triglycerides were observed in chickens fed supplemental spirulina.

CASE STUDY

MADE SMARTER HELP BUSINESSES TO CAPITALIZE ON THE MARKET WITH TECHNOLOGICAL INNOVATION

A food manufacturer, supported by Made Smarter, has launched an innovative product to capitalize on the personalized nutrition market.

abnormal.® is a new venture from the team behind The Protein Works, a fast-growing direct-to-consumer business that sells a wide range of protein powders, snacks, supplements, and foods, manufactured and distributed from its 75,000-square-foot facility in Cheshire.

With the global meal replacement market valued at $18 billion and forecast to grow, the business has invested over £1 m in cutting-edge technology to produce a personalized, nutritionally-complete meal.

With Made Smarter support it has invested in a bespoke software solution to fully automate the process from customer order through to product dispatch.

The data and systems infrastructure links a front-end web platform capturing data supplied by a customer to its back-office functions such as nutritional analysis, production, finance, warehouse management, data storage, and business intelligence.

- **The Challenge**

Since launching in 2012, Protein Works has been on a rapid growth journey using its digital platform to sell to consumers directly both in the UK and globally to over 150 countries.

While athletes and bodybuilders used to dominate the sports nutrition market, these days healthy living and fitness trends are driving growth in the sector.

Mark said: "With the global meal replacement market estimated to be over $18 billion in 2020 and forecast to continue to grow, complete meals have become a high-growth sector with consumers wanting the convenience of a nutritionally balanced meal in a convenient powder format."

The goal is to increase market share and capture a new generation of lifestyle customers who are looking for a healthy meal option by using digitalization to personalize not only the product but also the purchasing and consumer experience.

- **The Solution**

Made Smarter has supported the development of an integrated web platform, which has a marketing, subscription and commerce layer at the front-end and captures data supplied by a customer about their diet, lifestyle, health and habits, to create a unique profile.

The software then links to back-office functions such as nutritional analysis where algorithms process the data to create a meal containing the right amount of protein, essential fats, carbohydrates, vitamins and minerals for the customer. It even takes into account allergies and intolerances.

The meals are manufactured in-house using fully recyclable paper-based single-serve sachets and personalized packaging, which includes a bespoke nutrition guide.

Any changes the customer makes to their profile will resync to create a new meal recipe on their next order.

Mark said: "Digital is at the heart of everything that we do, and we aim to innovate and improve by using leading-edge technologies and real depth of data insight. abnormal. is about creating a great tasting, nutritionally balanced complete meal that is personalized to your own unique profile and lifestyle."

- **The Benefits**

The inherent automation that the new technology offers provides a competitive advantage in a new market and will give the business more opportunity to invest in achieving further efficiency and productivity gains as the business grows.

Mark said: "Having an affordable healthy meal, that is designed for the individual on so many levels, whether they are vegan, flexitarian or not, and has minimal impact on the planet, is a very powerful proposition. Personalized nutrition will be a game changer in the market. The technology we are creating will transform the way data is gathered and information is analyzed and how our business will operate. These new technologies enable faster, more responsive, and more efficient processes to deliver higher-quality personalized products at a reduced cost"

The project has created four new jobs and will upskill a further 10.

- **The Future**

The company aims to invest more in software and analytics to use data to empower decision-making across its range of brands.

Data will also drive continuous improvement and further investment in automation and improving its manufacturing operations to become a state of-the-art facility.

Its hardware and software solution will also enable the business to leverage its NPD team to continue driving innovation.

Mark said: "This is a transformational step for our business, as the data gathered will help to further develop the abnormal. nutritional proposition for our consumers. To create personalized nutritious food, which is backed up with a unique end-to-end experience."

CLASS ACTIVITY

PROTEIN DENATURING ACTIVITY

When a protein is exposed to conditions too far outside of a range it can tolerate, that protein's shape will come undone. This is called "denaturing" (basically, breaking) a protein. We denature proteins all the time when we cook food (think: eggs). In this activity, we will use common household products or processes to denature egg proteins in two main ways—by cooking them, and by exposing them to concentrated alcohol (ethanol). Do you think egg will look the same or different depending on how the proteins it holds are denatured?

1. **Materials**

- Stove or microwave
- Pot or microwave-safe container to boil water
- 1 fork
- 1 pair scissors
- 1 bowl
- 4 small glass containers of the same size
- 1 egg (split egg white into four parts); additional can be used
- 2/3 cup water (150 ml)
- 1/3 cup of rubbing alcohol (75 ml)

2. **Breaking Proteins in 8 Easy Steps**

1. Pour 1/3 cup rubbing alcohol into one glass container, room temperature water (1/3 cup) into another, and the rest of the water (1/3 cup) into a microwave-safe container (or into a pot).
2. Crack egg into the bowl, removing the yolks.
3. Cut the egg white into pieces so you can add ¼ into each glass container.
4. Heat up the water for your hot water treatment and pour it into one of the empty glass containers.

5. Quickly put ¼ of the egg yolk into the boiling water. Then put ¼ into the alcohol, ¼ into the room temperature water, and the remainder into the final, empty glass container.
6. Observe any immediate changes that occur in terms of egg white color and consistency. If you try stirring the different treatments, rinse your fork between stirs.
7. Wait for 30 minutes.
8. Use the fork to inspect the state of the egg whites in each treatment and note how they may have changed over time.

3. What Happened?

Why does denatured egg white turn from clear to white? If more than one treatment denatured egg whites, do you think the treatments denatured the egg whites in the same way?

Let's look at each of the treatments we used:

1. **Control**. Egg whites start out clear. They are almost 90% water, but the other 10% is packed with proteins. Egg whites contain more than 50% of the proteins found in the egg. The main protein in egg white is called albumin. The small, folded proteins in the egg white don't take up much space, and the gel-like egg white looks clear. The control egg showed us that, when left at room temperature, the egg whites stay clear, meaning the proteins maintain their original shape. These proteins were not denatured.
2. **Cooking (hot water)**.Whenever eggs are cooked with heat, the egg whites turn from clear to white, and the gel becomes rubberier. As heat denatured the proteins in the egg white, it broke apart some of the bonds (mostly hydrogen bonds) that were holding the proteins in their original shape. The proteins unfolded, taking up more space (turning the gel white) and hardening them in place next to one another.
3. **Alcohol**. Alcohol also denatures proteins. It does this the same way as heat, by breaking the bonds that hold parts of the protein in a folded shape. Sometimes the alcohol molecules bond directly to some of the parts of the protein, disrupting the normal way the protein would bond to itself. (So, alcohol is called a "bond disruptor.") The proteins again unfolded, taking up more space and hardening in place next to one another. This process took much longer with alcohol than it did with heat, however. The longer time for denaturation with alcohol is simply because it spreads more slowly than heat. The alcohol had to diffuse (or move through the fluid) into the egg in order to affect the proteins it touched.
4. **Room temperature water**. Sometimes in this experiment, room temperature water has a small denaturing effect on some of the egg white. It acts in the same way, by breaking bonds, but its effect isn't nearly as strong as alcohol or hot water.

4. **Questions to Think About After Denaturi**ng Egg Proteins

- Think about the effects of the two different water treatments. Do you think the water itself was denaturing proteins? If not, what was? If so, what was having the larger effect between the water treatments?
- Are there any other processes you know of that turn egg whites from clear to white? What is it and do you think the same processes are happening?
- Name another condition besides heat and exposure to a bond disruptor (like alcohol) that could affect the ability of a protein to maintain its shape.
- What other things change color when their proteins are denatured?
- Why might a living organism want to keep their proteins from denaturing?
- In this activity, why was it important to have egg whites that we did not cook or add alcohol to?

SUMMARY

- Physicochemical properties refers to the physical and chemical properties of the compound or molecular attributes that defines the intrinsic chemical reactivity of the compound. It includes stereochemistry, electrostatic characteristics, optical characteristics, absorption spectra, titration pattern, and isoelectric properties.
- The absolute configuration of the amino acid can be described based on the stereochemistry of the molecule. Since they are non-superimposable mirror images of each other, the two forms represent a class of stereoisomers known as enantiomers.
- Proteins are built by linking two or more amino acid residues together in different orientations or configurations. The amino acids in the proteins are joined together by peptide bonds formed through a condensation reaction between two amino acids, releasing water molecules in the process.
- Denaturation of proteins involves the alteration of their structure, causing them to lose their shape and function. This can occur when proteins are subjected to heat, acid, or other environmental factors. Denaturation of a protein leads to its inability to perform its original function, and this can result in changes to the texture and flavor of the food it is in.
- Various factors can cause protein denaturation. Some of these factors include heat, freezing, pH change, ionic strength change, and surface changes.
- Denaturing protein during cooking makes meat more palatable and easier to digest.
- Proteins can have surface properties such as the ability to form or stabilize emulsions (interfacial oil/water interface), the ability to create or stabilize foams (interfacial air/water interface), or solubility (combining the connections between water and proteins).
- The phenomenon of proteins stabilizing emulsions or foam is caused by their ability to adsorb at the interface, reduce surface tension, and create a coherent layer around oil droplets or air bubbles.
- Foams are created by dispersing air bubbles in the liquid phase. Adding protein increases the aqueous phase's viscosity, increasing the interfacial film's durability and producing foam. Proteins reduce the surface tension by interacting with the water molecules and air, allowing foam bubbles to form.
- Emulsions are dispersion systems consisting of two or more immiscible liquids in the form of a continuous and a dispersed phase (small droplets). When mixing oil and solutions with aqueous proteins, it is preferable to limit contact between them and phase separation. Initially, minimal contact is achieved due to spherical droplets' formation with energy input from the outside.
- Food proteins are essential nutrients required for maintaining various bodily functions and human health. Proteins, especially some traditional plant- and animal-derived protein sources, are essential food ingredients.

REVIEW QUESTIONS

1. What are the physicochemical properties of amino acids?
2. What is the function of protein in food chemistry?
3. Distinguish between the tertiary and quaternary levels of protein structure.
4. What are the functional properties of proteins?
5. Explain the levels of protein structure.
6. What are the properties of carbohydrates?
7. Discuss the classification and properties of edible fats and oils.
8. Describe the forces involved in protein folding.
9. What are the nutritional importance of protein?

REFERENCES

1. De Castro, R. J. S., Ohara, A., dos Santos Aguilar, J. G., & Domingues, M. A. F. (2018). Nutritional, functional and biological properties of insect proteins: Processes for obtaining, consumption and future challenges. *Trends in Food Science & Technology, 76*, 82–89. https://doi.org/10.1016/j.tifs.2018.04.006.
2. Duncan, T. V. (2011). Applications of nanotechnology in food packaging and food safety: Barrier materials, antimicrobials and sensors. *Journal of Colloid and Interface Science, 363*, 1–24. https://doi.org/10.1016/J.JCIS.2011.07.017.
3. Foegeding, E. A., & Davis, J. P. (2011). Food protein functionality: A comprehensive approach. *Food Hydrocolloids, 25*, 1853–1864. https://doi.org/10.1016/j.foodhyd.2011.05.008.
4. Huang, C., Feng, W., Xiong, J., Wang, T., Wang, W., Wang, C., & Yang, F. (2019). Impact of drying method on the nutritional value of the edible insect protein from black soldier fly (*Hermetia illucens* L.) larvae: Amino acid composition, nutritional value evaluation, in vitro digestibility, and thermal properties. *European Food Research and Technology, 245*, 11–21. https://doi.org/10.1007/s00217-018-3136-y.
5. Lam, R. S. H., & Nickerson, M. T. (2013). Food proteins: A review on their emulsifying properties using a structure-function approach. *Food Chemistry, 141*, 975–984. https://doi.org/10.1016/j.foodchem.2013.04.038.
6. Tadros, T., Izquierdo, P., Esquena, J., & Solans, C. (2004). Formation and stability of nano-emulsions. *Advances in Colloid and Interface Science, 108–109*, 303–318. https://doi.org/10.1016/j.cis.2003.10.023.
7. Udgaonkar, J. B. (2013). Polypeptide chain collapse and protein folding. *Archives of Biochemistry and Biophysics, 531*, 24–33. https://doi.org/10.1016/j.abb.2012.10.003.
8. Villarino, C. B. J., Jayasena, V., Coorey, R., Chakrabarti-Bell, S., & Johnson, S. K. (2016). Nutritional, health, and technological functionality of lupin flour addition to bread and other baked products: Benefits and challenges. *Critical Reviews in Food Science and Nutrition, 56*, 835–857. https://doi.org/10.1080/10408398.2013.814044.

9. Wijaya, W., Patel, A. R., Setiowati, A. D., & Van der Meeren, P. (2017). Functional colloids from proteins and polysaccharides for food applications. *Trends in Food Science & Technology, 68*, 56–69. https://doi.org/10.1016/j.tifs.2017.08.003.

CHAPTER 5

Enzymes

LEARNING OBJECTIVES

After studying this chapter, you will be able to:

- Examine the science of enzyme
- Identify the covalent modification of enzymes
- Access the enzyme regulation in metabolic pathways

KEY TERMS FROM THIS CHAPTER

Liver function	Chemical reactions
Metabolic pathways	Enzymology
Therapeutic drugs	Biochemical identity
Lysozyme	Biochemistry
Molecular biology	Enzyme denaturation
Bacteria	Stabilization

INTRODUCTION

Proteins called enzymes aid in accelerating chemical reactions within our bodies. Enzymes are necessary for many processes, including liver function and digestion. A particular enzyme's excess or deficiency can lead to health issues. Healthcare professionals can also use the enzymes in our blood to look for illnesses and injuries. Proteins called enzymes aid in accelerating our bodies' chemical reactions, or metabolism. They decompose some materials and construct others. Enzymes are present in all living things. Enzymes are naturally produced by our bodies. However, food and manufactured goods also contain enzymes. Digestive support is one of the most significant functions of enzymes. The process through which food is converted into energy is called digestion. Enzymes, for instance, are found in our stomach, intestines, pancreas, and saliva. They digest carbohydrates, proteins, and fats. These nutrients are used by enzymes for cell growth and repair.

Enzymes also help with:

- Breathing.
- Building muscle.
- Nerve function.
- Ridding our bodies of toxins.

Most enzymes are proteins with catalytic properties that are essential for carrying out various tasks. Enzymes are essential for the maintenance of life because they carry out various chemical reactions and metabolic processes within the cell. The enzymes that are involved in the first stage of the metabolic process react with a molecule known as the substrate. The substrates are changed into different molecules by enzymes, which are referred to as products. Because enzymes are essential to the maintenance of biological processes, their regulation has played a significant role in clinical diagnosis. All enzymes are composed of proteins as their macromolecular components, with the exception of ribozymes, a class of RNA catalysts. The ribonucleic acid enzyme is the source of the word "ribozyme.." Numerous ribozymes are ribonucleic acid molecules that catalyze reactions within or between other RNAs. Every bodily tissue and fluid contains enzymes. Enzymes found inside cells catalyze every reaction that occurs in metabolic pathways. Enzymes in the circulatory system control blood clotting, and enzymes in the plasma membrane control the catalysis that occurs in cells in response to cellular signals. The majority of vital biological processes rely on the actions of enzymes.

5.1. THE SCIENCE OF ENZYME

Proteins called enzymes (/ˈɛnzaɪmz/) speed up chemical reactions to function as biological catalysts. Enzymes can interact with molecules known as substrates, and they change the substrates into other molecules known as products. Enzyme catalysis is required for nearly all of the metabolic processes in a cell to proceed at rates quick enough to support life. Enzymes are required by metabolic pathways in order to catalyze specific steps. Enzymology is the study of enzymes, and pseudoenzyme analysis acknowledges that some enzymes have lost their capacity for biological catalysis during evolution. This loss is often reflected in the peculiar 'pseudocatalytic' properties that these enzymes possess in their amino acid sequences.

There are over 5,000 different types of biochemical reactions that enzymes can catalyze. Additional biocatalysts include ribozymes, which are catalytic RNA molecules. Although they are occasionally referred to as an enzyme type rather than being similar to an enzyme, the term "enzyme" still frequently refers to the particular type of protein even after ribozymes were discovered in 1980–1982. An enzyme's distinct three-dimensional structure determines its specificity.

Like all catalysts, enzymes work by reducing the activation energy of the reaction to speed it up. Certain enzymes have the ability to convert substrate to product millions of times more quickly. Orotidine 5'-phosphate decarboxylase is an extreme example; it can speed up a reaction that would normally take millions of years to occur in milliseconds. Enzymes don't change a reaction's equilibrium or be consumed in chemical reactions, just like any other catalyst does. Enzymes are far more specific than the majority of other catalysts. Enzyme activity can be affected by other molecules: inhibitors are molecules that decrease enzyme activity, and activators are molecules that increase activity. Enzyme inhibitors are found in many medicinal medications and toxins. When an enzyme is exposed to excessive heat, it becomes permanently denatured, losing its structure and catalytic properties. Additionally, an enzyme's activity significantly decreases outside of its ideal pH and temperature range. Commercial applications for certain enzymes include the synthesis of antibiotics. Enzymes are used in some household products to accelerate chemical reactions: proteins in biological washing powders are broken down by the enzymes, which also remove starch and fat stains from clothing. Similarly, meat tenderizer uses enzymes to break down proteins into smaller molecules, which facilitate easier chewing (Hammes et al., 2011).

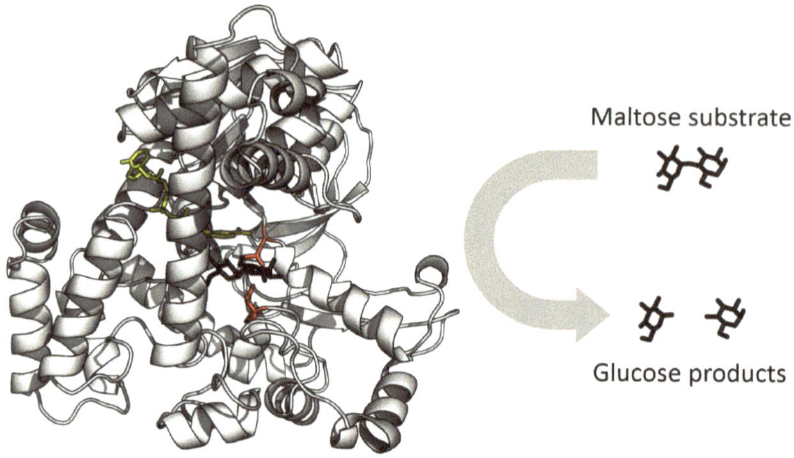

Figure 5.1. *The enzyme glucosidase converts the sugar maltose into two glucose sugars. Active site residues in red, maltose substrate in black, and NAD cofactor in yellow.*

Source: https://upload.wikimedia.org/wikipedia/commons/thumb/0/0a/Glucosidase_enzyme.png/400px-Glucosidase_enzyme.png.

5.1.1. Etymology and History

By the late 17th and early 18th centuries, it was known that saliva and plant extracts could convert starch to sugars and that stomach secretions could break down meat, but the exact mechanisms behind these processes were still unknown. Diastase was the first enzyme to be discovered in 1833 by French chemist Anselme Payen. A few decades later, Louis Pasteur discovered, while researching the fermentation of sugar into alcohol by yeast, that the fermentation was triggered by ferments, an essential substance thought to be found only in living things. According to what he wrote, the act of fermenting alcohol is connected to the organization and life of yeast cells rather than their decomposition or death. The term "enzyme" was first used in 1877 to describe this process by the German physiologist Wilhelm Kühne (1837–1900). The word is derived from the Ancient Greek word ἔνζυμον (énzymon), which means "leavened, in yeast." Later, the term "ferment" was used to describe the chemical activity produced by living things, while the word "enzyme" was used to describe nonliving substances like pepsin.

In 1897, Eduard Buchner turned in his first paper on the analysis of yeast extracts. Through a series of tests conducted at the University of Berlin, he discovered that yeast extracts could ferment sugar in the absence of live yeast cells. He gave the enzyme responsible for sucrose zymase fermentation a name. He was awarded the Nobel Prize in Chemistry in 1907 in recognition of his breakthrough in cell-free fermentation. In keeping with Buchner's example, the name of an enzyme is typically derived from the reaction it catalyzes. The suffix -ase is joined with the substrate's name (e.g., the lactase enzyme, which breaks down lactose), or to the kind of reaction (e.g., DNA polymers are created by DNA polymerase. Early in the 20th century, it was still unknown what enzymes' biochemical identities were. While many scientists noted that proteins were linked to enzymatic activity, others contended that proteins were only the actual enzymes' carriers and could not catalyze reactions on their own. In 1926, James B. Sumner crystallized the enzyme urease and demonstrated that it was a pure protein; in 1937, he achieved the same feat with the enzyme catalase. John Howard Northrop and Wendell Meredith Stanley, who studied the digestive enzymes pepsin (1930), trypsin, and chymotrypsin, conclusively proved that pure proteins can be enzymes. The 1946 Chemistry Nobel Prize went to these three scientists. Eventually, the ability to crystallize enzymes led to the solution of their structures through x-ray crystallography. This was first applied to lysozyme, an enzyme that breaks down the coating of some bacteria and is present in tears, saliva, and egg whites. David Chilton Phillips led a team that solved the structure of lysozyme, which was published in 1965. The field of structural biology and the pursuit of an atomic-level understanding of the functioning of enzymes began with the discovery of this high-resolution structure of lysozyme.

5.1.2. Classification and Nomenclature

Enzymatic activity and amino acid sequence similarity, which indicates evolutionary relationship, are the two primary classification criteria for enzymes. Activity of enzymes. The term "enzyme," which ends in -ase, is frequently derived from the substrate of the enzyme or the chemical reaction it catalyzes. DNA polymerase, lactase, and alcohol dehydrogenase are a few examples. Isomers are distinct enzymes that catalyze the same chemical reaction. The Enzyme Commission (EC) numbers are a system of enzyme nomenclature developed by the International Union of Biochemistry (Nelson & Cox, 2017) and Molecular Biology. The hierarchy of enzymatic activity is represented by a series of four numbers that follow the letter "EC" to describe each enzyme (from very general to very specific). That is, the enzyme is categorized in broad terms by the first number according to its mechanism, and then further specificity is added by the subsequent digits.

The top-level classification is:

- EC 1, Oxidoreductases: catalyze oxidation/reduction reactions
- EC 2, Transferases: transfer a functional group (e.g., a methyl or phosphate group)
- EC 3, Hydrolases: catalyze the hydrolysis of various bonds
- EC 4, Lyases: cleave various bonds by means other than hydrolysis and oxidation
- EC 5, Isomerases: catalyze isomerization changes within a single molecule
- EC 6, Ligases: join two molecules with covalent bonds.
- EC 7, Translocases: catalyze the movement of ions or molecules across membranes, or their separation within membranes.

These sections are subdivided by other features such as the substrate, products, and chemical mechanism. An enzyme is fully specified by four numerical designations. For example, hexokinase (EC 2.7.1.1) is a transferase (EC 2) that adds a phosphate group (EC 2.7) to a hexose sugar, a molecule containing an alcohol group (EC 2.7.1).

Sequence similarity. Sequence similarity is not reflected in EC categories. For example, two ligases with the same EC number but different sequences can catalyze the same reaction. Enzymes, like all other proteins, have been categorized into multiple families based on the similarity of their sequences, regardless of their intended use. Numerous databases, including Pfam, have records of these families for proteins and protein families.

Non-homologous isofunctional enzymes. Non-homologous isofunctional enzymes are unrelated enzymes that share the same enzymatic activity without any biological relationship. Homologous gene displacement can result from the spread of these genes through horizontal gene transfer to unrelated species, particularly bacteria, where they can supplant native genes that perform the same function.

5.1.3. Structure

Typically, enzymes are globular proteins that can function independently or in larger complexes. The structure of the enzyme is determined by its amino acid sequence, which also controls its catalytic activity. Even though structure dictates function, structure by itself is not yet able to predict a novel enzymatic activity. When heated or exposed to chemical denaturants, enzyme structures unfold, or denature, and this structural disruption usually results in a loss of activity. Since temperatures above a species' normal range are typically associated with enzyme denaturation, industrial users highly value the ability of bacteria found in volcanic environments, such as hot springs, to function at high temperatures. This enables the operation of enzyme-catalyzed reactions at extremely high rates.

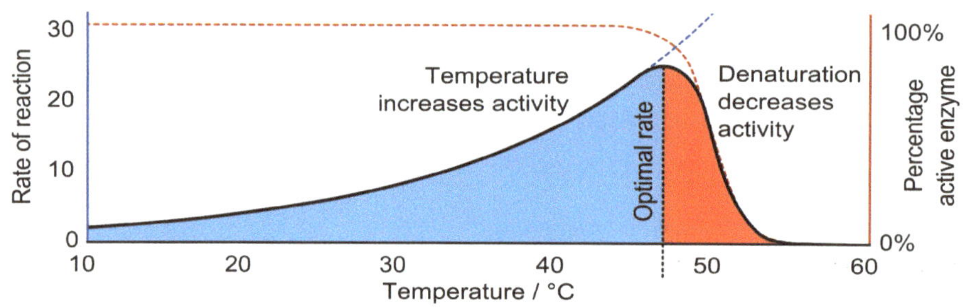

Figure 5.2. *Enzyme activity initially increases with temperature (Q10 coefficient) until the enzyme's structure unfolds (denaturation), leading to an optimal rate of reaction at an intermediate temperature.*

Source: https://upload.wikimedia.org/wikipedia/commons/thumb/d/d5/Q10_graph_c.svg/1920px-Q10_graph_c.svg.png.

Typically, enzymes are substantially bigger than their substrates. The monomer of 4-oxalocrotonate tautomerase has 62 amino acid residues, while animal fatty acid synthase has over 2,500 residues. The catalytic site, which makes up only 2–4 amino acids of their structure, is directly responsible for catalysis. This catalytic site is situated close to a binding site or sites where substrates are oriented by residues. The active site of the enzyme is made up of the binding and catalytic sites. The majority of the enzyme's remaining structure keeps the active site's exact orientation and dynamics in place. Certain enzymes have sites to bind and orient catalytic cofactors; amino acids are not directly involved in catalysis in these cases. Additionally, allosteric sites—where a small molecule binds and changes the structure of the enzyme to either increase or decrease activity—may be found in the structures of enzymes. There are a few RNA-based biological catalysts known as ribozymes that can function both independently and in conjunction with proteins. The most prevalent of these is the ribosome, a complex of catalytic RNA and protein components.

5.1.4. Mechanism

5.1.4.1. Substrate Binding

Before they can catalyze any chemical reaction, enzymes must bind to their substrates. When it comes to the substrates they bind and the chemical reactions they catalyze, enzymes are typically highly selective. By binding pockets with complementary charge, shape, and hydrophilic/hydrophobic properties to the substrates, specificity is attained. Therefore, enzymes are chemoselective, regioselective, and stereospecific because they can discriminate between very similar substrate molecules. Copying and expressing the genome involves some of the enzymes with the highest specificity and accuracy. There are proof-reading mechanisms in some of these enzymes. In this case, a reaction is first catalyzed by an enzyme like DNA polymerase, which is followed by a step where the product's accuracy is verified. For high-fidelity mammalian polymerases, this two-step procedure yields average error rates of less than one error in 100 million reactions. Ribosomes, aminoacyl tRNA synthetases, and RNA polymerase all have comparable proofreading processes. On the other hand, some enzymes exhibit broad specificity and exhibit enzyme promiscuity, meaning they can act on a variety of substrates that are physiologically relevant. Numerous enzymes have minor side effects that happened by accident (i.e., neutrally), which could serve as the foundation for a new function's evolutionary selection.

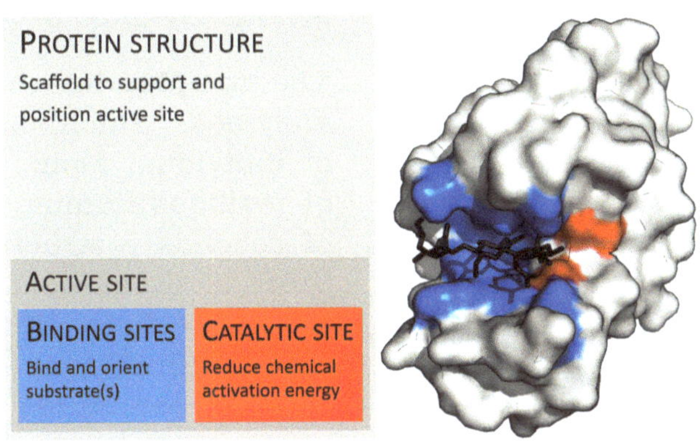

Figure 5.3. *Organization of enzyme structure and lysozyme example. Binding sites in blue, catalytic site in red and peptidoglycan substrate in black.*

Source: https://upload.wikimedia.org/wikipedia/commons/thumb/2/21/Enzyme_structure.svg/400px-Enzyme_structure.svg.png.

- **"Lock and Key" Model:** Emil Fischer postulated in 1894 that the enzyme and the substrate both have particular complementary geometric shapes that precisely fit into one another in order to explain the observed specificity of enzymes. This model is commonly known as the lock and key model. The stabilization of the transition state that enzymes reach is not explained by this early model, which only explains enzyme specificity.
- **Induced Fit Model:** Daniel Koshland proposed an alternative interpretation of the lock and key model in 1958. He proposed that because enzymes are relatively flexible structures, interactions between the enzyme and substrate constantly reshape the active site. Because of this, the substrate does not just attach to a rigid active site; rather, the side-chains of amino acids that comprise the active site are shaped into the exact locations necessary for the enzyme to carry out its catalytic activity. Certain molecules of the substrate, like glycosidases, also undergo a small shape change when they enter the active site. Until the substrate is fully bound, the active site keeps changing, and the ultimate shape and charge distribution are then established. Through the conformational proofreading mechanism, induced fit may improve the fidelity of molecular recognition in the presence of noise and competition.

5.1.4.2. Catalysis

Enzymes can accelerate reactions in several ways, all of which lower the activation energy (ΔG‡, Gibbs free energy)

- By stabilizing the transition state: Creating an environment with a charge distribution complementary to that of the transition state to lower its energy
- By providing an alternative reaction pathway: Temporarily reacting with the substrate, forming a covalent intermediate to provide a lower energy transition state
- By destabilizing the substrate ground state: Distorting bound substrate(s) into their transition state form to reduce the energy required to reach the transition state

Enzymes can use multiple of these mechanisms at once by arranging the substrates in a productive way to reduce the reaction entropy change (this mechanism's contribution to catalysis is relatively small). Proteases like trypsin, for instance, use a catalytic triad to carry out covalent catalysis, an oxyanion hole to stabilize charge accumulation on transition states, and an oriented water substrate to finish hydrolysis.

5.1.4.3. Dynamics

The complex internal dynamic motions of enzymes, which include the movements of individual amino acid residues, groups of residues forming a protein loop or unit of secondary structure, or even an entire protein domain, are what distinguish enzymes from stiff, static structures. A conformational ensemble of slightly different structures that interconvert with one another at equilibrium is produced by these motions. Aspects of an enzyme's activity may be linked to distinct states within this ensemble. Catalytic resonance theory is consistent, for instance, with the relationship between distinct conformations of the enzyme dihydrofolate reductase and the steps of the catalytic cycle that involve substrate binding, catalysis, cofactor release, and product release.

5.1.4.4. Substrate Presentation

The process of sequestering the enzyme from its substrate is known as substrate presentation. Enzymes can isolate themselves from a substrate in the cytosol or nucleus by adhering to the plasma membrane. Alternatively, an enzyme may be isolated from its substrate in the disordered area of the membrane by lipid rafts. The enzyme combines with its substrate when it is released. As an alternative, the enzyme can be activated by sequestering it close to its substrate. One type of enzyme is soluble, which acts on molecules in the plasma membrane by binding to a lipid in the membrane when it is activated.

5.1.4.5. Allosteric Modulation

Different from the active site, allostery sites are pockets on an enzyme that attach to molecules found in the surrounding cell. Following transduction to the active site, these molecules alter the dynamics or conformation of the enzyme, which in turn alters the enzyme's rate of reaction. Allosteric interactions have the ability to either activate or inhibit enzymes in this manner. Feedback regulation is the result of allostery with metabolites either upstream or downstream in an enzyme's metabolic pathway. This changes the enzyme's activity based on the flux through the remaining portion of the pathway.

5.1.5. Cofactors

Certain enzymes exhibit complete activity without the need for extra substances. Others need cofactors, which are non-protein molecules, to bind in order to function. One can have inorganic cofactors (e.g., giron-sulfur clusters and metal ions) or organic substances (e.g., flavin and heme). Metal ions, for example, can aid in stabilizing nucleophilic species within the active site. These cofactors have a variety of uses. Coenzymes, which are released from the active site of an enzyme during a reaction, and prosthetic groups, which are firmly bound to an enzyme, are two types of organic cofactors. Covalent bonds can form between organic prosthetic groups (e.g., biotin in pyruvate carboxylase and other enzymes). Carbonic anhydrase, an example of an enzyme containing a cofactor, has a zinc cofactor bound to its active site. Usually located in the active site, these firmly bound ions or molecules play a role in catalysis. Heme and flavin cofactors, for instance, are frequently engaged in redox reactions. Apoenzymes, also known as apoproteins, are enzymes that need a cofactor but do not have one bound to them. An enzyme is referred to as a holoenzyme (or haloenzyme) when it combines with the cofactor(s) needed for activity. The term "holoenzyme" can also be used to describe enzymes that have more than one protein subunit, like DNA polymerases, in which case the entire complex with all the subunits required for activity is referred to as the holoenzyme.

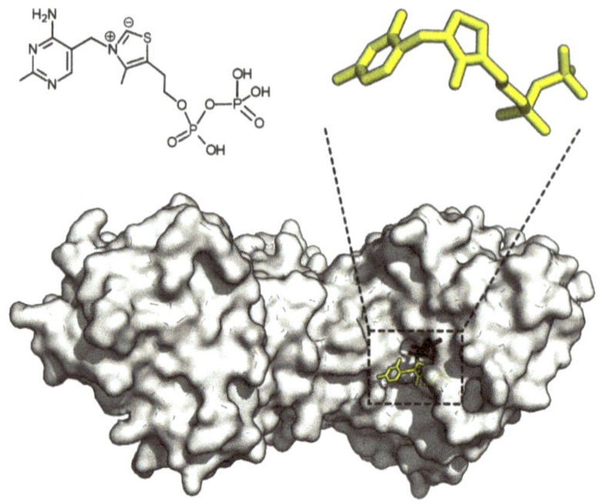

Figure 5.4. *Chemical structure for thiamine pyrophosphate and protein structure of transketolase. Thiamine pyrophosphate cofactor in yellow and xylulose 5-phosphate substrate in black.*

Source: https://upload.wikimedia.org/wikipedia/commons/thumb/f/f4/Transketolase_%2B_TPP.png/400px-Transketolase_%2B_TPP.png.

5.1.5.1. Coenzymes

Coenzymes are organic molecules that are attached to enzymes either tightly or loosely. Chemical groups are transferred between enzymes by coenzymes. Adenosine triphosphate (ATP), NADH, and NADPH are a few examples. Vitamins are the source of several coenzymes, including thiamine pyrophosphate (TPP), tetrahydrofolate (THF), flavin mononucleotide (FMN), and flavin adenine dinucleotide (FAD). The body is unable to synthesize these coenzymes from scratch, so closely related substances like vitamins must be obtained through diet.

The chemical groups carried include:

- the hydride ion (H^-), carried by NAD or $NADP^+$
- the phosphate group, carried by adenosine triphosphate
- the acetyl group, carried by coenzyme A
- formyl, methenyl or methyl groups, carried by folic acid and
- the methyl group, carried by S-adenosylmethionine

Coenzymes are a unique class of substrates, or second substrates, that are shared by many different enzymes because they undergo chemical changes as a result of enzyme activity. For instance, the coenzyme NADH is known to be used by roughly 1,000 enzymes. Within the cell, coenzymes are typically constantly renewed and their concentrations are kept constant. For instance, methionine adenosyltransferase generates S-adenosylmethionine and the pentose phosphate pathway regenerates NADPH. Because of their constant regeneration, coenzymes can be utilized in very small quantities and with great intensity. The human body, for instance, uses up its own weight in ATP every day.

5.1.6. Biological Function

Inside living things, enzymes have a wide range of uses. They play a crucial role in cell regulation and signal transduction, frequently through kinases and phosphatases. They also produce movement, as myosin, a component of the cytoskeleton, transports cargo throughout the cell and hydrolyzes adenosine triphosphate (ATP) to produce muscular contraction. Ion pumps involved in active transport are other ATPases found in the cell membrane. More unusual roles for enzymes include the production of light in fireflies by luciferase. Enzymes for viral release from cells, such as the influenza virus neuraminidase, or for infecting cells, such as the HIV integrase and reverse transcriptase, can also be found in viruses. Animal digestive systems utilize enzymes for a variety of purposes. Large molecules (proteins or carbohydrates, respectively) are broken down into smaller ones by enzymes like amylases and proteases so the intestines can absorb them. For example, starch molecules are too big to be absorbed from the intestine. However, starch chains are hydrolyzed by enzymes to produce smaller molecules like maltose and eventually glucose, which can be absorbed. Distinct enzymes break down various food ingredients. Because ruminants eat only plants, their gut microbes produce the enzyme cellulase, which helps break down the cellulose cell walls of plant fiber.

5.1.6.1. Metabolism

Enzymes can collaborate in a particular order to form metabolic pathways. One enzyme uses the byproduct of another enzyme as a substrate in a metabolic pathway. Another enzyme receives the product after the catalytic reaction. Occasionally, multiple enzymes can catalyze the same reaction simultaneously, enabling more intricate regulation. For instance, one enzyme may provide a low, constant activity while another may induce a high activity. The sequence in which these pathways proceed is set by enzymes. Without enzymes, metabolism could not be controlled to meet the demands of the cell or proceed through the same stages. A few critical steps in most central metabolic pathways are regulated, usually by enzymes that catalyze the hydrolysis of ATP. Due to the large amount of energy released by this reaction, thermodynamically unfavorable reactions can be coupled to ATP hydrolysis, which in turn drives the entire chain of linked metabolic reactions.

5.1.6.2. Control of Activity

There are five main ways that enzyme activity is controlled in the cell.

1. **Regulation:** Other molecules have the ability to either activate or inhibit enzymes. As an illustration, the final product(s) of a metabolic pathway frequently function as inhibitors for one of the pathway's initial enzymes (typically the first irreversible step, or committed step), controlling the quantity of end product produced by the pathways. Because its own concentration controls the amount of the final product produced, this type of regulatory mechanism is known as a negative feedback mechanism. The rate at which intermediate metabolites are synthesized can be efficiently adjusted by a negative feedback mechanism to meet the needs of the cells. This minimizes the overproduction of final goods and aids in the efficient use of resources and energy. Enzymatic action regulation aids in the

preservation of a steady internal environment in living things, much like other homeostatic mechanisms.

2. **Post-translational modification:** Phosphorylation, myristoylation, and glycosylation are a few instances of post-translational modification. For instance, the phosphorylation of several enzymes, such as glycogen synthase, aids in the regulation of glycogen synthesis or degradation in response to insulin and enables the cell to react to variations in blood sugar. The cleavage of the polypeptide chain is an additional instance of post-translational modification. Digestive protease chymotrypsin is made in the pancreas as chymotrypsinogen, an inactive form that is then carried to the stomach where it is activated. This prevents the pancreas or other tissues from being broken down by the enzyme before it reaches the gut. Zymogens, also referred to as proenzymes, are this kind of dormant precursor to an enzyme.
3. **Quantity:** A cell's ability to produce enzymes (the transcription and translation of enzyme genes) can be altered in response to environmental changes. Enzyme induction is the term for this type of gene regulation. For instance, because the penicillin molecule's essential beta-lactam ring is hydrolyzed by enzymes known as beta-lactamases, bacteria may develop resistance to antibiotics like penicillin. Another illustration comes from the liver's cytochrome P450 oxidases, which are crucial for the metabolism of drugs. Drug interactions may result from these enzymes being induced or inhibited. The rate at which an enzyme degrades can also be altered to control the amount of that enzyme. Enzyme repression is the antithesis of enzyme induction.
4. **Subcellular Distribution:** Different metabolic pathways can occur in different cellular compartments due to the ability of enzymes to be compartmentalized. For instance, fatty acids are produced in the cytosol, endoplasmic reticulum, and Golgi by one group of enzymes, and then utilized by another group of enzymes in the mitochondrion through the process of β-oxidation as an energy source. Furthermore, the degree of protonation may vary due to the trafficking of the enzyme to distinct compartments (e.g., the acidic lysosome and the neutral cytoplasm) or oxidative state (e.g., decreasing cytoplasm or oxidizing periplasm), all of which have an impact on the activity of enzymes. Enzyme subcellular localization can also be changed by polymerization of the enzymes into macromolecular cytoplasmic filaments, as opposed to partitioning into membrane-bound organelles.
5. **Organ specialization:** Different organs and tissues in multicellular eukaryotes have distinct gene expression patterns, which results in distinct sets of enzymes (referred to as isozymes) available for metabolic reactions. This offers a method for controlling the organism's overall metabolism. Hexokinase, the initial enzyme in the glycolysis pathway, for instance, possesses a distinct variant known as glucokinase, which is expressed in the pancreas and liver and has a lower affinity for glucose but a higher sensitivity to

glucose concentration. This enzyme is involved in controlling the synthesis of insulin and measuring blood sugar.

5.1.6.3. Involvement in Disease

A genetic disease can result from any malfunction (mutation, overproduction, underproduction, or deletion) of a single critical enzyme because homeostasis depends on the precise regulation of enzyme activity. Among the thousands of different types of enzymes found in the human body, a single malfunction can be lethal. Tay-Sachs disease, in which patients are deficient in the enzyme hexosaminidase, is an illustration of a fatal genetic disease brought on by an insufficient enzyme. The most prevalent kind of phenylketonuria is one instance of an enzyme deficiency. Phenylalanine and its related products accumulate as a result of numerous single amino acid mutations in the enzyme phenylalanine hydroxylase, which catalyzes the initial stage of phenylalanine degradation. While many mutations are located outside of the active site and lower activity by altering proper oligomerization or destabilizing the protein structure, some mutations are located in the active site and directly disrupt binding and catalysis. If the condition is not treated, this may result in intellectual disability. Pseudocholemetrase deficiency is an additional instance wherein the body's capacity to hydrolyze choline ester medications is compromised. Certain functional enzyme deficiencies, such as pancreatic insufficiency and lactose intolerance, can be treated by oral enzyme administration. Germline mutations in the genes encoding DNA repair enzymes are another way that malfunctioning enzymes can lead to illness. Because cells are less able to repair mutations in their genomes, defects in these enzymes lead to cancer. Cancers develop as a result of the gradual accumulation of mutations brought on by this. Xeroderma pigmentosum is one such hereditary cancer syndrome that results in skin cancer development even after very little exposure to UV light.

5.2. COVALENT MODIFICATION OF ENZYMES

Covalent modification of enzymes is a basic regulatory mechanism in which chemical groups covalently attached to particular amino acid residues are added or removed, thereby reversibly altering the activity of the enzyme. In response to different environmental cues and metabolic demands, cells use this dynamic and finely tuned regulatory strategy, known as post-translational modification, to modulate enzyme function. One of the most frequent covalent modifications is phosphorylation, which involves adding a phosphate group to an enzyme's side chain containing serine, threonine, or tyrosine residues. Protein kinases, which move the phosphate from ATP to the desired amino acid, catalyze this process. Phosphorylation frequently causes an enzyme's conformation to change, which can have an impact on the enzyme's catalytic activity, affinity for binding substrates, or interactions with other biological elements. Protein phosphatases, on the other hand, catalyze the process of dephosphorylation, which eliminates the phosphate group and returns the enzyme to either its active or inactive state. This alteration is noteworthy because it plays a role in controlling various biological functions such as metabolism, cell cycle progression, and signal transduction.

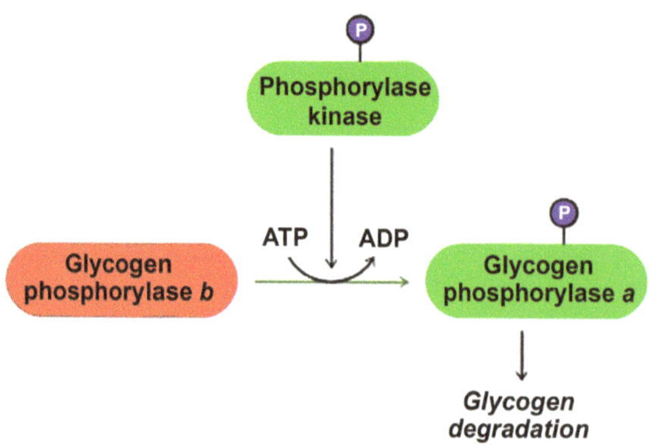

Figure 5.5. *State-of-the-art in phosphoproteomics.*

Source: https://3.bp.blogspot.com/_DZH2cmCoois/R03GjBbPZeI/AAAAAAAAD7g/cipwasQod7g/s400/tmp.jpg.

Another common covalent modification that usually affects lysine residues is acetylation. This alteration entails supplementing the ε-amino group of lysine with an acetyl group, which is commonly obtained from acetyl coenzyme A (acetyl-CoA). By changing the charge distribution within the protein and affecting how it interacts with substrates or other proteins, acetylation can have an effect on the activity of enzymes. One well-researched example is the regulation of gene expression through chromatin structure modulation, namely histone acetylation. Ubiquitination, a covalent modification mainly linked to protein degradation, is the process by which ubiquitin molecules bind to lysine residues. A polyubiquitin chain is created when the small protein ubiquitin is successively added to a lysine residue in the target protein. The ubiquitinated protein is recognized and broken down by the proteasome, a cellular protease complex, thanks to the signal provided by this chain. Substratum specificity is conferred by ubiquitination-related enzymes, such as E3 ubiquitin ligases, which dictate which proteins are broken down. The control of biological processes, such as the advancement of the cell cycle and the removal of misfolded or damaged proteins, depends on this covalent modification. The covalent modification known as "S-nitrosylation" involves combining a nitric oxide (NO) group with cysteine thiol groups. This alteration acts as a redox-dependent regulatory mechanism, modifying the activity of enzymes in response to alterations in the redox status of cells. Enzymes involved in cellular metabolism, apoptosis, and signal transduction can have their functions altered by s-nitrosylation.

5.2.1. Purpose

Reversible inhibitors play a crucial role in controlling the activity of enzymes. They have the ability to activate or inhibit enzymes, respectively. Furthermore, post-translational modification and covalent modification are two more ways that enzymes can be controlled. This implies that the structure of the enzyme can be further altered by adding unique groups to particular sites once it has been assembled in the cell. In the context of regulation, the addition of these groups is reversible. Although the group is bonded to the protein and added covalently, there is a reaction pathway that allows the group to be removed again. We'll examine three of these modifications in this instance. In every instance, alterations in the intermolecular interactions within the protein (or between the protein and another molecule) cause modifications to the protein's behavior. One extremely frequent modification is phosphorylation. A phosphate group is joined to the side chain of an amino acid during phosphorylation. Serine side chains are phosphorylated most frequently. Tyrosine is also frequently phosphorylated. The strong P-O bond that forms at those sites during the reaction makes them preferred.

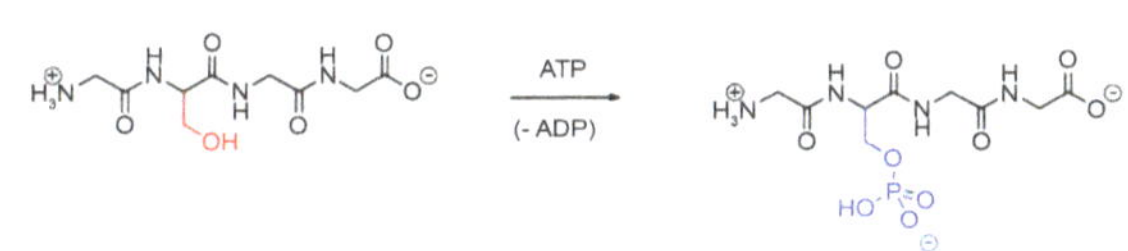

Figure 5.6. *Phosphorylation.*

Source: https://employees.csbsju.edu/cschaller/Reactivity/enzyme/EZphosphorylate.png.

Usually, phosphorylation is controlled by an additional enzyme known as a kinase. Yes, one enzyme will attach itself to another, attaching a phosphate group, and then freeing the attached enzyme. Phosphatase is another type of enzyme that can be used to remove phosphate groups again. The specificity of these modifications can be explained by the fact that they are carried out by particular enzymes. Some kinases may be less selective than others, but a given kinase may bind to its target protein in a specific location and phosphorylate it only there. An anionic phosphate group abruptly masks a neutral serine, which is, of course, the main effect of phosphorylation. Because of the sudden appearance of both repulsive and attractive forces where none previously existed, the negative charge drastically changes the intermlecular attractions. Acetylation is also fairly typical. The process of adding an acetyl or ethanoyl group —typically to a lysine — is known as acetylation. The resulting strong amide bond drives the forward reaction.

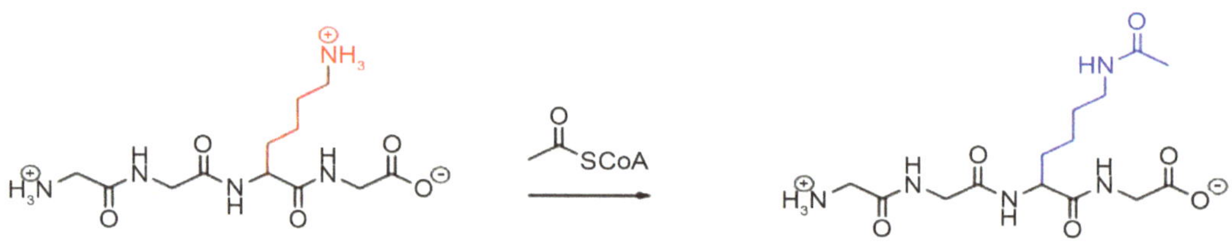

Figure 5.7. *Acylation.*

Source: https://employees.csbsju.edu/cschaller/Reactivity/enzyme/EZacetylate.png.

An acyltransferase or an acetylase performs acetylation. These are the same enzymes as phosphatases, binding to their target proteins to change their structures. Since lysines are typically positively charged at biological pH, acetylation causes charged species to abruptly vanish because neutral acyl groups cover them up. As a result, intermolecular attractions may change significantly. The process of prenylation involves adding a hydrocarbon side chain—most frequently, a cysteine side chain. For these kinds of carbon chains, sulfur is an especially effective nucleophile, improving selectivity for cysteine. However, as in the other instances, the reaction is regulated by an enzyme (a farnesyltransferase or geranyltransferase, for instance), so alternative side chains may be the focus.

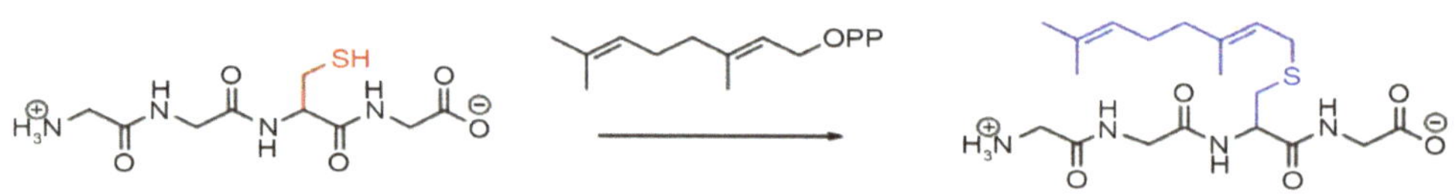

Figure 5.8. *Prenylation.*

Source: https://employees.csbsju.edu/cschaller/Reactivity/enzyme/EZprenylate.png.

The geranyl group and the related farnesyl group are two common groups that are added in these situations, and those enzyme types highlight them. Both of these groups belong to the terpenoid family of natural products and have structural similarities. The five-carbon building blocks that underpin terpenes are readily identifiable in the structures of geranyl and farnesyl.

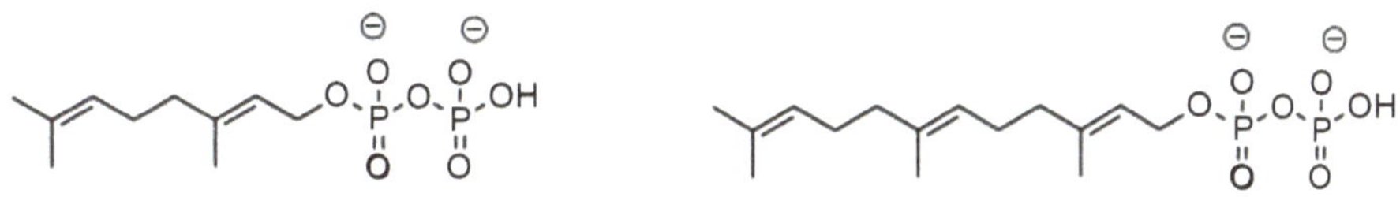

Figure 5.9. *Two common prenyl groups.*

Source: https://employees.csbsju.edu/cschaller/Reactivity/enzyme/EZterpene.png.

Because prenylation in these situations does not result in a complete alteration of a side chain's charge, as it did in the other scenarios, changes in intermolecular attractions are more subtle. Rather, the protein's hydrophilicity is increased in that region upon the addition of this group. A hydrophobic group is one that lacks hydrogen bonding sites, ions, or dipoles and is totally incapable of interacting with water molecules. Consequently, the surrounding water molecules tend to push it aside or displace it. The overall result of the water's pushing away is the stacking of hydrophobic groups. Thus, what we typically see is hydrophobic groups adhering to one another despite having relatively little mutual attraction (London forces). Naturally, the amount of London forces, or the relative area of a molecule that can interact with another via London forces, becomes very important because London forces are relatively weak. Due to their moderate length, geranyl and farnesyl groups exert a considerable amount of London forces. Consequently, a protein's two prenylated regions may cling to one another, or a single newly prenylated region may cluster with hydrophobic side chains like valines, leucines, and isoleucines. Additionally, a protein that was previously more water soluble may occasionally migrate to the cell membrane due to prenylation, where it may interact with the lipid bilayer. Therefore, prenylation can result in significant behavioral changes in the protein even though it only seems to slightly alter intermolecular attractions.

5.2.2. Feedback Inhibition of Metabolic Pathways

The final product of a metabolic pathway acts on the key enzyme controlling entry to that pathway, preventing the production of more of the end product, a process known as feedback inhibition. Though this may seem strange—why would a molecule want to block its own pathway?—the cell uses this as a clever mechanism to produce precisely the right amount of the product. When the product concentration is low, the pathway will work nonstop to replenish the supply because the enzyme won't be inhibited. A large amount of the product will clog the enzyme when it is accumulated, halting the synthesis of new product until the supply runs out. Diagram demonstrating inhibition of feedback. A multi-step metabolic pathway's final product reduces the activity of the enzyme by binding to an allosteric site on the enzyme that catalyzes the committed step of the pathway. When the final product levels are already high, this regulation aids in slowing down the pathway (when more is not needed). Feedback inhibition usually operates at the pathway's first committed step, or the first step that is essentially irreversible. Feedback inhibition, however, can occasionally occur at several locations along a pathway, especially if the pathway contains a large number of branch points. Allosteric enzymes are frequently used as catalysts for the pathway steps that feedback inhibition regulates.

For instance, some of the enzymes involved in cellular respiration—the process by which ATP is produced to power cellular reactions—are allosteric inhibitors of the energy carrier molecule ATP. This feedback inhibition prevents more ATP from being produced when there is an abundance of ATP. Given that ATP is an unstable molecule, this is helpful. A large amount of ATP could be wasted if it were produced in excess, disintegrating spontaneously back into its constituent parts (ADP and P_i). Conversely, for some of the same enzymes that ATP inhibits, ADP acts as a positive allosteric regulator, or an allosteric activator. To increase an enzyme's activity, for example, ADP may bind to it and alter its structure. Because of this pattern of regulation, cellular respiration enzymes become extremely active and increase ATP production through cellular respiration when ADP levels are high relative to ATP levels.

5.2.3. Phosphorylation in Cell Signaling

In signal transduction pathways, phosphorylation is a crucial regulatory mechanism that converts extracellular signals into intracellular reactions. This post-translational modification occurs when a phosphate group is added to a particular amino acid residue on proteins, including enzymes. These residues are usually serine, threonine, or tyrosine. The protein undergoes conformational changes upon the addition of the phosphate group, which can impact its activity, localization, or interactions with other cellular components. In this section, we will examine how phosphorylation functions in signal transduction pathways while focusing on particular enzymes and how they are regulated (Hedstrom, 2002).

- **Protein Kinases and Phosphorylation Cascades:** A key component of phosphorylation-mediated signal transduction is protein kinases. The cascade of phosphorylation events in the mitogen-activated protein kinase (MAPK) pathway is a well-known example. Extracellular signals in this pathway cause a sequence of protein kinases to become active, which phosphorylates MAP kinase (ERK). To control the expression of genes, ERK phosphorylates downstream substrates, such as transcription factors. The transmission of signals from the cell surface to the nucleus, which controls activities like cell proliferation, differentiation, and survival, depends on this phosphorylation cascade.
- **Phosphorylation in G Protein-Coupled Receptor (GPCR) Signaling:** The vast family of cell surface receptors known as GPCRs is essential to signal transduction. Intracellular heterotrimeric G proteins are activated by GPCRs upon ligand binding, which in turn activates downstream effectors. One well-known instance is the pathway involving adenylate cyclase. The process of converting ATP to cAMP is carried out by G protein-regulated adenylate cyclase. The subsequent binding of cAMP activates protein kinase A (PKA), which phosphorylates the target proteins. Numerous cellular responses, including

neurotransmission and hormone regulation, are mediated by this pathway.

- **Phosphorylation of Receptor Tyrosine Kinases (RTKs):** Cell surface receptors known as RTKs are essential for growth factor signaling. RTKs autophosphorylate on tyrosine residues upon ligand binding. The receptor's kinase activity is activated by this autophosphorylation, which starts subsequent signaling cascades. For example, tyrosine phosphorylation of the epidermal growth factor receptor (EGFR) initiates the Ras-MAPK pathway, which affects cell survival, proliferation, and differentiation.
- **Regulation of Glycogen Metabolism by Phosphorylation:** Phosphorylation controls glycogen metabolism-related enzymes very closely. When protein kinase A and other kinases phosphorylate glycogen, glycogen phosphorylase, which breaks down glycogen, gets activated. On the other hand, phosphorylation inhibits glycogen synthase, which is involved in the synthesis of glycogen. The equilibrium of these phosphorylation processes guarantees that the metabolism of glycogen is suitably controlled in reaction to energy requirements.
- **Phosphorylation in Cell Cycle Control:** Phosphorylation events tightly regulate the progression of the cell cycle. CDKs, or cyclin-dependent kinases, are important enzymes that regulate the cell cycle. The binding of cyclins and phosphorylation control their activity. For instance, the cyclin D-CDK4/6 complex regulates the G1-to-S transition by phosphorylating the retinoblastoma protein (Rb), which releases E2F transcription factors and encourages the advancement of the cell cycle.

5.2.4. Acetylation and Epigenetic Regulation

One important epigenetic alteration that significantly affects gene expression and cellular differentiation is histone acetylation, which is essential for controlling chromatin structure. The fundamental structural components of chromatin, nucleosomes, are formed by winding DNA around histone proteins. Histone acetyltransferases (HATs) catalyze the addition of acetyl groups to lysine residues on the histone tails, a process known as histone acetylation. By neutralizing lysines' positive charge, this process reduces the electrostatic interactions that hold histones and DNA together. Histone acetylation's effects on chromatin structure have a profound effect on the control of gene expression and the determination of cell fate. Histone acetylation primarily causes the chromatin structure to relax, making the underlying DNA easier for transcriptional machinery to access. The chromatin environment made more hospitable by acetylated histones makes it easier for transcription factors and RNA polymerase to bind to gene promoters. Because of its greater accessibility, genes can be more easily transcribed, which raises mRNA levels and eventually protein synthesis. Histone acetylation functions as a strong activator of gene expression in this manner. In the context of cellular differentiation, the relationship between histone acetylation and gene expression is particularly clear. Specific gene expression programs are carried out by stem cells during cellular differentiation, resulting in the emergence of specialized cell types with unique functions. During this process, histone acetylation patterns dynamically shift, designating genes for activation or repression. For instance, the transcription of genes related to cellular identity and function is aided by the acetylation of those histones, which causes cells to differentiate into distinct lineages. On the other hand, gene repression is caused by histone deacetylation, which is catalyzed by histone deacetylases (HDACs) and results in a more compact chromatin structure. One of the key components of epigenetic regulation is the dynamic interaction between histone acetylation and deacetylation. The equilibrium between these conflicting processes controls gene transcription and is essential for preserving cellular homeostasis.

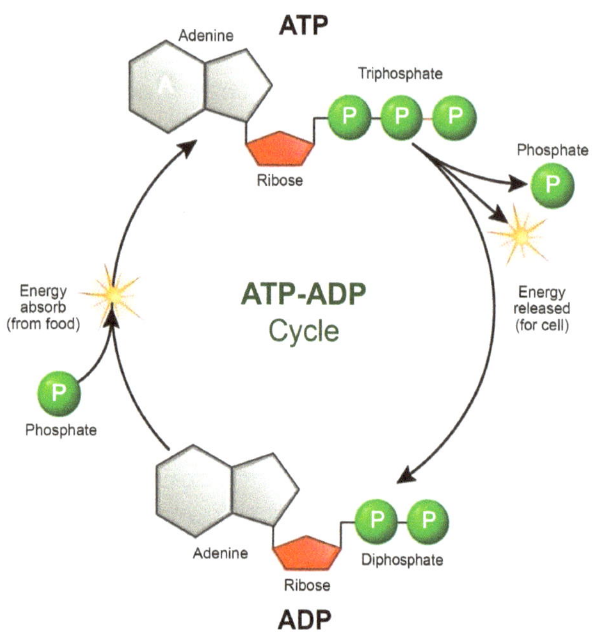

Figure 5.10. *ATP ADP cycle. Adenosine triphosphate (ATP) is an organic chemical that provides energy for cell.*

Source: https://d2jx2rerrg6sh3.cloudfront.net/image-handler/ts/20190520013010/ri/673/picture/2019/5/shutterstock_1238619880.jpg.

Histone acetylation affects global chromatin states and higher-order chromatin structures in addition to individual genes. In contrast to heterochromatin, which is more condensed and transcriptionally repressed, it aids in the formation of euchromatin, a more open and transcriptionally active chromatin state. The intricate gene expression programs that are orchestrated throughout development, tissue maintenance, and response to environmental cues are made possible by this epigenetic regulation. Furthermore, cancer and neurodegenerative illnesses are among the diseases linked to the dysregulation of histone acetylation. Genes may be improperly activated or repressed as a result of aberrant acetylation patterns, which can contribute to the development and progression of certain illnesses. As a result, histone deacetylase inhibitors are being investigated as possible treatments for a number of diseases, and histone acetylation has become a desirable target for therapeutic interventions.

5.2.5. Ubiquitination in Protein Degradation

Addition of ubiquitin, a small protein with 76 amino acids, to target proteins is known as ubiquitination, a highly regulated post-translational modification. This alteration, which marks proteins for proteasome destruction, is essential to the degradation of proteins. The UPS plays a pivotal role in maintaining cellular homeostasis by monitoring and controlling multiple physiological processes and guaranteeing the timely turnover of proteins. Examining the ubiquitination mechanisms reveals information about the enzymes involved, substrate selectivity, and the UPS's importance in preserving cellular equilibrium. Three primary classes of enzymes play a major role in the process of ubiquitination: ubiquitin-activating enzymes (E1), ubiquitin-conjugating enzymes (E2), and ubiquitin ligases (E3). These enzymes work together to orchestrate a cascade of enzymatic reactions. By adenylating ubiquitin and creating a high-energy thioester bond with its C-terminus, the E1 enzyme starts the ubiquitination process. The activated ubiquitin is then moved to an E1 enzyme cysteine residue after that.

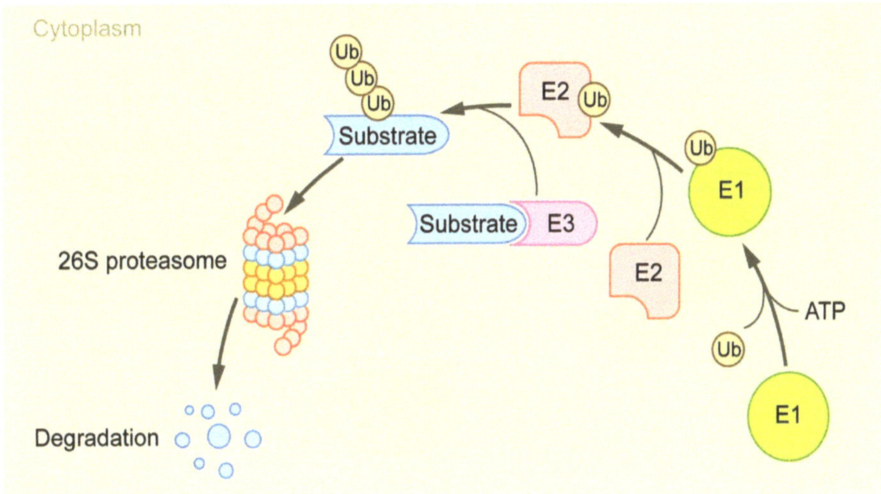

Figure 5.11. *Protein degradation via the ubiquitin-proteasome system (UPS).*

Source: https://media.springernature.com/m685/springer-static/image/art%3A10.1038%2Fs41392-022-00966-4/MediaObjects/41392_2022_966_Fig1_HTML.png.

Subsequently, ubiquitin is transferred to an ubiquitin-conjugating enzyme (E2) via a trans-thiolation reaction by the ubiquitin-charged E1 enzyme. In order to transport and transfer ubiquitin to the intended substrate, E2 enzymes are essential. Lastly, ubiquitin is transferred from the E2 enzyme to the substrate protein with the help of an E3 ubiquitin ligase. When it comes to granting substrate specificity during the ubiquitination process, the E3 ligase is very significant. Homologous to E6-AP C-terminus (HECT) ligases and RING (a really intriguing new gene) ligases are the two primary forms of E3 ligases. While RING ligases aid in the direct transfer of ubiquitin from the E2 enzyme to the substrate, HECT ligases first form an intermediate thioester bond with ubiquitin. The specificity of ubiquitin conjugation to distinct substrates is influenced by the variety of E3 ligases. Different target protein attachment strategies for ubiquitin can result in different functional outcomes. A single ubiquitin molecule is attached to a target protein through a process known as monoubiquitination, which is frequently connected to the control of endocytosis or DNA repair. However, in polyubiquitination, ubiquitin chains are added to the substrate through one of its seven lysine residues (K6, K11, K27, K29, K33, K48, K63). The most well-known polyubiquitin chains connected by K48 act as a cue for the 26S proteasome to break down proteins.

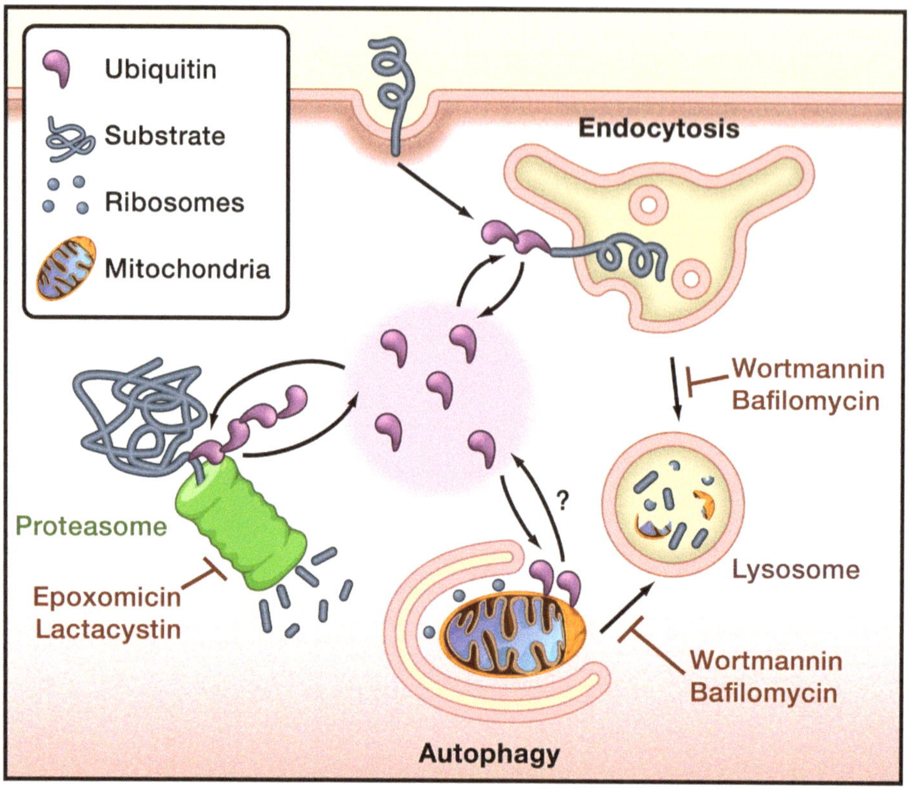

Figure 5.12. *Ubiquitin Is a common denominator of protein degradation pathways.*

Source: https://www.cell.com/cms/asset/206b5977–94bd-435f-a6fa-cdf690bc83d8/gr1.jpg.

The large, multicatalytic protease complex known as the 26S proteasome is in charge of breaking down ubiquitinated proteins. The topology of the polyubiquitin chain largely dictates the proteasome's substrate selectivity. The proteasome recognizes K48-linked polyubiquitin chains, which causes the substrate to unfold and move into the proteasomal core particle for degradation. The ubiquitin-proteasome system plays a critical role in maintaining cellular homeostasis. It controls the turnover of many different proteins, such as those involved in signal transduction, DNA repair, immune response, and cell cycle progression. UPS dysregulation has been linked to a number of illnesses, including cancer and neurological conditions, underscoring the UPS's significance in preserving cellular health.

5.3. ENZYME REGULATION IN METABOLIC PATHWAYS

In order to ensure that the output of the pathways satisfies biological demand and that energy in the form of ATP is not wasted by having opposing pathways run concurrently in the same cell, exquisite mechanisms that regulate the flux of metabolites through metabolic pathways have evolved. Enzymes can be controlled by varying their quantity or by altering the activity of an already-existing enzyme.

5.3.1. Changing the Activity of a Pre-Existing Enzyme

The quickest way to modulate the activity of an enzyme is to alter the activity of an enzyme that already exists in the cell. The list below, illustrated in the following figure, gives common ways to regulate enzyme activity

1. **Substrate Availability:** Reactants, or substrates, have a unique affinity for enzymes that is indicated by a dissociation constant and a kinetic parameter known as Km, or units of molarity. The activity of the enzyme is very low if the actual concentration of a substrate in a cell is significantly less than the Km. the enzyme is at its most active when the concentration of substrate in the active site is significantly higher than Km.
2. **Product Inhibition:** It should be obvious that the product of an enzyme-catalyzed reaction should bind to the activity site as well, albeit most likely with less affinity, since the product of such a reaction often resembles the starting reactant. When a reaction's product is present in high concentration, it would be beneficial for the cell's energy balance if no additional product was produced. Thus, product inhibition is frequently seen. Similarly, a cell would benefit energetically if the final product of a pathway could also bind to the first enzyme in the pathway and inhibit it, thereby inhibiting the pathway as a whole. Feedback inhibition of this kind is frequently seen. Inhibition of the product and end product is shown in Figure 5.13.

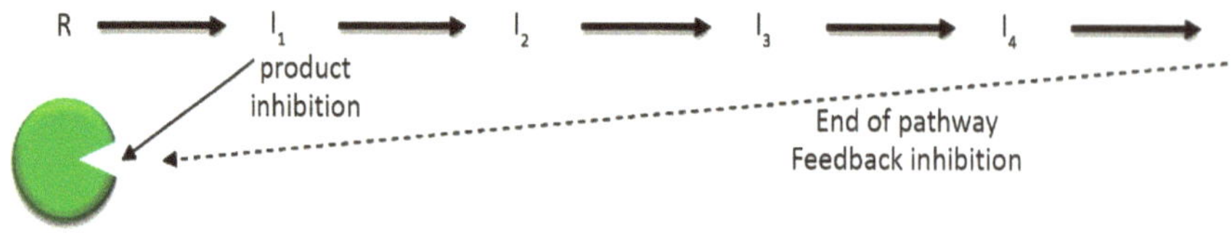

Figure 5.13. *Product and end product inhibition of an enzyme.*

Source: https://bio.libretexts.org/@api/deki/files/49574/ProductEnzPathwayFeedbackInhib.gif?revision=1.

3. **Allosteric Regulation:** Given the number of interconnected pathways, it would be ideal if molecules in one pathway had an impact on the activity of enzymes in another, even in cases where the molecules in one pathway differ structurally from reactants or products in another. Target enzyme activity can be controlled by molecules that bind to allosteric sites, which are locations other than the active site. These molecules may differ from those that bind at the active site in terms of structure. By causing conformational changes, they are able to either stimulate or inhibit the activity of the target enzyme.
4. **pH and Enzyme Conformation:** The pH can fluctuate during metabolic processes like respiration (aerobic glycolysis, for instance), which can change an enzyme's conformation and, consequently, its activity. The first alterations are covalent (a change in the protein's protonation state), which may change the delicate balance of forces that determines how the protein is arranged.
5. **pH and Active Site Protonation State:** Without influencing the local or global conformation of the protein, pH changes can have an impact on the protonation state of important amino acid side chains in the active site of proteins. If, for example, the catalytic mechanism requires the deprotonation of an active site nucleophile in order for catalysis to occur, then catalysis may be impacted.
6. **Covalent Modification:** Post-translational modifications affect most proteins, if not all of them. These modifications can alter the shape of the protein locally or globally, promote or inhibit the binding interaction between substrates and allosteric regulators, or even change the protein's location within the cell. Proteins can undergo a variety of reversible modifications, including phosphorylation, acetylation, methylation, sulfation, glycosylation, amidation, hydroxylation, prenylation, and myristoylation. A portion of these changes are reversible. Dephosphorylations by phosphates and phosphorylations by kinases are very common methods of regulation. Protein kinases and phosphatases within the cell are activated or inhibited as a result of signal transduction pathways that begin at the cell membrane and regulate the phosphorylation state.

Figure 5.14 shows ways to regulate the activity of pre-existing enzymes.

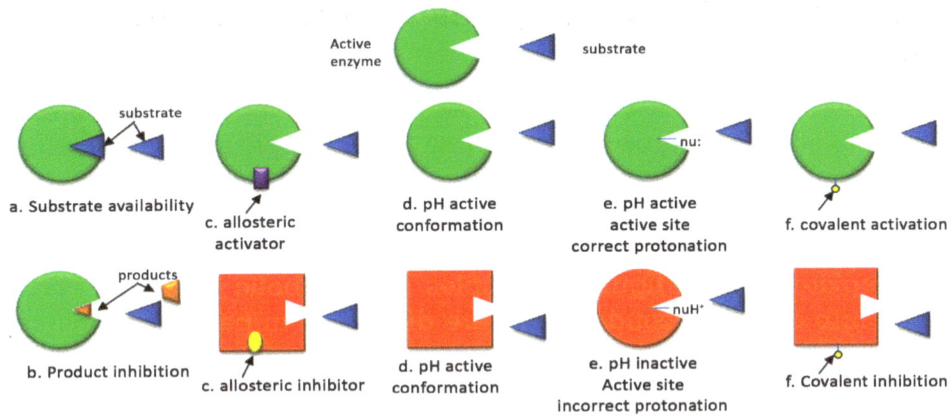

Figure 5.14. *Ways to regulate the activity of pre-existing enzymes.*

Source: https://bio.libretexts.org/@api/deki/files/49578/RegExistEnzActivity.gif?revision=1.

Mitogen-activated protein kinase 2 (MAPK2), another name for extracellular regulated kinase 2 (ERK2), is a protein that is essential for cell signaling across the cell membrane. The protein's structure and activity are altered when ERK2 is phosphorylated on tyrosine 185 (Tyr185) and threonine 183 (Thr153). An interactive iCn3D model is presented in Figure 5.15, which illustrates the structural alignment of ERK2 in the phosphorylated (pY185) and dephosphorylated (5UMO) forms (2ERK). Use the key to switch back and forth between the two structures.

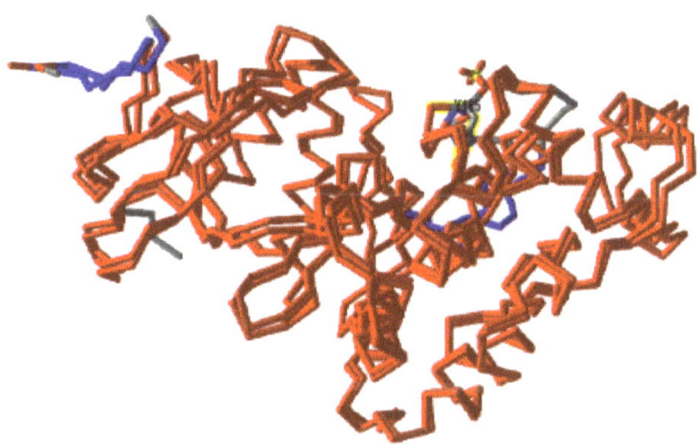

Figure 5.15. *Structural alignment of ERK2 in the dephosphorylated (5UMO) and phosphorylated (pY185) forms (2ERK).*

Source: https://bio.libretexts.org/@api/deki/files/60908/structaligERK2__dephospho_(5UMO)_and_phospho_(pY185)_forms_(2ERK).png?revision=1&size=bestfit&width=407&height=252.

The residues that change significantly in conformation on phosphorylation are shown in blue. The side chain of tyrosine 185 in the unphosphorylated form is shown in CPK-colored sticks and labeled.

5.3.2. Regulation of Single Enzymes or Entire Pathways: Enzyme Condensates

By assembling the enzymes in a single, sizable complex made of soluble enzymes that generate a condensate via a procedure akin to phase separation, individual enzymes or all of the enzymes in a particular pathway can be co-ordinatedly controlled to optimize end-product output. Figure 5.16 displays these condensates for a number of different enzymes. Enzymes linked to metabolism that form polymers in a variety of organisms are depicted in the figure. Examples of metabolic enzymes that have been seen to combine into cytosolic condensates are displayed in panel (a). The structures of metabolic enzymes that polymerize into filaments are displayed in panel (b) above. The polymer's protomer is positioned within the filament below and is visible above. These are P-Fructo-Kinase (4XYJ), mTORC1 (5FLC), glutamine synthetase (3FKY), acetyl-CoA carboxylase (6G2D), and cytidine triphosphate synthase (5U03) from PDB files.

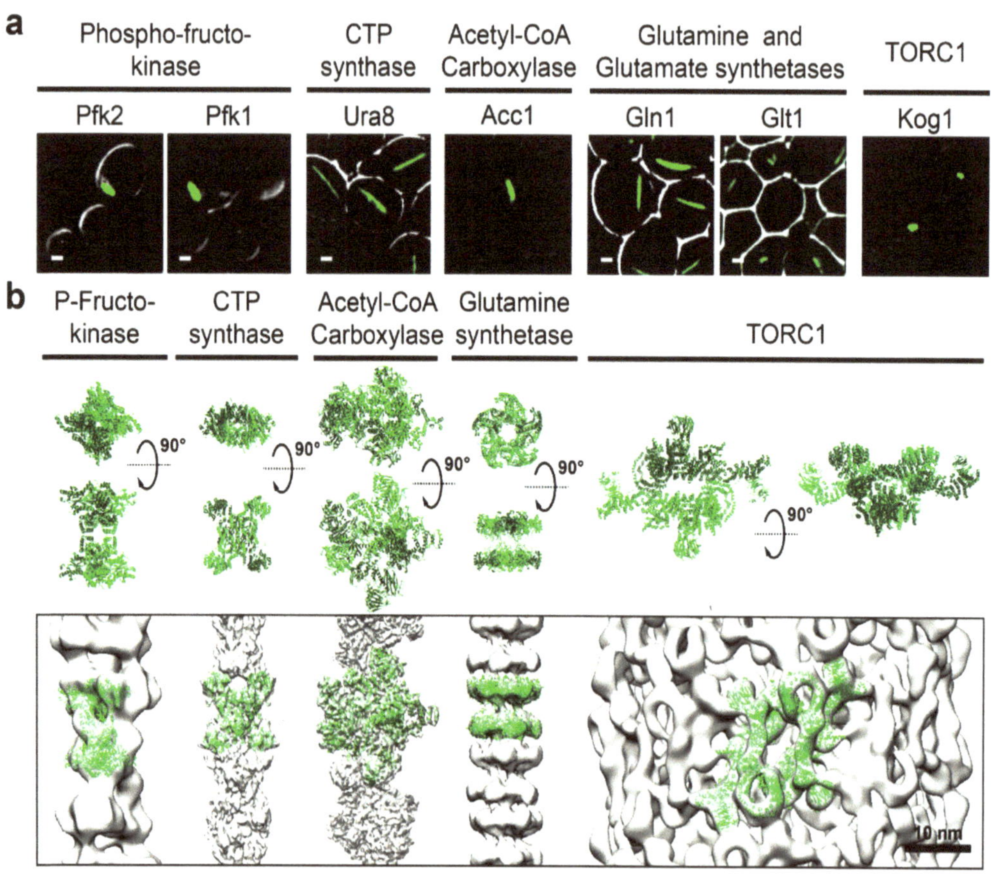

Figure 5.16. *Supramolecular assembly of enzyme condensates.*

Source: https://bio.libretexts.org/@api/deki/files/60830/Regulation_of_Cellular_Metabolism_through_Phase_Separation_of_EnzymesFig2.svg?revision=1&size=bestfit&width=820&height=651.

The same structures from the Electron Microscopy Data Bank are displayed in panel (b) bottom, including the yeast glutamine synthetase filament, human CTP synthase filament (and-8474), human acetyl-CoA carboxylase with citrate (emd-4342), and P-Fructo-Kinase filament (emd-8542).

5.3.3. Changing the Amount of an Enzyme

A longer-term strategy for regulating an enzyme's activity is to change the activity of an enzyme that is already present in the cell. The mechanisms for controlling enzyme concentrations are depicted in Figure 5.17.

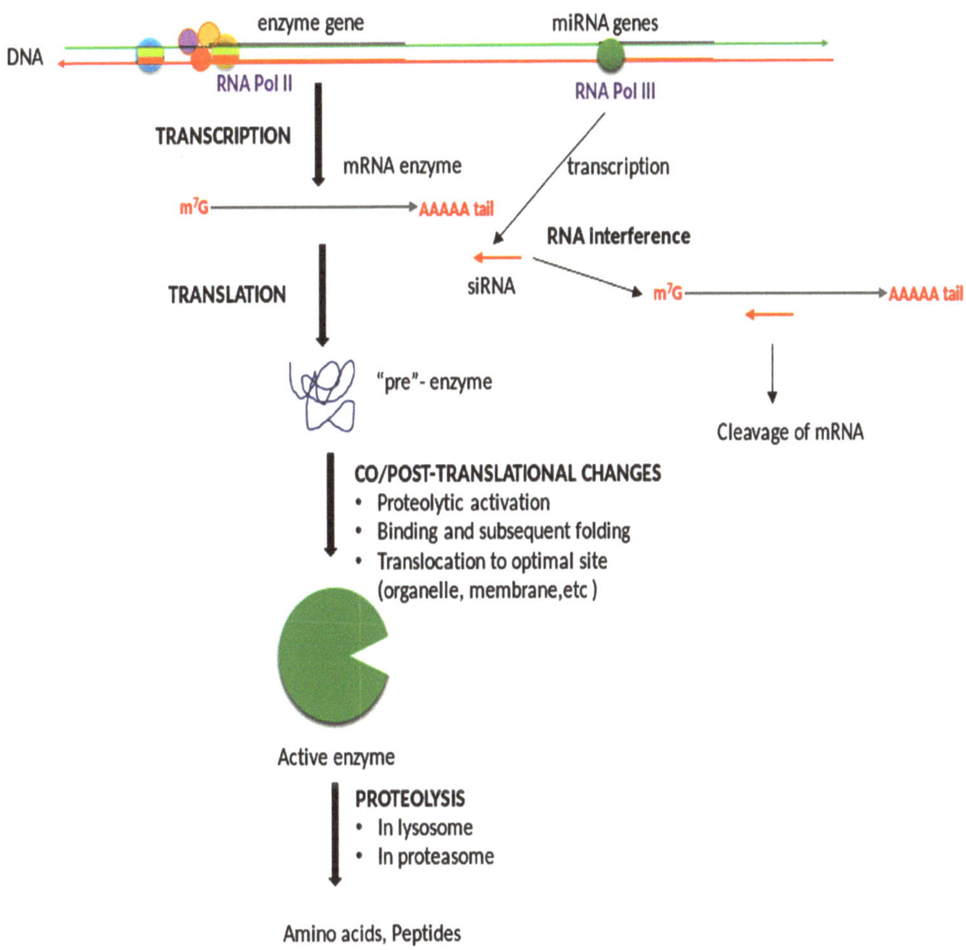

Figure 5.17. *Ways to regulate the concentration of an enzyme.*

Source: https://bio.libretexts.org/@api/deki/files/49576/RegExistEnzAmount.gif?revision=1.

Methods Include:

1. **Alterations in the Transcription of Enzyme's Gene:** Signs from outside the cell (hormones, neurotransmitters, etc., may cause signal transduction reactions, which in turn may activate or inhibit a protein enzyme's gene's transcription. Transcription factors, or proteins, bind to specific DNA sequences to control the transcription of the enzyme gene, causing these alterations.
2. **Degradation of Messenger RNA for the Enzyme:** The amount of a protein that is synthesized is directly correlated with its messenger RNA levels. Target enzyme mRNA can have certain sequences that small inhibitor RNAs, which are produced from microRNA molecules transcribed from cellular DNA, can bind to. An enzyme (Dicer) is drawn to the resulting double-stranded RNA complex, and it cleaves the complex, reducing the amount of the protein enzyme that is translated from its mRNA.
3. **Co/Post-Translational Changes:** A protein enzyme can experience a variety of modifications that control its activity after it is translated from its mRNA. Certain proteins are synthesized in a pre-formed state that requires selective and limited cleavage by proteases in order to release the protein enzyme. In order to take on a catalytically active form, certain proteins that are not fully folded must attach to other components within the cell. Lastly, the proteasome, a complex found in cells, or lysosomes, which are organelles that contain proteolytic enzymes, can fully proteolyze fully active protein.

In the end, all proteins are controlled, even if it's just by adjusting how quickly they synthesize and degrade. Nonetheless, because their activity can impact the outcome of entire pathways, certain enzymes that are strategically located within metabolic pathways make excellent candidates for regulation. These enzymes usually share two properties: they catalyze early committed steps in pathways and reactions that are far from equilibrium.

5.3.4. Enzymes to Regulate: Reactions not at Equilibrium

The enzymes at the start of pathways carrying out thermodynamically favored reactions are the best candidates for regulation. Because these enzymes regulate the flow of metabolites through pathways, we can use the analogy of water flowing (or fluxing) from one container to another, as shown in Figure 5.18, to understand their regulation and understand why the latter is so crucial.

Assume you want to fill a swimming pool to any height you choose. There are two ways to accomplish this (see Figure 5.19). You could open a valve that regulates the water flow from the water tower in your town to the swimming area. Given the variation in the water levels and the potential energy difference between them, the reaction (water flow) is energetically (thermodynamically) favored in this situation. Even though flow (or flux) is flavorless, you can control it by opening and closing the valve, which works similarly to activating and inhibiting an enzyme, to go from no flow to maximum flow. In the second scenario, where you fill the pool from a lake, you have fewer options. Reaching the ideal water level would be challenging (especially if it was an above-ground pool). It would be challenging to control the flow. Comparatively speaking, the most promising candidates for regulation are those enzymes whose reactions can be regulated by the mechanisms covered in the preceding section but which are thermodynamically favored (not at equilibrium).

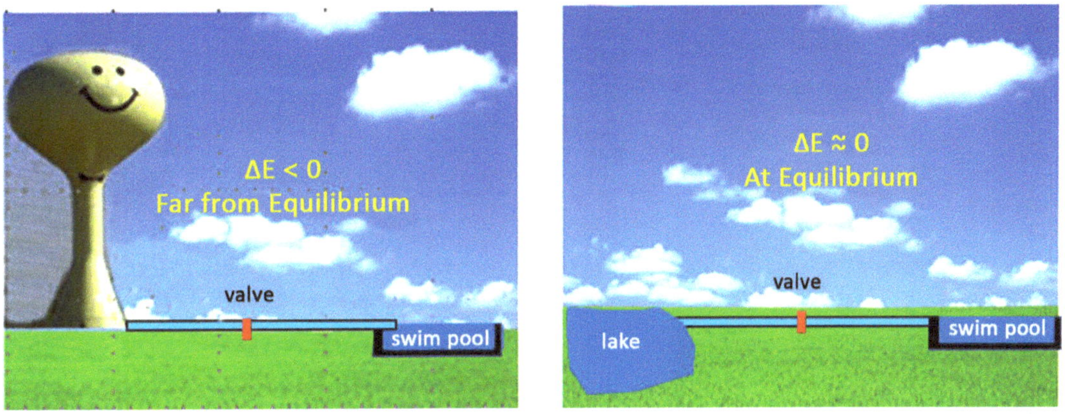

Figure 5.18. *Regulation of water flow in pipe.*

Source: https://bio.libretexts.org/@api/deki/files/49577/EnzRegulationWaterTower.jpg?revision=1.

Which reactions are commonly not at equilibrium (i.e., $\Delta G<0$ and usually also $\Delta G^0 <0$ if the ratio of products to reactants is not too high)? The answer is those that have reactants that are thermodynamically unstable compared to their reaction products. Several types of reactions often fit these criteria:Hydrolysis (or similar reactions) of anhydride or analogous motifs: The Figure 5.19 shows molecules with similar "anhydride" motifs and the ΔG^0 for hydrolysis of the molecules. Those with more negative ΔG^0 values can transfer their phosphate group to ADP to make ATP, which is necessary to drive unfavorable biological reactions. Metabolic reactions that involve hydrolysis (or other types of transfer reaction of these groups) usually proceed with a negative ΔG^0 and ΔG, making them prime candidates for pathway regulation. Many textbooks label these types of molecules as having "high energy" bonds. Many students find this confusing because bonds between atoms reduce energy in comparison to an unbonded state. The high energy phosphoanhydride covalent bond requires energy to break. Because more energy is released during the bond formation of the new products than was needed to break the bonds in the reactants, the hydrolysis of the molecules below is said to be exergonic. The hydrolysis of the reactants is thermodynamically advantageous in addition to other effects like the products' preferential hydration, reduced charge density, and fewer competing resonances. Thermodynamically unstable molecules are depicted in Figure 5.19 (in comparison to their reaction products in aqueous solutions).

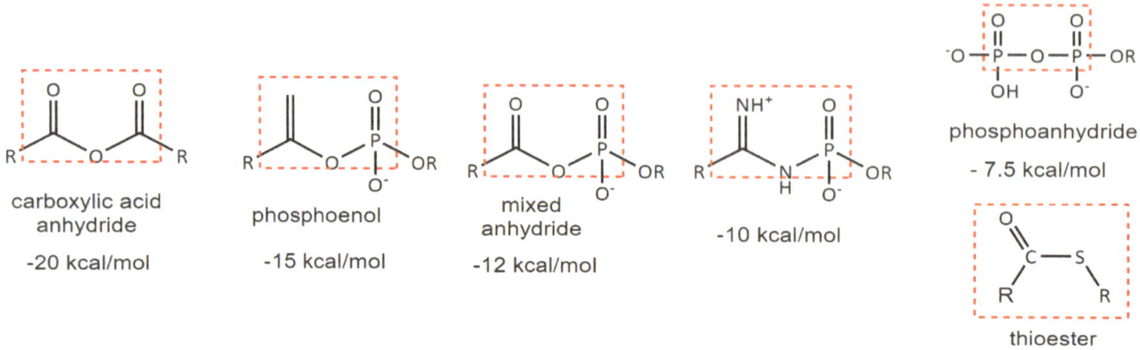

Figure 5.19. *Molecules that are thermodynamically unstable compared to their hydrolysis products.*

Source: https://bio.libretexts.org/@api/deki/files/60909/metabolismWP_HighNrgMotiff.svg?revision=1&size=bestfit&width=808&height=295.

Even though they don't have an anhydride motif, thioesters like Acetyl-SCoA are included because they have the same negative ΔG^0 of hydrolysis as ATP. Due to the weaker C-S bond, thioesters are destabilized when compared to both their hydrolysis products and alcohol-based esters. Why? Redox reactions: Redox reactions are thermodynamically preferred when a potent oxidizing agent is used. This much is known. Dioxygen's oxidation reactions of hydrocarbons, sugars, and fats are unquestionably exergonic—after all, we refer to these processes as combustion reactions. When it comes to redox reactions involving weaker oxidants, NAD+ is a commonly used biological oxidizing agent. Hardly speaking, are all of these reactions as preferred as combustion. Recall that an oxidizing and reducing agent reacts to form another oxidizing and reducing agent in every redox reaction.

Consider the following reaction:

Pyruvate + NADH ↔ Lactate + NAD^+.

This is a reversible reaction that can go either way. When both Pyr and NADH levels are high, the above form is written in the preferred direction for anaerobic metabolism. Despite the fact that ΔG^0 favors lactate oxidation, the reaction is driven in the opposite direction and proceeds as shown because of the high concentration of Pyr and NADH. The ΔG^0 for a redox reaction should be calculated from standard reduction potentials using the formula $\Delta G^0 = -nFE^0$ in order to ascertain whether the reaction is favored and likely to occur (and possibly regulated).

5.3.5. Enzymes to Regulate: Those Catalyzing Committed Steps in Pathways

The enzymes that catalyze the reaction pathway's first committed step are the most suitable for regulation. The committed step is essentially irreversible and proceeds with a $\Delta G^0 < 0$. These reactions frequently start from important metabolic intermediates that are close to or right before reaction pathway branches.

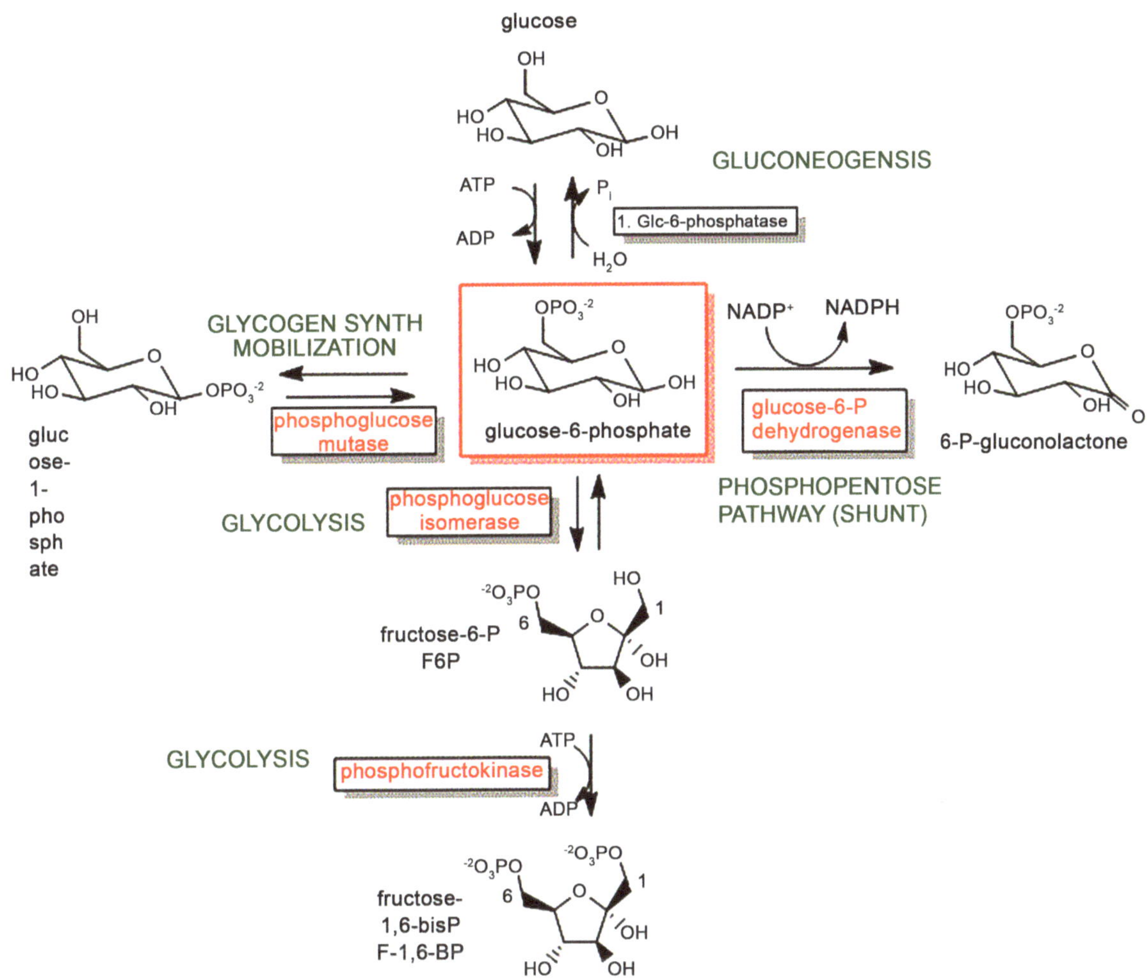

Figure 5.20. *Reactions for the production and use of the intermediate glucose-6-phosphate.*

Source: https://bio.libretexts.org/@api/deki/files/49579/metabolismWP_CommIntermG6P.gif?revision=1.

Figure 5.20, which depicts the reactions for the synthesis and utilization of the intermediate glucose-6-phosphate, provides two instances of important intermediates at metabolic pathway branch points.

Figure 5.21 shows reactions for the production and use of the intermediate acetyl-CoA.

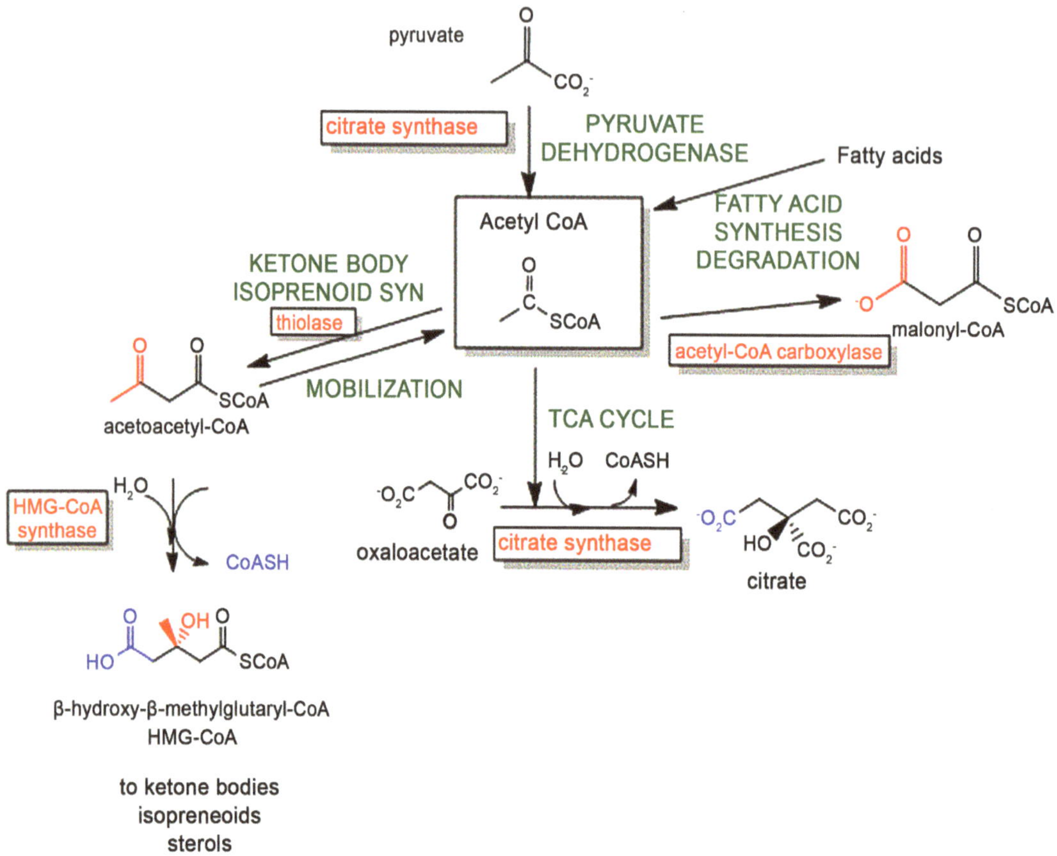

Figure 5.21. *Reactions for the production and use of the intermediate acetyl-CoA.*

Source: https://bio.libretexts.org/@api/deki/files/49585/metabolismWP_CommIntermAcetylCoA.gif?revision=1.

In reality, metabolic regulation is more complex and is distributed to many steps in a reaction pathway in ways that might not be evident without detailed mathematical analyzes. We will discuss that in the next sections on metabolic control analysis.

5.3.6. Control of Metabolism through Enzyme Regulation

The requirements and circumstances of individual cells fluctuate over time and differ from one cell to the next. A stomach cell, for instance, uses a different quantity of energy than a skin, blood, fat-storing, or nerve cell. Additionally, the same stomach cell might require less energy in between meals and more energy right after a meal.

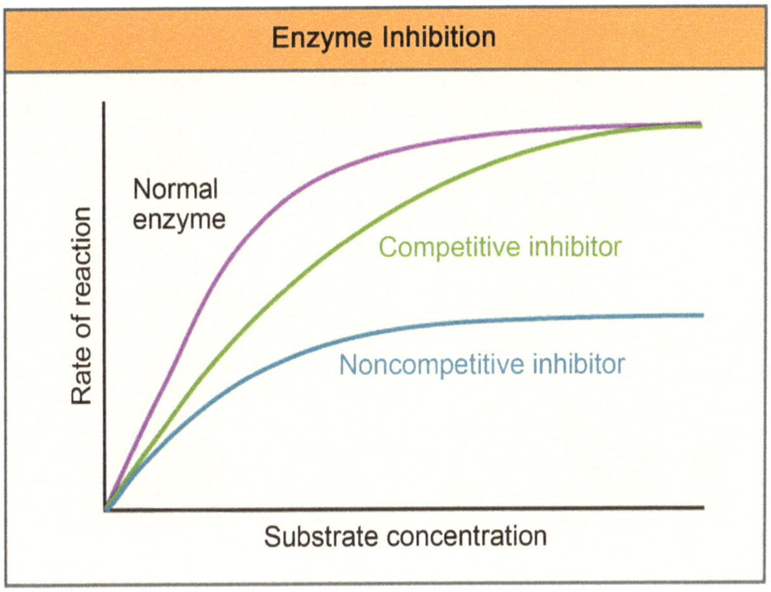

Figure 5.22. *Enzyme inhibition: Competitive and noncompetitive inhibition affect the rate of reaction differently. Competitive inhibitors affect the initial rate, but do not affect the maximal rate, whereas noncompetitive inhibitors affect the maximal rate.*

Source: https://s3-us-west-2.amazonaws.com/courses-images/wp-content/uploads/sites/1950/2017/05/31183105/figure-06-05-04.jpeg.

The chemical reactions that a cell is capable of performing capture its function. Chemical reactions that are specific to a cell's function are promoted by enzymes, which also lower the activation energies of chemical reactions. Enzymes are essential to cells because they ultimately dictate the kinds of chemical reactions that a cell can perform and how quickly they can do so.

5.3.6.1. Competitive and Noncompetitive Inhibition

The cell controls enzymes to either stimulate or impede particular chemical reactions using particular molecules. There are several ways in which an enzyme can be inhibited, which is sometimes required to lower a reaction rate. An inhibitor molecule can bind to the active site of the enzyme in competitive inhibition, preventing the enzyme from binding to the substrate, because it resembles a substrate sufficiently. In order to bind to the enzyme, it "competes" with the substrate.

An inhibitor molecule binds to the enzyme at an allosteric site, which is different from the active site, in noncompetitive inhibition. The inhibitor modifies the shape of the enzyme, making it less suitable to catalyze the reaction, but the substrate can still bind to the enzyme.

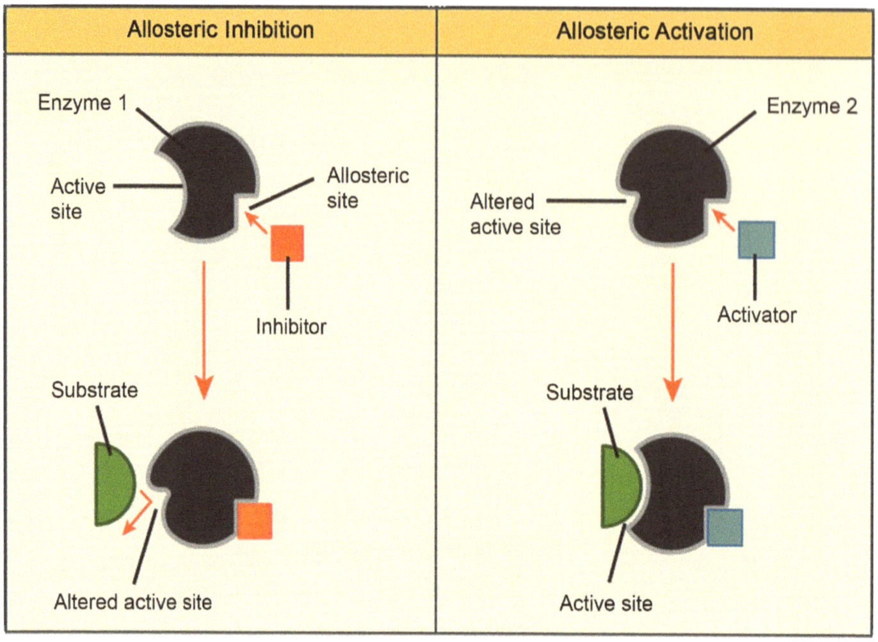

Figure 5.23. *Allosteric inhibitors and activators: Allosteric inhibitors modify the active site of the enzyme so that substrate binding is reduced or prevented. In contrast, allosteric activators modify the active site of the enzyme so that the affinity for the substrate increases.*

Source: https://s3-us-west-2.amazonaws.com/courses-images/wp-content/uploads/sites/1950/2017/05/31183107/figure-06-05-05.jpeg.

5.3.6.2. Allosteric Inhibition and Activation

Inhibitor molecules attach to an enzyme at the allosteric site in noncompetitive allosteric inhibition. Their binding causes a conformational shift in the enzyme that lowers the active site's affinity for its substrate. The enzyme's active site and conformation are altered by the binding of this allosteric inhibitor, making it impossible for the substrate to bind. As a result, the reaction rate is lowered and the enzyme is unable to lower the reaction's activation energy. Allosteric sites can be bound by other molecules besides allosteric inhibitors. Reaction rates can be accelerated by allostery activators. When they attach to an allosteric site, the enzyme undergoes a conformational shift that improves the active site's affinity for its substrate. The reaction rate rises as a result.

5.3.6.3. Cofactors and Coenzymes

Cofactors and coenzymes are non-protein helper molecules that many enzymes require to function. For each of these molecules, binding results in the best possible conformation and activity for the corresponding enzyme. These molecules can bind together permanently through stronger covalent bonds or momentarily through ionic or hydrogen bonds. Inorganic ions like iron (Fe^{2+}) and magnesium (Mg^{2+}) are cofactors. For instance, zinc ions (Zn^{2+}) are needed by DNA polymerase in order to construct DNA molecules. Coenzymes are organic aid molecules consisting of hydrogen and carbon atoms in their basic atomic structure. Dietary vitamins are the most prevalent type of coenzyme. Vitamin C functions as a coenzyme for several enzymes involved in the synthesis of collagen, a crucial component of connective tissue. Pyruvate dehydrogenase is an enzyme complex that needs five distinct organic coenzymes in addition to one cofactor to catalyze its chemical reaction. The availability of different cofactors and coenzymes controls how well enzymes work.

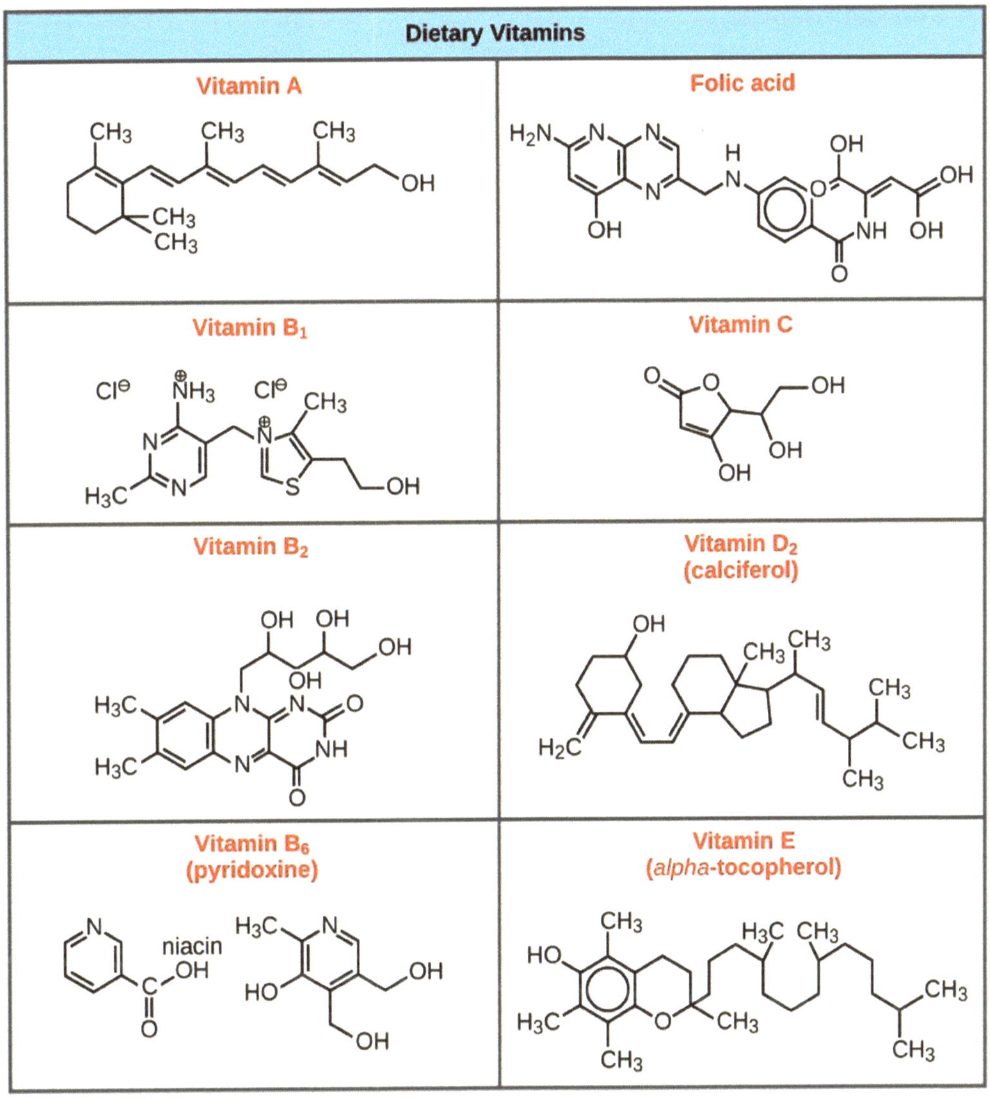

Figure 5.24. *Vitamins: Vitamins are important coenzymes or precursors of coenzymes and are required for enzymes to function properly. Multivitamin capsules usually contain mixtures of all the vitamins at different percentages.*

Source: https://assets.coursehero.com/study-guides/lumen/images/sanjacinto-biology1/enzymes/Figure_06_05_067.jpg.

5.3.6.4. Enzyme Compartmentalization

Enzymes and other molecules are typically divided into distinct organelles in eukaryotic cells. Because certain biological processes are housed in distinct organelles, this organization plays a role in the regulation of enzymes. For instance, only mitochondria are used by the enzymes involved in the later phases of cellular respiration to carry out reactions. Lysosomes contain the enzymes that aid in the breakdown of foreign substances and cell debris.

5.3.6.5. Feedback Inhibition in Metabolic Pathways

When a reaction product controls its own subsequent production, this is known as feedback inhibition. Through the use of the byproducts of enzymatic reactions to prevent additional enzyme activity, cells have evolved to use feedback inhibition to control enzyme activity in metabolism. Anabolic and catabolic processes are examples of metabolic reactions that must proceed in accordance with the needs of the cell. Certain metabolic products inhibit the enzymes in the chemical pathway while certain reactants activate them in order to maintain chemical equilibrium and meet the needs of the cell.

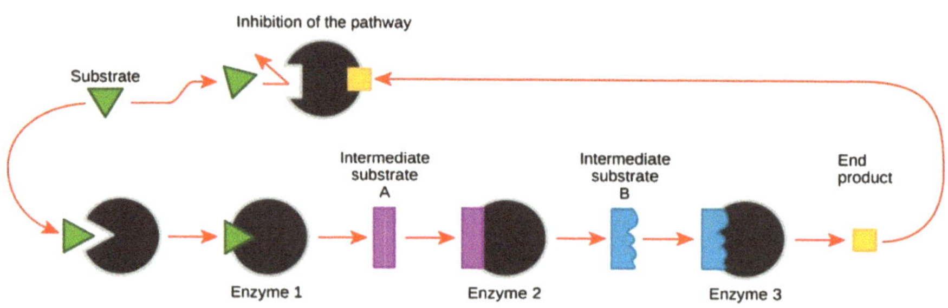

Figure 5.25. *Feedback inhibition: Metabolic pathways are a series of reactions catalyzed by multiple enzymes. Feedback inhibition, where the end product of the pathway inhibits an earlier step, is an important regulatory mechanism in cells.*

Source: https://slcc.pressbooks.pub/app/uploads/sites/20/2020/11/Feedback-inhibition-1024x346.jpeg.

Feedback inhibition regulates the synthesis of nucleotides and amino acids. Take ATP as an illustration of feedback inhibition. It is a byproduct of the catabolic metabolism of sugar (cellular respiration), and the same enzymes that produced it also function as allosteric regulators. Since ATP is an unstable molecule that can split spontaneously into ADP, most of it would be wasted if there was an excess of it. If there is already an abundance of ATP, this feedback inhibition stops it from being produced. ADP is an allosteric activator, but ATP is an inhibitor. When ADP is present in greater quantities than ATP, it causes the breakdown of sugar, which results in the production of more ATP.

CASE STUDY

ENZYMES ARE VITAL FOR EVERY PART OF LIFE

With the limited evidence, the daily, monotonous life of ancient humans is hard to predict with precision, yet one thing known for sure is that they also required various food sources to sustain life. Based on the preexisting knowledge in addition to newly emerging data, we are putting pieces together and trying to predict and clarify which enzyme-bound products they used.

What was the role of fermentation in the emergence of enzyme products in the ancients' life? In fact, was the first enzyme product a fermented one or not? Which came first? Bread? Wine? Vinegar? Beer? Soy sauce? Kumis? Sake? Koji? Kefir? Sour cream? Pickles? Sauerkraut? Sourdough? Yoghurt? Boza? Kimchi? Miso? Tempeh? Or perhaps a completely distinct, unknown enzyme product was consumed that unfortunately left the scene with no trace for us to follow.

Neanderthal man's food culture began to emerge approximately in 30,000 BC with the introduction of bread and, accordingly, it is very hard to find any documented evidence on this subject. Bakery technology begins with the cultivation of wheat in Göbekli Tepe that is, with its 11,600 years old existence, also the world's oldest temple. Sourdough starters or yeast was used as a food additive in bread baking. However, the exact or approximate date is not clear. Dairy products were another important food source in ancient societies. Cheese is an effortless candidate to claim the role of being the first enzymatic product. And the use of milk dates back to 8,000 years ago and it evolved in the 'Fertile Crescent' between the Tigris and Euphrates rivers.

Koumiss was an alcoholic beverage (V century BC) of ancient times in central Asia and was used in the treatment of various diseases such as phthisis. Yogurt and kefir are also among the ancient traditional dairy products and were used extensively as preventive compounds against diseases, and for curative purposes. The nomads carried fresh milk in bags, probably crudely made from animals' stomach, and the milk, in this relatively perfect condition, fermented into yogurt or kefir. The origins of wine and vinegar go back to as far as 2,500 BC. Hammurabi tablets, casted in stone, are perhaps the first documented artifacts on fermentation of grapes. These tablets date back to 2,100 BC and again this was probably the first document on the commercial use of enzymes. The second commercial product of the enzyme origin might have been vinegar, which was widely used in those days both for medicinal purposes as a painkiller, and in disinfection of wounds and finally in food storage. However, C. Wang and his colleagues, using five analytical methods to identify the chemical constituents of the potteries originating from China, found out that the emergence of various fermented products dates back to as early as 7,000 BC. On the other hand, the origin of sake is an oblique one. Sake may be as old as wine. It goes back as far as 2,500 years when the cultivation of rice became widespread. The most mysterious and acutely necessary component of sake is koji which contains alpha-amylase, beta-amylase proteases, peptidases, sulfatases enzymes. Traditionally, koji is also used in soy sauce and in miso. Natto is a traditional, dateless Japanese meal and the enzyme found in natto has been named nattokinase.

Smell or the occurrence of various odors, again an interesting façade of enzyme catalyzed reactions, is another, crucial factor in survival. The first study on enzymatic catalysis and the related occurrence of odors was performed approximately four decades ago. Investigators, for instance, studied the effect of γ-irradiation on cystein sulfoxide lyase and odor of onions. Odors originating from the catalytic activities of the enzymes are potent regulators of biological functions. Jasmonic acid, a highly volatile terpene compound, shows its effect on protein kinases or transcription factors. Odor evoked behaviors or social interactions show their effect on the enzymes such as flavin containing enzyme monooxygenase-3. In the determination and approximate diagnosis of various health problems, dogs and cats smell some of the diseases such as cancer, epileptic seizures, and hypoglycemia.

Meat tenderizer, which contains the enzyme papain, has been recommended for the treatment of bee and fire ant stings. It was also established that these meat tenderizers were safe and effective for patients with a phytobezoar.

SUMMARY

- Enzymes are proteins that help speed up chemical reactions in our bodies. Enzymes are essential for digestion, liver function and much more. Too much or too little of a certain enzyme can cause health problems.
- The majority of enzymes are proteins with catalytic capabilities crucial to perform different processes. Metabolic processes and other chemical reactions in the cell are carried out by a set of enzymes that are necessary to sustain life.
- Enzymes are generally globular proteins, acting alone or in larger complexes. The sequence of the amino acids specifies the structure which in turn determines the catalytic activity of the enzyme. Although structure determines function, a novel enzymatic activity cannot yet be predicted from structure alone.
- Enzymes are usually much larger than their substrates. Sizes range from just 62 amino acid residues, for the monomer of 4-oxalocrotonate tautomerase, to over 2,500 residues in the animal fatty acid synthase.
- Despite of enzymes have complex internal dynamic motions, they are not rigid and static structures. Indeed, movements of parts of the enzyme's structure such as individual amino acid residues, groups of residues forming a protein loop or unit of secondary structure, or even an entire protein domain.
- Enzymes serve a wide variety of functions inside living organisms. They are indispensable for signal transduction and cell regulation, often via kinases and phosphatases.
- Covalent modification of enzymes is a fundamental regulatory mechanism that involves the reversible alteration of an enzyme's activity through the addition or removal of chemical groups covalently attached to specific amino acid residues.
- Phosphorylation is typically carried out under the control of another enzyme called a *kinase*. That's right, one enzyme will bind another, tying a phosphate group onto it before releasing it again. A phosphate group can be removed again via another enzyme called a *phosphatase*.

REVIEW QUESTIONS

1. How are enzymes systematically classified and named according to their catalytic functions?
2. What specific roles do enzymes play in biological processes and how do they enhance reaction rates?
3. How does feedback inhibition regulate metabolic pathways and maintain homeostasis within cells?
4. In what ways does acetylation influence gene expression through epigenetic regulation?
5. What mechanisms can alter the activity of pre-existing enzymes within a cell?
6. Why is it crucial to regulate enzymes that catalyze committed steps in metabolic pathways?

REFERENCES

1. Berg, J. M., Tymoczko, J. L., & Gatto, G. J. (2018). *Stryer's Biochemistry*. W.H. Freeman and Company.
2. Cornish-Bowden, A. (2014). *Fundamentals of Enzyme Kinetics*. John Wiley & Sons.
3. Cornish-Bowden, A. (1997). *New Beer in an Old Bottle: Eduard Buchner and the Growth of Biochemical Knowledge*. Universitat de València. (ISBN: 84-370-3328-4).
4. Fenton, A. W. (2014). Allostery: An illustrated definition for the 'second secret of life.' *Trends in Biochemical Sciences, 39*(8), 556–569. https://doi.org/10.1016/j.tibs.2014.09.001.
5. Hammes, G. G., Benkovic, S. J., & Hammes-Schiffer, S. (2011). Flexibility, diversity, and cooperativity: Pillars of enzyme catalysis. *Biochemistry, 50*(46), 10422–10430. https://doi.org/10.1021/bi201312u.
6. Hedstrom, L. (2002). Serine protease mechanism and specificity. *Chemical Reviews, 102*(12), 4501–4524. https://doi.org/10.1021/cr0101777.
7. Nelson, D. L., & Cox, M. M. (2017). *Lehninger Principles of Biochemistry*. W. H. Freeman and Company.

CHAPTER

6 Vitamins

LEARNING OBJECTIVES

After studying this chapter, you will be able to:

- Discuss the importance of vitamins
- Explain the effects of processing and storage of food

KEY TERMS FROM THIS CHAPTER

Organic compounds
Coenzyme
Molecules
Metabolism
Metabolic energy
Pyridoxine
Tricarboxylic acid cycle
Fat-soluble vitamins
Citric acid

INTRODUCTION

Our bodies need vitamins, which are necessary organic compounds, in tiny amounts for a variety of physiological processes. They are essential for immunity, metabolism, and general health maintenance. There are 13 known vitamins that fall into two categories: water-soluble and fat-soluble. Each vitamin has a unique source and function. The B-complex vitamins (B1, B2, B3, B6, B12, folate, and biotin) and vitamin C are examples of water-soluble vitamins. With the exception of vitamin B12, these vitamins dissolve in water and are not significantly stored in the body. They are essential for the synthesis of collagen, red blood cell formation, energy production, and nerve function. Whole grains, meat, fish, dairy products, and leafy greens are good sources of B vitamins, and fruits and vegetables like citrus fruits, bell peppers, and broccoli are good sources of vitamin C.

The vitamins A, D, E, and K, on the other hand, dissolve in fat and are stored in the body's fatty tissues and liver. They are involved in blood clotting (vitamin K), antioxidant defense (vitamin E), bone health (vitamin D), and vision (vitamin A). Vitamins A and D from animal products, vitamin E from vegetable oils and nuts, and vitamin K from leafy greens and dairy are examples of fat-soluble vitamin sources.

Vitamin deficiencies can result in a number of health issues, which is why they are so important. For example, a vitamin D deficiency can result in weakened bones (osteomalacia) or softening of the bones in children (rickets), while a vitamin C deficiency can cause scurvy, which is characterized by bleeding gums and skin issues. Anemia, fatigue, and nerve damage can all be caused by B vitamin deficiencies. However, taking too much of some vitamins, particularly fat-soluble vitamins like vitamin A, can be toxic and cause health problems. A balanced diet full of fruits, vegetables, whole grains, lean meats, and dairy products is the best way to get vitamins. Nonetheless, dietary supplements might be required in some circumstances, particularly for people following restricted diets or dealing with particular medical conditions.

Comprehending the function of vitamins in preserving well-being emphasizes how crucial a balanced diet is. Supplements can be useful in bridging gaps, but a balanced diet should always come first. Because each vitamin has a distinct role in physiological processes, diet and general health are closely related. Therefore, maintaining optimal health and vitality requires making sure that one consumes a sufficient amount of vitamins through a varied and nutritious diet.

6.1. IMPORTANCE OF VITAMINS

Organic compounds known as vitamins, or a group of closely related molecules known as vitamers, are necessary for an organism's correct metabolism when present in small amounts. Essential nutrients must be obtained through diet because the body is unable to synthesize them in sufficient amounts for survival. For instance, some species can synthesize vitamin C, but others cannot; in the former case, vitamin C is not regarded as a vitamin, but in the latter, it is. Most vitamins are actually collections of related molecules called vitamers rather than single molecules. For instance, vitamin E is composed of eight different forms: four tocopherols and four tocotrienols.

6.1.1. Water-Soluble Vitamins

Water-soluble vitamins are easily absorbed by the body and are typically eliminated by the kidneys in excess. Water-soluble vitamins are required by the body often and in small amounts. The B vitamins and vitamin C are examples of water-soluble vitamins. Since they are easily eliminated in urine, with the exception of vitamins B6 and B12, and do not store well, frequent consumption is required. When present in excess of needs, they are usually not harmful, though megadoses of pyridoxine (vitamin B6), vitamin C, or niacin have been linked to symptoms. As coenzymes or cofactors, all B vitamins support the normal functioning of key enzymes and the proper progression of reactions that produce energy. Therefore, tissues that are growing or metabolizing quickly, such as skin, blood, the digestive system, and the nervous system, are primarily affected by any deficiency in water-soluble vitamins.

6.1.1.1. Functions Water-Soluble Vitamins

Certain proteins, fats, and carbohydrates are created as a result of the B-vitamin coenzymes' involvement in enzyme systems that transfer specific groups between molecules. These molecules can then be used to build bodily tissues or to store and release energy. The pantothenic acid coenzyme, also known as coenzyme A, is a key player in regulating the interconversion of fats, proteins, and carbohydrates into metabolic energy. It is involved in the tricarboxylic acid cycle, also known as the Krebs or citric acid cycle, which links the metabolism of carbohydrates, fats, and proteins. During the citric acid cycle, the conversion of proteins and carbohydrates into metabolic energy is regulated by the coenzymes thiamin and vitamin B6. Coenzymes niacin and riboflavin help transfer hydrogen ions or electrons (particles with a negative charge) during the tricarboxylic acid cycle reactions. Not related to the tricarboxylic acid cycle, all of these coenzymes are also active in transfer reactions that are necessary for the synthesis of structural compounds.

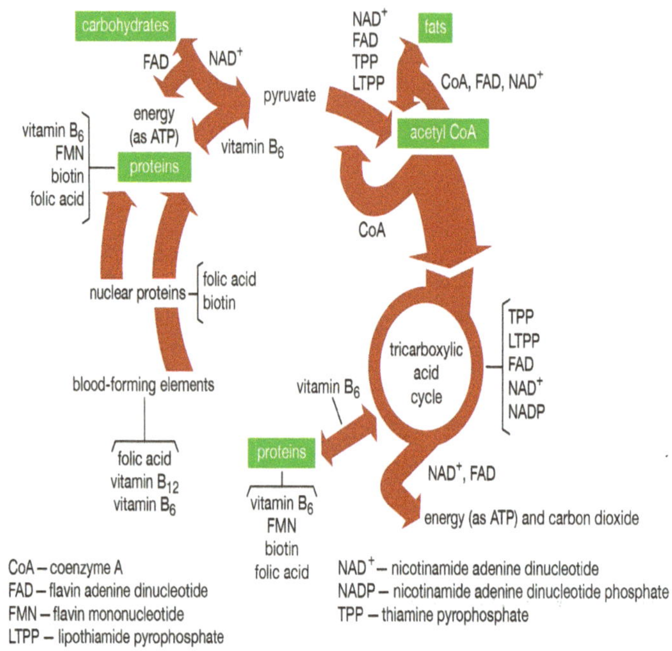

Figure 6.1. *Functions of B-vitamin coenzymes in metabolism.*

https://cdn.britannica.com/22/6522–050–729662DF/Functions-B-vitamin-coenzymes-metabolism.jpg.

While vitamin C has been shown to take part in certain enzyme-catalyzed reactions, its status as a coenzyme has not yet been determined. Its role most likely stems from its potent reducing agent qualities (i.e., it transfers electrons to other molecules with ease).

6.1.1.2. Metabolism

The animal intestine absorbs the water-soluble vitamins, which then travel straight to the blood and the tissues where they are needed. To be absorbed, vitamin B12 needs a material called intrinsic factor. There are some B vitamins that can exist in forms that animals cannot utilize. For example, the majority of the niacin in certain cereal grains (wheat, corn, rice, barley, and bran) is combined with another material to form a complex known as niacytin, which is indigestible by animals. The protein avidin, which is present in raw egg white, has the ability to bind biotin. Because digestive-tract enzymes are unable to absorb or degrade this complex, biotin cannot be used. In items made from animals (e.g., Biotin, vitamin B6, folic acid, and meat) are bound to other molecules to form conjugated molecules or complexes. While none of the three vitamins are active in their complex form, they are typically released from their bound forms by intestinal tract enzymes (for biotin and vitamin B6) or tissues (for folic acid) and can therefore be used. The majority of plant and animal tissues that are involved in metabolism contain B vitamins.

Human urine typically contains the excretion of water-soluble vitamins. Urine contains free vitamins such as thiamin, riboflavin, pantothenic acid, vitamin B6, vitamin C, and biotin instead of coenzymes; free niacin is not much excreted in the urine. Urine also contains byproducts (also known as metabolites) that are created during the metabolism of thiamin, niacin, and vitamin B6. Additionally, pantothenic acid, riboflavin, and biotin metabolites are produced in the urine. When intake of these vitamins is adequate for normal bodily function, excretion of these vitamins (or their metabolites) is low. Vitamin excess is stored in the tissues if intake starts to surpass minimum requirements. However, the tissue's capacity for storage is finite, and the rate of excretion increases dramatically as the tissues become saturated. In contrast to the other water-soluble vitamins, vitamin B12 is eliminated exclusively through the feces. This is also how some biotin and folic acid are typically eliminated. Water-soluble vitamins (apart from vitamin B12, folic acid, and biotin) are excreted in the feces, but the source of these vitamins is most likely the intestinal bacteria that produce them, not the vitamins the animal has eaten and used.

If taken in excess, the water-soluble vitamins are usually not thought to be harmful. There is one exception, though, and that is in humans. High doses of niacin (50–100 mg; 1 mg = 0.001 gram) cause blood vessels to dilate; the effects are more severe and can even impair liver function. Animals given thiamin in quantities 100 times higher than necessary (i.e., roughly 100 mg) can result in respiratory failure and death. Humans have not been shown to be toxically affected by therapeutic doses of thiamin (100–500 mg), with the possible exception of rare cases of anaphylactic shock in susceptible people. Other B vitamins are not known to be toxic.

6.1.2. Fat-Soluble Vitamins

Unlike water-soluble vitamins, fat-soluble vitamins are stored in the body's cells and are not as readily eliminated. They must be taken in sufficient quantities, but not as frequently as water-soluble vitamins. A fat-soluble vitamin may turn toxic if taken in excess. A balanced diet typically supplies adequate fat-soluble vitamins, but your body is particularly sensitive to excesses of vitamin D and vitamin A from animal sources (retinol).

Lipids (fats) are soluble in fat-soluble vitamins. Typically, these vitamins are absorbed as fat globules, or chylomicrons, which pass through the small intestine's lymphatic system and enter the body's general blood circulation. Because fat-soluble vitamins are stored by the body for later use, the body does not require them on a daily basis like it does water-soluble vitamins, which must be replenished regularly. They are primarily kept in fat tissues and the liver. Although it is beneficial for us to accumulate these vitamins in our bodies so that they are available when we need them, overconsumption can result in toxicity.

The body may be lacking in fat-soluble vitamins if fat intake is excessively low or if fat absorption is impaired, as in the case of some medications that prevent fat from being absorbed from the intestines or certain illnesses like cystic fibrosis, which causes a pancreatic enzyme deficiency that also prevents fat from being absorbed from the intestines. The majority of fat-soluble vitamins can be found in fatty foods like dairy products, oily fish, liver, vegetable oils, and animal fats like butter and lard. It's also worth noting that cooking does not destroy fat-soluble vitamins, in contrast to water-soluble vitamins.

6.1.2.1. Functions of Fat-Soluble Vitamins

For the linings of the body's surfaces (e.g., the skin) to remain intact, the vitamin A group is crucial. g. skin, respiratory system, and cornea), for the development of sperm, and for the healthy operation of the immune system. Retinal and opsin, a protein found in the retina of the eye, combine to form complex molecules known as rhodopsin, or visual purple, which are involved in dark vision. The vitamin D group is necessary for growth, particularly the calcification or growth of bones. In addition, the vitamin E group is required for healthy animal growth; in the absence of vitamin E, animals become infertile and exhibit abnormalities in their muscles, organs, central nervous system, and liver, among other areas. Vitamin K is essential for healthy metabolism, which includes converting food into cellular energy in some biological membranes. Vitamin K is also needed for healthy blood clotting.

6.1.2.2. Metabolism

From the intestines to the circulating blood, lymph is the main means of transporting the fat-soluble vitamins. Because bile salts are necessary for the intestinal system to efficiently absorb fat-soluble metabolites, anything that hinders fat absorption also prevents the body from absorbing fat-soluble vitamins. Retinol (vitamin A alcohol) is primarily found in the bloodstream in its ester form during digestion because, prior to being transported by the lymph, a fatty acid, most commonly palmitic acid, is added to the molecule. Fatty acid molecules are not necessary for the absorption of vitamins D, E, or K. While vitamins A and D are both bound to proteins during bloodstream transit, small amounts of vitamin A (and possibly vitamin K) may be absorbed directly into the bloodstream.

The body has the capacity to store more fat-soluble vitamins than water-soluble ones. The majority of stored vitamin E is found in body fat, though significant amounts are also found in the uterus of females and the testis of males.

Vitamins A, D, and K are primarily stored in the liver, with smaller amounts also being stored in other soft body tissues. Alpha-tocopherol is stored in tissues at higher concentrations than the other forms of vitamin E. The different forms of vitamin E are stored in tissues in different amounts. Vitamin A is the fat-soluble vitamin that is stored the most.

Overconsumption of either vitamin A or D can result in toxicity, also known as hypervitaminosis A or D. However, excessive use of pharmaceutical vitamin preparations can easily result in vitamin D and A toxicity. Vitamin A at toxic levels is 100 times higher than what is normally needed. e. over the course of several months, around 150,000 micrograms (μg; 1 μg = 0.000001 gram) every day.

Infants may experience toxicity from much lower dosages. Large doses of vitamin K3 are toxic to newborn infants, but excessive doses of the natural vitamins K1 and K2 have no apparent effects other than the development of resistance to anticoagulant drug therapy. Even when taken in significant excess of the recommended amount, vitamin E does not appear to have any obvious negative effects.

Quinones are a class of organic compounds that includes vitamin groups E and K. These chemicals are converted into alpha-lactones, which are sugar-like compounds eliminated through the urine. Additionally, some vitamin K1 is eliminated in the bile and so shows up in the feces. Urine and bile, which are excreted along with feces, are the breakdown products of vitamin A. Only in the feces are vitamin D and the products of its breakdown excreted.

6.1.3. Chemical Structure of Vitamins

Although vitamins play a significant role in our diets, you have probably not given much thought to how they are structured chemically. All 13 vitamins have chemical structures, though there may be some variation depending on the vitamin's source.

- **Calcium:** The most prevalent mineral in the human body is calcium (Ca). The bones and teeth contain over 99% of the calcium in the body. Moreover, calcium is present in bodily fluids, where it controls how tightly muscles and blood vessels contract. Because the body grows so quickly during puberty and maturity, this is the time when calcium is most needed. Dairy products and milk are excellent sources of calcium.

Age	Calcium DRI (mg/day)
0–6 months	210
7–12 months	270
1–3 years	500
4–8 years	800
9–18 years	1300
19–50 years	1000
51+ years	1200

- **Fluorine:** Teeth and bones hold the majority of the fluorine (F) in the body. Fluoride is mostly found in drinking water. Fluorine effectively prevents dental caries and hardens tooth enamel. Fluorosis can result from too much fluorine in drinking water building up in teeth and bones. Permanent teeth that form during periods of high fluorine consumption have chalky patches on the enamel's surface that are unevenly distributed and eventually turn brown or yellow, giving the teeth their distinctive mottled look.
- **Iodine:** iodine (I) is primarily involved in the synthesis of two thyroid hormones, thyroxine and triiodothyronine. In adults, about 80% of the iodide absorbed is trapped by the thyroid gland.

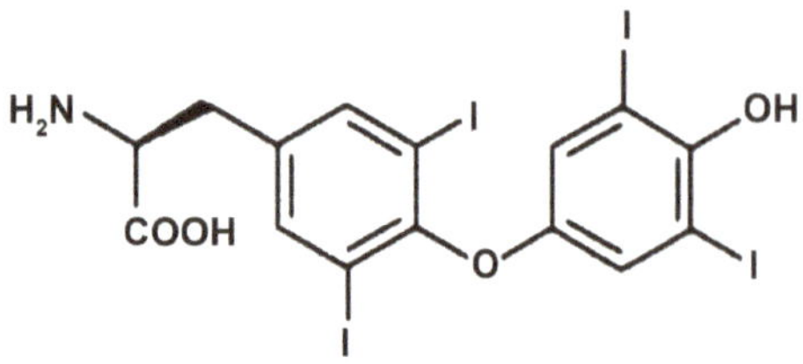

Thyroxine

Seawater is where most environmental iodine is found. Those who live far from the sea are especially vulnerable to deficiencies. Iodide fortified salt (usually 70 µg/g) helps guarantee sufficient intake (150 µg/day). Deficiency is common worldwide but rare in places where iodized salt is used. When daily iodide intake is less than 20 µg, iodine deficiency occurs. A goiter, or enlargement of the thyroid gland that manifests as swelling in the front of the neck, is the result of the thyroid gland hypertrophying in mild to moderate deficiency in order to concentrate iodide within itself. Eating foods high in iodine, like seaweed or kombu-type kelp, can cause excessive iodine consumption, which can cause thyrotoxicosis, a condition caused by high concentrations of thyroid hormones in the body.

- **Iron:** Hemoglobin, myoglobin, and numerous enzymes in the body are all made of iron (Fe). More than 85% of iron in the average diet comes from nonheme iron, which is absorbed considerably less efficiently than heme iron, which is mostly found in animal products. On the other hand, nonheme iron absorbs more readily when combined with vitamin C and animal protein. For men and postmenopausal women, the Recommended Daily Allowance (RDA) for iron is 8 mg; for premenopausal women, it is 18 mg. Globally, iron deficiency is the most prevalent nutritional deficiency, leading to anemia. It could be brought on by malabsorption or insufficient consumption of iron. The best food sources of dietary iron are meat products. Diets that are too vegan or ovo-lacto vegetarian can lead to iron deficiency. Iron deficiency can also result from chronic bleeding. When a person receives repeated blood transfusions or consumes excessive amounts of iron supplements, iron may build up in their body. Being toxic, too much iron can harm the intestines and other organs and induce diarrhea and vomiting.

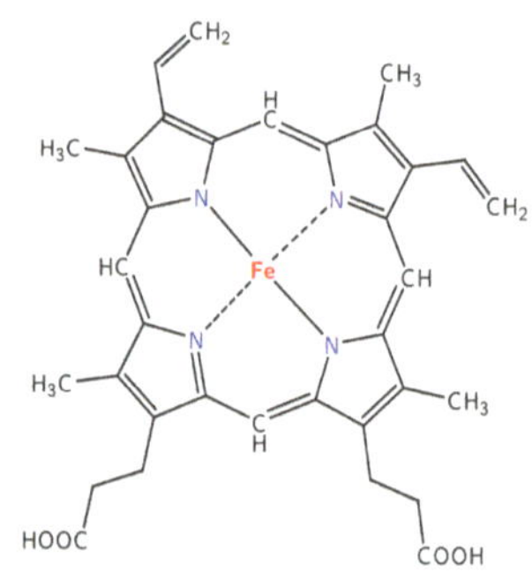

- **Magnesium:** Magnesium (Mg) plays several crucial metabolic roles in the creation and movement of energy. It also has an impact on how muscles contract and relax. Magnesium helps certain enzymes function and is involved in the synthesis of proteins. Chlorophyll-rich dark green leafy vegetables, cereals, and nuts provide the majority of the magnesium in the diet.

Age	Magnesium DRI (mg/day)
0–6 months	30
7–12 months	75

1–3 years	80
4–8 years	130
9–13 years	240
14–18 years	boys: 410, girls: 360
19–30 years	men: 400, women: 310
31+ years	men: 420, women: 320

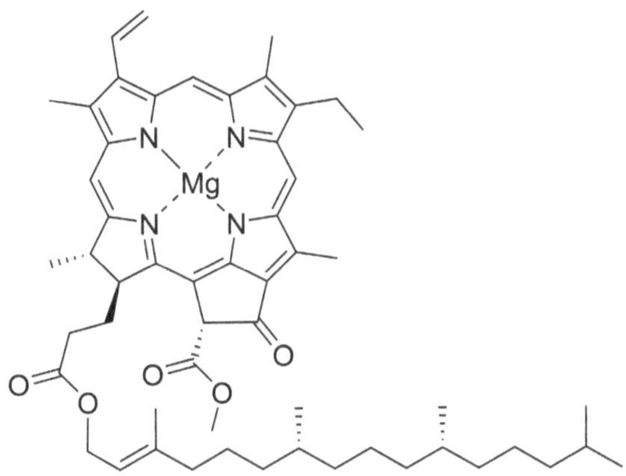

Chlorophyll A

- **Manganese:** In addition to being an essential component of several enzyme systems, such as phosphoenolpyruvatecarboxykinase and manganese-specific glycosyltransferases, manganese (Mn) is also required for a healthy bone structure. Nuts and cereals contain manganese. Manganese needs to be consumed in amounts of 2 to 5 mg daily.
- **Molybdenum:** Coenzymes that are required for the activity of xanthine oxidase, sulfite oxidase, and aldehyde oxidase contain molybdenum (Mo). The conversion of sulfite to sulfate, which is required for the metabolism of sulfur-containing amino acids like cysteine, is catalyzed by sulfite oxidase. Peas, beans, and lentils are examples of legumes that are high in molybdenum. 45 micrograms per day is the Recommended Dietary Allowance (RDA) for adult men and women. It is acceptable to take up to 2 milligrams per day.
- **Potassium:** Potassium (K) helps to prevent blood pressure from rising too high when too much sodium is consumed, and it also maintains fluid volume both inside and outside of cells. For children aged 9 to 13, 4 points 5 grams of potassium per day is the recommended intake, while for older adults 4 points 7 grams per day is the recommended intake. Generally speaking, foods high in potassium include fruits, vegetables, dairy, meats, nuts, and dried peas. People with diabetes or chronic renal insufficiency may experience hyperkalemia and possibly even die suddenly if they consume too much potassium from supplements or salt substitutes.
- **Selenium:** The enzyme glutathione peroxidase, which breaks down hydroperoxides made from polyunsaturated fatty acids, contains selenium (Se). The enzymes that deiodinate thyroid hormones also contain selenium. Selenium typically functions as an antioxidant in conjunction with vitamin E. Keshan disease is a kind of congestive cardiomyopathy that is brought on by a selenium deficiency. Selenium has an RDA of 70 micrograms (mcg). Based on the avoidance of brittle nails and hair as well as the early indicators of chronic selenium toxicity, an adult's

tolerable upper level of selenium is 400 mcg/day. Blood selenium concentrations that match an 850 mcg/day intake have been shown to have toxic effects. Nuts, meats, fish, and poultry are good sources of selenium.

- **Sodium:** Table salt (Sodium Chloride, NaCl) is the most common form of sodium (Na) consumption. The recommended daily intake of 1.5 grams, with a maximum of 2.3 grams, is calculated to fulfill the requirements for sweat losses for individuals 8 years of age or older who participate in recommended levels of physical activity. People who are physically active and sweat a lot in humid climates might require more than the Adequate Intake. The upper limit may be excessive for people with hypertension, but it applies to healthy people without the condition.
- **Zinc:** The primary tissues that contain zinc (Zn) are the testes, liver, muscle, teeth, hair, and skin. Numerous nicotinamide adenine dinucleotide (NADH) dehydrogenases, RNA and DNA polymerases, DNA transcription factors, alkaline phosphatase, superoxide dismutase, and carbonic anhydrase are among the several 100 enzymes that contain zinc. For men and women, respectively, the Recommended Daily Allowance (RDA) for zinc is 11 milligrams and 8 milligrams. Cereals and mollusks like oysters are good dietary sources of zinc.

6.1.4. Types of Vitamins

Vitamin deficiencies in developed nations are primarily caused by food trends, drug abuse, alcoholism, and poverty. Taking excessive amounts of vitamins A, D, B6, or niacin can lead to vitamin toxicity, also known as hypervitaminosis.

Vitamins can be water soluble (vitamin C) or fat soluble (vitamins A, D, E, and K). Biotin, folate, niacin, pantothenic acid, riboflavin, thiamin, pyridoxine, and B12 are among the B vitamins. Vegans who don't eat fermented foods or yeast extracts may become vitamin B12 deficient. Strictly vegetarian diets also frequently lack zinc, iron, and calcium.

- **Biotin (Vitamin B7):** Biotin (Vitamin B7) acts as a coenzyme for carboxylation reactions essential to fat and carbohydrate metabolism. Adequate intake for adults is 30 µg/day. Liver, egg yolks, green vegetables, and whole grains are rich sources of biotin.
- **Folic Acid (Vitamin B9):** Folate, also known as vitamin B9, is necessary for the development of the fetal nervous system and plays a role in the maturation of red blood cells as well as the synthesis of purines and pyrimidines. A healthy diet rich in folic acid both preconception and during the first trimester of pregnancy helps prevent some brain and spinal cord abnormalities, including spina bifida. The upper jejunum and duodenum absorb folate. In the United States, 400 µg of folate is advised daily, with an upper limit of 1000 µg. Folate is basically safe. Megaloblastic anemia caused by deficiency is identical to that caused by vitamin B12 deficiency. Dementia risk is greatly increased in the elderly when there is a folate deficiency. Dried peas, dried beans, yeast, and

leafy greens like mustard greens, spinach, endive, and lettuce are good sources of folic acid.

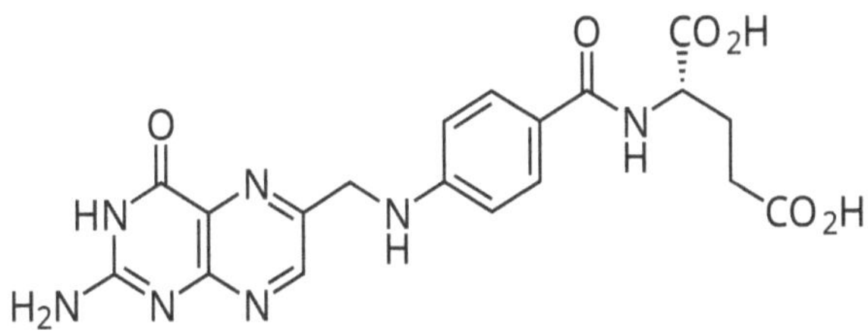

- **Niacin (Vitamin B3):** Nicotinamide adenine dinucleotide (NAD) and nicotinamide adenine dinucleotide phosphate (NADP) are derivatives of niacin (also known as vitamin B3 or nicotinic acid), and they are coenzymes in oxidation-reduction reactions that are essential to cell metabolism. Pellagra is a condition marked by dermatitis, gastrointestinal issues, and mental disturbances that is brought on by dietary niacin deficiency. Primary deficiency, which typically arises in regions where maize (Indian corn) makes up a significant portion of the diet, is caused by an extremely low intake of both the amino acid tryptophan and niacin. Fish and mushrooms are excellent providers of niacin.

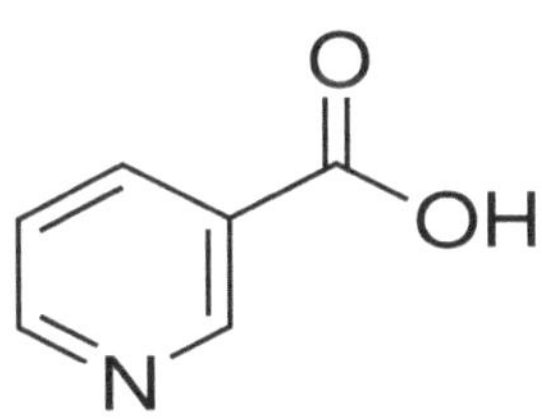

Niacin

- **Pantothenic Acid (Vitamin B5):** Foods are a great source of pantothenic acid (Vitamin B5), with whole grain cereals, legumes, eggs, and meat having the highest concentrations. Pantothenic acid plays a crucial role in the metabolism and synthesis of proteins, fats, and carbohydrates. It is also required to form coenzyme-A (CoA). Adults need roughly 5 mg daily. You can find pantothenic acid in yeast, liver, and mushrooms.

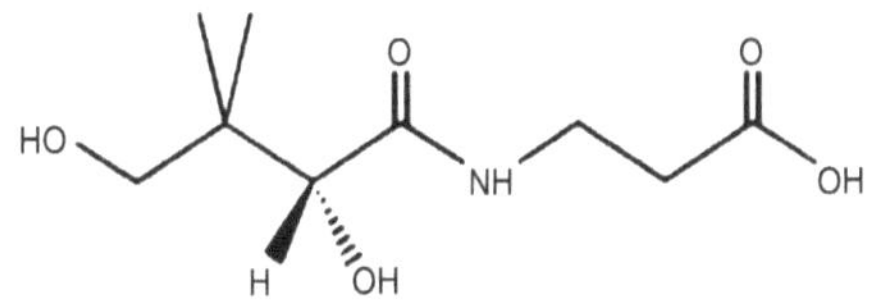

Pantothenic acid (vitamin B5)

- **Riboflavin (Vitamin B2):** In numerous oxidation-reduction reactions, riboflavin, or vitamin B2, is a necessary coenzyme that plays a role in the metabolism of carbohydrates. In essence, riboflavin is nontoxic. Deficiency in riboflavin typically coexists with deficiencies in other B-vitamins. A sore throat, lesions of the lips and oral mucosa, glossitis, conjunctivitis, seborrheic dermatitis, and normochromic-normocytic anemia are some of the symptoms and indicators. In addition to meats like beef, pork, and lamb, riboflavin can be found in mushrooms and yeast.

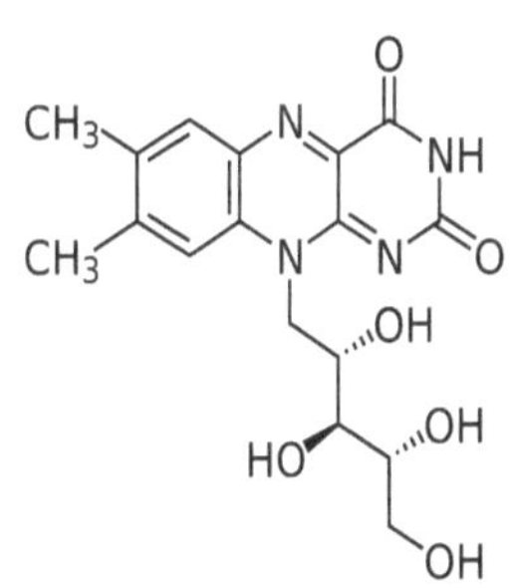

- **Thiamin (Vitamin B1):** Vitamin B1, also known as thiamin or thiamine, is commonly found in food. Thiamin plays a role in the metabolism of fat, glucose, amino acids, carbohydrates, and alcohol. In general, thiamin is nontoxic. People in developing nations who live primarily on highly refined rice or other carbohydrates are most likely to suffer from thiamin deficiency, which can lead to beriberi. Brewer's yeast, fortified cereals, and bean sprouts are excellent sources of thiamin.

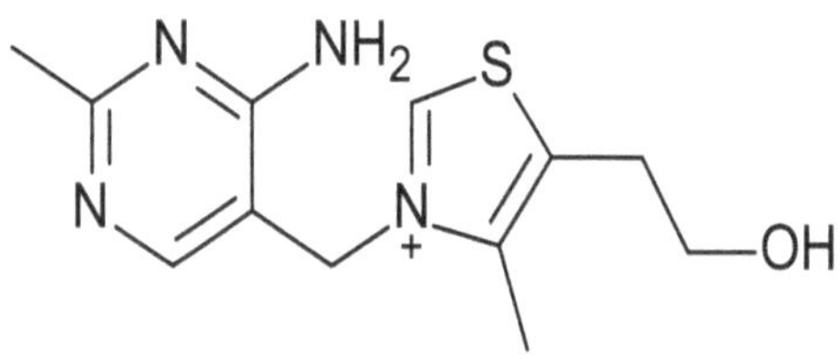

Thiamin (Vitamin B1)

- **Vitamin A:** The synthesis of rhodopsin, a pigment that is found in photoreceptors in the retina, depends on vitamin A (retinol). Vitamin A supports the upkeep of epithelial tissues. Normally, the body stores 90% of its Vitamin A in the liver. The body releases vitamin A into the bloodstream bound to retinol-binding protein and prealbumin (transthyretin) in order to use it. Converted to Vitamin A are β-carotene and other provitamin carotenoids found in brightly colored fruits and leafy green vegetables. Vegetables that have been cooked or homogenized and served with some fats or oils absorb carotenoids more readily. For men and women, respectively, the Recommended Dietary Allowance (RDA) for vitamin A is 900 micrograms and 700 micrograms. Deficiency leads to rashes on the skin, immune system impairment, and common ocular symptoms like dry eyes and night blindness.

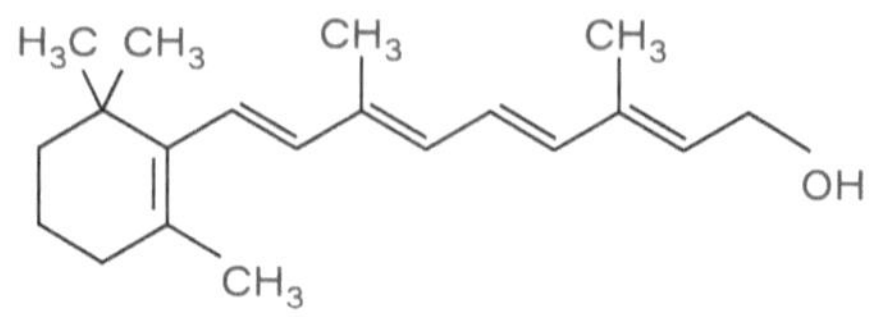

Retinol (Vitamin A)

- **Vitamin B12:** The term "cobalamin" refers to any substance having biologic vitamin B12 activity. These substances play a role in the synthesis and repair of myelin, methyl transfer, and nucleic acid metabolism. They are required for healthy red blood cell formation. In the acidic environment of the stomach, vitamin B12 is released and bound to R protein. The B12-R protein complex in the small intestine is broken down by pancreatic enzymes. Intrinsic factor, which is released by the parietal cells in the stomach mucosa, binds to vitamin B12 following cleavage. Vitamin B12 is absorbed in the terminal ileum, where intrinsic factor is necessary. Three ounces (85 grams) of meat contains 2.4 micrograms, which is the Recommended Dietary Allowance (RDA). Clams, oysters, turkey, chicken, beef, and pork are foods high in vitamin B12. Although insufficient absorption is typically the cause of dietary vitamin B12 deficiency, vegans who do not take vitamin supplements may also become deficient. Deficiency results in peripheral neuropathy, which is typified by tingling or numbness in the hands and feet, megaloblastic anemia, and damage to the white matter of the brain and spinal cord.

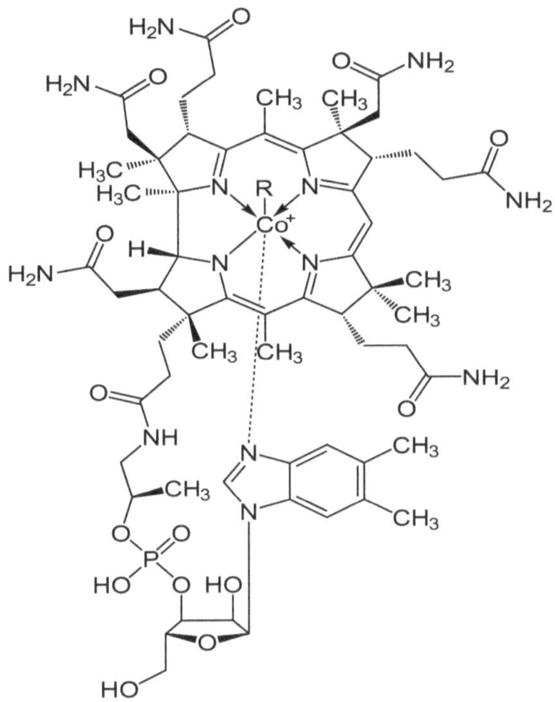

Cyanocobalamin (Vitamin B12)

- **Vitamin B6:** Pyridoxine, pyridoxal, and pyridoxamine are a trio of chemical compounds that are closely related and comprise vitamin B6. The body breaks them down into pyridoxal phosphate, which is a coenzyme in numerous vital processes in the blood, brain, and skin metabolism. The metabolism of lipids, carbohydrates, and amino acids, as well as the biosynthesis of heme and nucleic acid, depend on vitamin B6. Many different types of vegetables and meats contain vitamin B6. A lot of cereals for breakfast are enhanced with B6. Red and green peppers, brewer's yeast, and Chinese cabbage (pak-choi) are a few foods that naturally contain vitamin B6.

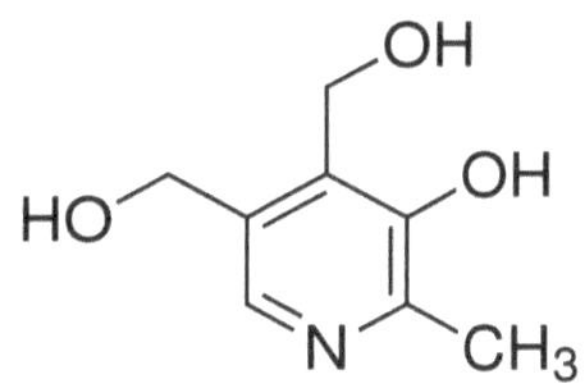

Pyridoxine (Vitamin B6)

- **Vitamin C:** The synthesis of collagen, carnitine, hormones, and amino acids is influenced by vitamin C, also known as ascorbic acid. It is necessary for the healing of wounds and speeds up burn recovery. In addition to being an antioxidant, vitamin C boosts immunity and makes iron absorption easier. While general undernutrition can result in deficiency in developed nations, severe deficiency leading to scurvy is rare. Fatigue, depression, and connective tissue abnormalities like gingivitis, rash, internal bleeding, or poor wound healing are signs of deficiency. For women and men, respectively, the Recommended Dietary Allowance (RDA) is 75 milligrams and 90 milligrams. About two grams (2,000 mg) of vitamin C per day is the maximum amount that can be consumed safely. Increased dosages may result in diarrhea and upset stomach. Vegetables and fruits that are fresh contain vitamin C. Vitamin C is found in abundance in citrus fruits, such as lemons and oranges.

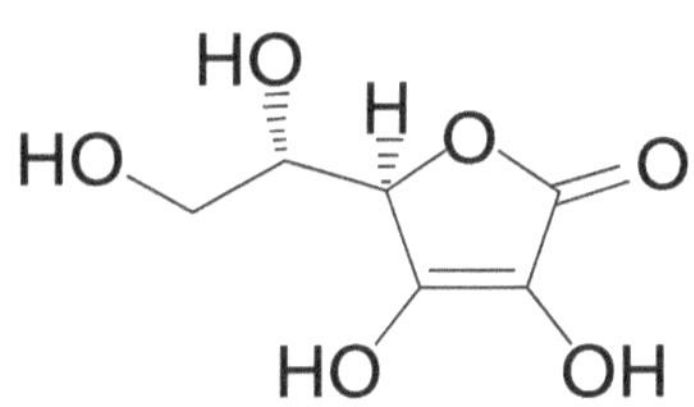

Ascorbic Acid (Vitamin C)

- **Vitamin D:** The two primary forms of vitamin D are D2 (ergocalciferol) and D3 (cholecalciferol). Ultraviolet radiation from the sun causes the skin to produce vitamin D3, which is mostly found in fish liver oils and egg yolks. Human breast milk has an average of only 10% of the vitamin D found in fortified cow's milk, whereas some developed countries fortify milk and other foods with the vitamin. As we age, our need for vitamin D rise. With multiple active metabolites that function as hormones, vitamin D is a prohormone. The kidneys transform vitamin D3 into 1,25(OH)2D, also known as 1,25-dihydroxycholecalciferol, calcitriol, or active vitamin D hormone, after the liver metabolizes vitamin D3 into 25(OH)D. The most metabolically active form is 1,25(OH)2D, although the main circulating form, 25(OH)D, also has some metabolic activity. Vitamin D deficiency may result from insufficient sun exposure. Deficiency hinders the process of bone mineralization, which can lead to osteomalacia in adults and rickets in children. It can also exacerbate osteoporosis.

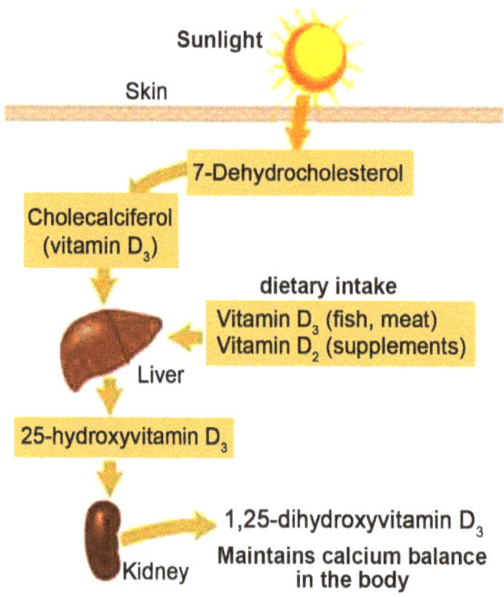

Vitamin D metabolism

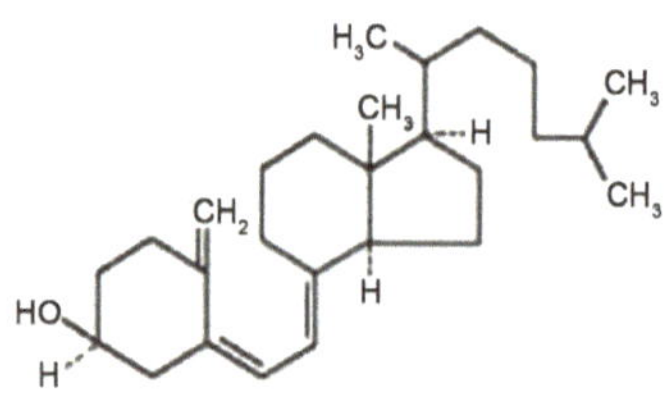

Cholecalciferol (Vitamin D_3)

As of right now, the Institute of Medicine recommends 200 IU per day for people under 50, 400 IU for people between 51 and 70, and 600 IU for people over 70. These guidelines were developed by figuring out how much vitamin D was necessary to stop rickets or bone demineralization. The maximum amount of vitamin D that can be safely consumed each day is 10,000 IU. There has been no discernible decrease in the risk of fractures in randomized trials that used the 400 IU of vitamin D daily that is currently advised. Trials using 700–800 IU of vitamin D per day, however, revealed a lower incidence of fractures. For adults, maintaining blood serum levels of Vitamin D that effectively strengthen bones requires consuming at least 1000 IU per day.

- **Vitamin E:** Tocopherols and tocotrienols are among the compounds that make up the vitamin E group and share similar biologic activities. While α-tocopherol is the most biologically active, β-, γ-, and δ-tocopherols are also highly biologically active.

By acting as antioxidants, these substances stop polyunsaturated fatty acids in cellular membranes from being peroxidized. The levels of total plasma lipids influence the levels of plasma tocopherol. The normal range for plasma α-tocopherol is 5–20 μg/mL. In developing nations, dietary vitamin E deficiency is prevalent. A lack of vitamin E results in the axons of neurons, or nerve cells, degenerating, leading to neurologic deficiencies and red blood cell fragility, which is commonly diagnosed as hemolytic anemia. It is not advised to take vitamin E supplements due to research showing a higher risk of heart failure and overall death. Spinach, watercress, mustard greens, and other green leafy vegetables are good sources of vitamin E. Nuts and sunflower kernels, as well as other oily plant seeds, are good providers of Vitamin E.

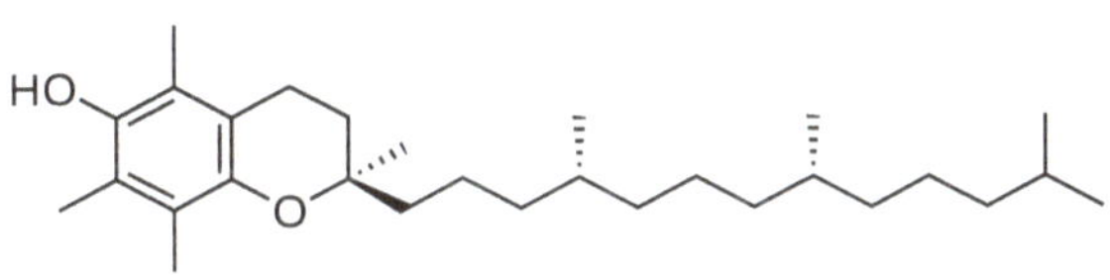

Alpha-tocopherol (Vitamin E)

- **Vitamin K:** Dietary vitamin K is called phytoquinone, or vitamin K1. Dietary fat improves the absorption of it. Infant formulas have extra vitamin K in them. Menaquinones, a class of compounds produced by intestinal tract bacteria, are referred to as vitamin K2. However, the quantity of these compounds synthesized is insufficient to meet the body's need for vitamin K. For men and women, respectively, the Recommended Dietary Allowance (RDA) is 120 micrograms and 90 micrograms. The hepatic production of coagulation factors VII, IX, X, and II (prothrombin) is regulated by vitamin K. Because vitamin K is abundant in green vegetables like kale, spinach, and mustard greens, dietary vitamin K deficiency is rare in healthy adults. Menaquinones are also produced by the bacteria found in the normal gut.

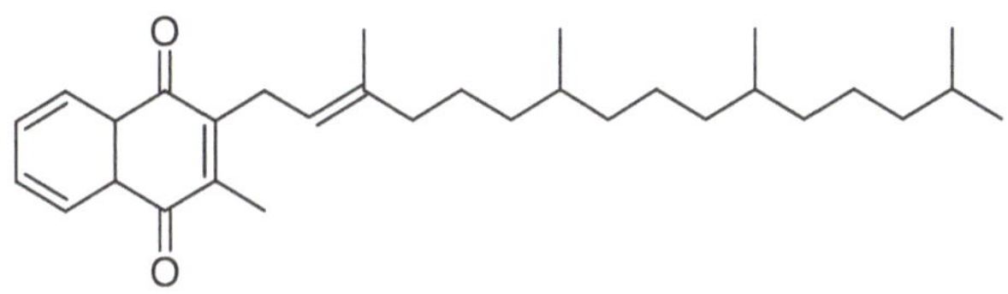

Vitamin K_1 (phylloquinone)

6.1.5. Pro-Vitamin A

A potent antioxidant, vitamin A is a fat-soluble vitamin. In addition to its many other benefits, vitamin A is essential for maintaining neurological function, healthy skin, and good vision. Like all antioxidants, vitamin A helps to fight the damage caused by free radicals, which in turn helps to reduce inflammation. One approach to halt aging naturally is to eat a diet rich in antioxidants. Vitamin A and other antioxidants are also important for healthy, clear skin, strong bones, immune system support, and gene regulation. They also help with cell differentiation and differentiation. Eggs, milk, liver, carrots, and yellow or orange vegetables like spinach, squash, and other leafy greens are some of the best foods to eat when looking for vitamin A.

6.1.5.1. Vitamin A Sources

There are two main forms of vitamin A in the body: beta carotene and active vitamin

A. Retinol, or active vitamin A, is derived from foods that contain animal products. Without requiring conversion, the body can utilize this "pre-formed" form of Vitamin A. The body transforms "pro Vitamin A" carotenoids, which are found in vibrant fruits and vegetables, into retinol once the food is consumed. This is the other form of Vitamin A. To be used by the body, beta carotene, a kind of carotenoid that is mostly found in plants, must first be transformed into active Vitamin A. Numerous studies have demonstrated the importance of antioxidants, such as vitamin A, for long-term health and well-being. They support cell growth, strengthen immunity, and improve eye health. When it comes to getting antioxidants like Vitamin A, doctors and nutrition experts advise consuming a well-balanced diet rich in fruits, vegetables, and whole foods instead of taking supplements.

6.1.5.2. Symptoms of Vitamin A Deficiency

Normal vision, healthy skin, normal bone growth, and the defense of infection-resistant mucous membranes in the respiratory, digestive, and urinary tracts are all dependent on vitamin A. Individuals who have chronically impaired fat metabolism are particularly vulnerable to becoming vitamin A deficient.

The most frequent medical conditions that result in malabsorption of vitamin A are pancreatic disorders, inflammatory bowel disease, leaky gut syndrome, auto immune responses, and gluten sensitivity. Alcoholics are also far more likely to experience a vitamin A deficiency due to their excess toxicity, which lowers vitamin A levels.

In more than half of all countries, particularly in Africa and South-East Asia, vitamin A deficiency has emerged as a public health concern. In low-income countries, it particularly affects pregnant women and young children. This can be a major issue for kids because deficiency in vitamin A raises the risk of serious illness, including death, as well as severe visual impairment and blindness. Common childhood infections like diarrheal disease and measles can pose a threat to children.

6.1.5.3. Poor Eye Health

A lack of vitamin A can cause the cornea to thicken and eventually cause blindness. Keratomalacia is a bilateral condition that typically affects both eyes and is caused by a severe deficiency of vitamin A. This kind of deficiency could be metabolic, referring to your body's capacity to absorb the vitamin, or dietary, referring to your regular consumption of the vitamin. Extreme dry eyes and night blindness are two possible early signs of Keratomalacia. Wrinkles, cloudiness, and a softening of the corneas may occur after your vision. Without proper care and attention, the corneas may continue to soften, which could result in infections, tears, or degenerative tissue changes that could all result in blindness.

6.1.5.4. Premature Skin Damage

Vitamin A deficiency will lead to the drying, scaling, and follicular thickening of the skin. Keratinization of the skin, when the epithelial cells lose their moisture and become hard and dry, can occur in the mucous membranes of the respiratory, gastrointestinal tract, and urinary tract.

6.1.5.5. Respiratory Infections

Respiratory infections can occur because the body's immunity is impaired by the lack of vitamin A. The younger the patient, the more severe the effects can be. Growth retardation and infections are common among children, and the mortality rate can reportedly exceed 50% in children with severe vitamin A deficiency.

6.1.5.6. At Risk Pregnancy

For pregnant women, the vitamin A demand is the highest during the last trimester; most often, women suffer from vitamin A deficiencies during this time. A pregnant woman can suffer from night blindness if her vitamin A intake is not sufficient.

6.1.6. Pro-Vitamin D

Fat-soluble vitamin D can be found as a dietary supplement and is only found naturally in a small number of foods. It can also be added to other foods. Additionally, it is produced internally when sunlight's UV rays strike the skin and start the synthesis of vitamin D. Sunlight, food, and supplement-derived vitamin D is physiologically inert and needs to go through two hydroxylation processes in the body to become active. The first one, called calcidiol, is the result of the liver converting vitamin D to 25-hydroxyvitamin D [25(OH)D]. The second, which mainly takes place in the kidney, produces calcitriol, or 1,25(OH)2D, the physiologically active form of vitamin D.

In order to allow for normal bone mineralization and prevent hypocalcemic tetany, vitamin D maintains sufficient serum calcium and phosphate concentrations and encourages calcium absorption in the gut. Additionally, it is necessary for osteoblasts and osteoclasts to grow and remodel bone. Inadequate levels of vitamin D can lead to thinning, fragile, or malformed bones. Inadequate levels of vitamin D shield adults from osteomalacia and prevent rickets in children. In addition to calcium, vitamin D also aids in the prevention of osteoporosis in older adults. In addition, vitamin D regulates immunological, neuromuscular, and cell growth in addition to lowering inflammation. Vitamin D modulates the expression of many genes that code for proteins that control cell division, proliferation, and apoptosis. Many cells contain vitamin D receptors, some of which convert 25(OH)D to 1,25(OH)2D.

The most accurate measure of vitamin D status is the serum concentration of 25(OH) D. With a fairly long circulating half-life of 15 days, it reflects both vitamin D produced cutaneously and that obtained from food and supplements. While 25(OH)D functions as a biomarker of exposure, it is unclear to what extent levels of 25(OH)D also function as a biomarker of effect (i.e., concerning health status or results). Vitamin D stores in bodily tissues are not indicated by serum 25(OH) D levels. Unlike 25(OH)D, which has a short half-life of 15 hours, circulating 1,25(OH)2D is generally not a good indicator of vitamin D status because serum concentrations are tightly regulated by calcium, phosphate, and parathyroid hormone. Usually, 1,25(OH)2D levels do not drop until a severe case of vitamin D deficiency.

The levels of 25(OH)D in the serum linked to deficiencies (e.g., rickets), sufficiency for bone health, ideal general health, and cut points have not been established through a process of scientific consensus. A committee of the Institute of Medicine concluded that people are at risk of vitamin D deficiency at serum 25(OH)D concentrations based on its review of data regarding vitamin D needs. <30 nmol/L (<12 ng/mL). Some are potentially at risk for inadequacy at levels ranging from 30–50 nmol/L (12–20 ng/mL). Practically all people are sufficient at levels ≥ 50 nmol/L (≥ 20 ng/mL); the committee stated that 50 nmol/L is the serum 25(OH) D level that covers the needs of 97.5% of the population. Serum concentrations > 125 nmol/L (>50 ng/mL) are linked to possible negative consequences.

6.2. EFFECTS OF PROCESSING AND STORAGE OF FOOD

The amount of vitamins in food and its overall nutritional value can be greatly impacted by how it is processed and stored. To ensure optimal intake of essential vitamins and to make informed dietary choices, it is imperative to comprehend these effects. Certain vitamins are less impacted by processing and therefore more stable than others. When food is processed and stored, water-soluble vitamins (B-group and C) are more unstable than fat-soluble vitamins (K, A, D, and E).

The most unstable vitamins include:

- folate
- thiamine
- vitamin C.

More stable vitamins include:

- niacin (vitamin B3)
- vitamin K
- vitamin D
- biotin (vitamin B7)
- pantothenic acid (vitamin B5).

Processing Effects

1. **Heat Exposure:** Many vitamins, particularly water-soluble ones like vitamin C and B vitamins, are sensitive to heat. Cooking, boiling, baking, and other forms of heat processing can lead to significant losses of these vitamins. For example, boiling vegetables can cause leaching of water-soluble vitamins into the cooking water, reducing their content in the food itself.
2. **Freezing and Canning:** Freezing can preserve some vitamins better than other methods, but it still leads to some nutrient loss, especially if foods are stored for extended periods. Canning involves heat treatment, which can cause varying degrees of vitamin degradation depending on the type of vitamin and the intensity of the heat applied during processing.

3. **Milling and Refining:** Processing grains to make refined products like white flour removes the outer layers where many vitamins, such as B vitamins and vitamin E, are concentrated. As a result, refined grains are often lower in these essential nutrients compared to whole grains.
4. **Exposure to Light and Air:** Vitamins such as vitamin A and vitamin C are sensitive to exposure to light and air. Packaging foods in opaque containers and minimizing exposure to light during storage can help preserve these vitamins.

Storage Effects

1. **Temperature:** Proper storage temperature is crucial for maintaining vitamin content. Refrigeration slows down the degradation of vitamins, whereas higher temperatures can accelerate it. For example, storing fruits and vegetables at room temperature can lead to vitamin C losses over time.
2. **Duration:** The longer food is stored, the more likely it is to experience nutrient losses. This applies particularly to water-soluble vitamins like vitamin C and B vitamins. Fresh produce loses nutrients gradually from the moment of harvest, so consuming it soon after purchase can help maximize nutrient intake.
3. **Packaging:** Packaging can affect vitamin stability. Oxygen and moisture can promote vitamin degradation, so vacuum-sealed or tightly sealed packaging can help preserve nutrient content better than loosely packaged foods.

Implications for Nutrition

The impact of storage and processing on vitamin content emphasizes the need of eating a varied diet rich in fresh, minimally processed foods. Food processing frequently results in the loss of nutrients even though it can make foods safer and more convenient to eat.

To mitigate these losses:

- **Choose Fresh Foods:** Opt for fresh fruits, vegetables, and whole grains whenever possible to maximize vitamin intake.
- **Minimize Heat Exposure:** Use gentle cooking methods such as steaming or stir-frying to preserve water-soluble vitamins.
- **Store Properly:** Store foods in appropriate conditions (e.g., cool and dark for certain produce, refrigerated for perishables) to minimize nutrient degradation.
- **Consider Supplementation:** In cases where dietary intake may be insufficient, such as for individuals with limited access to fresh foods or specific dietary restrictions, vitamin supplements can help fill nutrient gaps.

While processing and storage play essential roles in food safety and convenience, they also affect the vitamin content of foods. Being mindful of these effects and making informed choices about food preparation and storage can help maintain a nutrient-rich diet essential for overall health and well-being.

6.2.1. Processes Affecting Food Nutrient Content

The processes used to grow, harvest, store, and prepare food can all have an impact on the nutritional value of the food. Foods that undergo high-temperature, high-light, or high-oxygen processing lose the most nutrients.

- **Fertilizers:** Most plant crops are produced with the aid of fertilized soils. High use of nitrogen fertilizers tends to reduce the vitamin C content in many fruit and vegetable crops. It does not seem to make any difference to the plant's nutrient value whether the fertilizer is organic or not.
- **Milling**: Grind cereals, like wheat, to get rid of their fibrous husks. The majority of the plant's dietary fiber, B-group vitamins, phytochemicals, and certain minerals are found in the husks. Because of this, even if whole meal breads have been artificially fortified with some of the nutrients that were lost during milling, they are still less nutritious than white breads.

Everything that is removed cannot be replaced, especially the phytochemicals. In certain products, "fiber" is reintroduced as resistant starch, which might not have the same health benefits as the original fiber.

- **Blanching:** Before a food is canned or frozen, it is usually heated very quickly with steam or water. The water-soluble vitamins, including vitamin C and B-complex, are sensitive and easily destroyed by blanching.
- **Canning**: To extend its shelf life and eradicate any potentially harmful microorganisms, food is heated inside the can. Certain microorganisms need to be heated to extremely high temperatures, which can change the food's texture and flavor and make it less appetizing. Foods that are canned typically don't require or use preservatives. Vitamins that are soluble in water are especially vulnerable to heat. Although fresh food often spoils faster than canned food, many people mistakenly think that canned foods are less nutrient-dense than their fresh counterparts.
- **Freezing:** The nutrient value of a food is retained when it is frozen. Any nutrient losses are due to the processing prior to freezing and the cooking once the frozen food is thawed.
- **Pasteurization:** Pasteurization involves heating liquid foods such as milk and fruit juices to specific temperatures to destroy micro-organisms. The nutrient value of milk is generally unaffected. In the case of pasteurized fruit juices, some losses of vitamin C can occur.
- **High Pressure Processing:** This alternative preservation method subjects a food to elevated pressures, with or without the use of heat to kill micro-organisms. This method has been used in foods such as fruit juices. As heat is not required, this process impacts less on the vitamin content, flavor and color of foods.
- **Dehydrating:** Fruits and other foods can lose some of their vitamin C content when dried, but other nutrients—especially fiber from plants—can be concentrated. Food items become more energy dense after being dried, which could lead to weight gain. Additional nutrients are lost in the cooking liquid when dehydrated food is reconstituted and cooked with water.

6.2.2. Free Radicals and Antioxidants

Atoms or molecules that have one or more unpaired electrons make them extremely "reactive," and these are known as free radicals. Activated atoms or groups of atoms with an odd (unpaired) number of electrons—typically comprising oxygen or nitrogen—are considered biologically significant free radicals. Every orbit in a non-radical compound has two electrons occupying it. Free radicals are created when a chemical reaction breaks the bonds that hold paired electrons together. Consequently, the outer orbit of a "free radical" compound contains a single unpaired electron.

An excited single electron is looking to join a paired set, and to do so, it will steal an electron from an adjacent atom. The neighboring atom loses an electron during this theft, turning it into a free radical, while the original free radical becomes stable. Then, to start a chain reaction, this new free radical will look for another atom to steal from.

Their ability to interact with and eventually damage tissue stems from their extreme reactivity, which is driven by a desire to acquire another electron. Damage is done to molecules that are important to biology, like proteins, lipids, carbohydrates, and DNA. Reactive oxygen species (ROS) and reactive nitrogen species (RNS) are collective terms for free radicals that encompass not only the radicals of oxygen and nitrogen but also certain non-radical reactive derivatives of these elements.

During a tissue's regular physiological metabolism, free radicals (ROS/RNS) are continuously produced. In mitochondria, oxygen is used to produce ATP. Most of the oxygen is reduced to form water, but only a small percentage (between 2 and 5%) is not completely reduced as a result of the formation of oxygen intermediates. The electron transport chain in the mitochondria is thought to be the source of the super oxide radical, which is the primary free radical produced in living cells. If there is an excess of free radicals produced naturally during oxidative metabolism, they react with fatty acids to form fatty acid hydroperoxides, which can set off a chain reaction that produces more free radicals and hydroperoxides, ultimately causing damage to cells if the chain reaction is not stopped. Another significant source of free radical generation is the activation of macrophages under stressful conditions. Immune cells create ROS and RNS, which are vital weapons in the fight against infections. Under normal circumstances, the body's antioxidant defenses counteract the harmful effects of ROS/RNS, and these defenses are bolstered by dietary intake of essential nutrients (e.g., trace minerals and vitamins). By stabilizing these extremely reactive free radicals, antioxidants help to preserve the structural and functional integrity of cells. Antioxidants are therefore crucial for both human and animal health and immune system protection. When the body's antioxidant system cannot handle the increased production of reactive oxygen metabolites, oxidative stress results in damage to the cell, tissue, or organ.

The body has evolved a complex antioxidant system that is dependent on antioxidant nutrients because free radicals are harmful to cells (Table 6.1). The main sources of vitamin antioxidants in tissue defense mechanisms against free-radical damage are generally found in vitamin C, vitamin E, and β-carotene (along with other carotenoids). Furthermore, a number of metalloenzymes, such as superoxide dismutase (Cu, Zn, and Mn), glutathione peroxidase (Se), and catalase (Fe), are essential for preventing oxidative damage to the internal components of cells. The animal body can only produce these antioxidant enzymes when adequate amounts of these metals are provided through diet. On the other hand, oxidative stress and damage to biological molecules and membranes result from a lack of those elements.

The quantity of free radicals produced by the body and the antioxidants required to defend against them are delicately matched. This equilibrium can be thrown off, leading to oxidative stress, by an abundance of free radicals or a deficiency in antioxidant defense. All of these nutrients—vitamins and trace elements—need to be balanced in the diet and in the body's tissues in order for tissues to be protected from free radical damage and to support immune function.

Table 6.1. Vitamins and Minerals in Antioxidant Systems

Nutrients	Component (location in cell)	Function
Vitamin C	Ascorbic acid (cytosol)	Reacts with several types of ROS/RNS
Vitamin E	α-tocopherol (membranes)	Breaks fatty acid peroxidation chain reactions
β-carotene	β-carotene (membranes)	Prevents initiation of fatty acids peroxidation chain reaction
Selenium	Glutathione peroxidase (cytosol)	An enzyme that converts hydrogen peroxide to water
Copper and zinc	Superoxide dismutase (cytosol)	An enzyme that converts superoxide to hydrogen peroxide
Manganese and zinc	Superoxide dismutase (mitochondria)	An enzyme that converts superoxide to hydrogen peroxide
Copper	Ceruloplasmin (water phase)	An antioxidant protein, may prevent copper and iron from participating in oxidation reactions
Iron	Catalase (cytosol)	An enzyme (primarily in liver) that converts hydrogen peroxide to water

6.2.3. Antioxidant Vitamins and Selenium

The main antioxidant vitamins that protect tissues from damage caused by free radicals are β-carotene, C, and E. The mineral most directly linked to antioxidant function would be selenium. Selenoprotein P, three thioredoxin reducases (TrXR1), five glutathione peroxidases, and at least 35 other antioxidant Se proteins are known to exist. According to the most recent studies, TrXR1 lowers damaging ROS and promotes the expression of other cytoprotective antioxidant genes. As a membrane-bound antioxidant, vitamin E traps lipid peroxyl free radicals that are generated during "oxidative stress" from unsaturated fatty acids.

It seems that vitamin E's orientation within cell membranes is essential to its activity. As a chain-breaking antioxidant, vitamin E keeps lipids in membranes from oxidizing and neutralizes free radicals. The first line of defense against phospholipid peroxidation is vitamin E. The second line of defense is glutathione peroxidase (GSH-Px), which is composed of selenium and breaks down peroxides and hydroperoxides. The different GSH-Px enzymes are expressed from distinct genes and have varying tissue specificities. Generally speaking, various GSH-Px forms work together to fulfill their protective roles, each offering antioxidant defense at various body locations.

α-tocopherol is the main form of vitamin E that has immune and antioxidant properties. Nonetheless, tocotrienols and non-α-tocopherol have significant roles, despite the paucity of research on the subject. It has been demonstrated that γ-tocopherol is a more potent inhibitor of lipid peroxidation induced by peroxy nitrite. Furthermore, γ-tocopherol has a greater ability to prevent inflammatory responses. Tocotrienols have been shown to suppress ROS more effectively than tocopherols and to exhibit superior antioxidant activity *in vitro*. It was discovered that tocotrienols were more successful in slowing down the aging process and age-related illnesses.

A combination of carotenoids and other antioxidants, particularly tocopherols, may offer superior antioxidant protection in the lipid environment of biological membranes than tocopherols alone. Carotenoids have antioxidant qualities that protect lipids from superoxide and hydroxyl radical damage by scavenging singlet oxygen and peroxyl radicals as well as sulfur, thiyl, sulfonyl, and NO2 radicals. Carotenoids can therefore actively quench singlet oxygen (O_2), stop O_2 from causing lipid peroxidation, and stop the spread of lipid peroxidation. One β-carotene molecule can effectively quench 1,000 O_2 molecules before they undergo chemical reactions and produce products. While β-carotene is the main antioxidant carotenoid, other carotenoids (e.g., Zeaxanthin, lutein, and lycopene) have potent antioxidant properties.

The most significant antioxidant found in extracellular fluids, vitamin C can prevent lipid peroxidation damage to biomembranes by removing peroxyl radicals from the aqueous phase before they have a chance to cause peroxidation. Vitamin C is found in cells' aqueous phase, where it aids in scavenging free radicals. Being a strong antioxidant, vitamin C readily gives up electrons to stabilize reactive species like ROS.

CASE STUDY

MANAGING VITAMIN D DEFICIENCY IN AN ELDERLY PATIENT

- **Patient Background:** John, a 70-year-old male, presents with complaints of chronic joint pain, fatigue, and frequent falls over the past year. He lives alone and spends most of his time indoors due to limited mobility and fears of falling.
- **Initial Assessment:** Physical examination reveals tenderness in his joints, particularly around the knees and hips. John's medical history includes osteoarthritis and hypertension. Blood tests confirm a severe deficiency in vitamin D.
- **Case Discussion:** John's symptoms—joint pain, fatigue, and increased falls—are indicative of vitamin D deficiency, which can exacerbate bone health issues like osteoarthritis and increase the risk of falls and fractures in the elderly.

1. **Causes:**

1. **Limited Sun Exposure:** John's avoidance of outdoor activities due to mobility issues and fear of falling has reduced his exposure to sunlight, a primary source of vitamin D synthesis.
2. **Dietary Insufficiency:** His diet lacks foods rich in vitamin D, such as fatty fish, fortified dairy products, and egg yolks.

2. **Treatment Plan:**

1. **Supplementation:** John is prescribed high-dose vitamin D supplements to rapidly correct the deficiency. This may include weekly or monthly doses depending on his specific needs and response to treatment.
2. **Fall Prevention Strategies:** Occupational therapy is recommended to improve John's mobility and reduce the risk of falls at home.

- **Education and Follow-up:** John and his caregiver receive education on the importance of vitamin D for bone health and strategies to increase dietary intake and sunlight exposure safely. Regular follow-up visits are scheduled to monitor his vitamin D levels and assess improvements in symptoms.
- **Outcome:** With supplementation and lifestyle modifications, John experiences reduced joint pain, improved energy levels, and fewer falls. Monitoring continues to ensure his vitamin D levels are maintained within the optimal range.

CLASS ACTIVITY

For a class activity on vitamins, students can engage in a group discussion where each group is assigned a specific vitamin (e.g., vitamin C, vitamin D, vitamin B12) to research. They should explore its functions, sources, recommended daily intake, and the consequences of deficiency or excess. Each group presents their findings to the class, emphasizing the importance of their assigned vitamins in maintaining health. This activity promotes research skills, teamwork, and understanding of the vital roles that vitamins play in human nutrition and overall well-being.

SUMMARY

- Vitamins are essential organic compounds that our bodies require in small amounts for various physiological functions. They play crucial roles in metabolism, immunity, and overall health maintenance.
- Water-soluble vitamins travel freely through the body, and excess amounts usually are excreted by the kidneys. The body needs water-soluble vitamins in frequent, small doses.
- The B-vitamin coenzymes function in enzyme systems that transfer certain groups between molecules; as a result, specific proteins, fats, and carbohydrates are formed and may be utilized to produce body tissues or to store or release energy.
- Fat-soluble vitamins are stored in the body's cells and are not excreted as easily as water-soluble vitamins. They do not need to be consumed as often as water-soluble vitamins, although adequate amounts are needed.
- Vitamins are an important part of our diet, but you probably have not given a great deal of thought to their chemical structures. Chemical structures for all 13 vitamins, though there can be some variability in these structures in sources of the vitamins.
- Riboflavin (Vitamin B2) is involved in carbohydrate metabolism as an essential coenzyme in many oxidation-reduction reactions. Riboflavin is essentially nontoxic. Riboflavin deficiency usually occurs with other B-vitamin deficiencies.
- Vitamin A is a fat soluble vitamin that is also a powerful antioxidant. Vitamin A plays a critical role in maintaining healthy vision, neurological function, healthy skin, and more. Vitamin A- like all antioxidants- is involved in reducing inflammation through fighting free radical damage.
- Vitamin D is a fat-soluble vitamin that is naturally present in very few foods, added to others, and available as a dietary supplement.

- The effects of processing and storage on vitamin content highlight the importance of consuming a balanced diet that includes a variety of fresh, minimally processed foods.

REVIEW QUESTIONS

1. What are vitamins, and why are they essential for human health?
2. Differentiate between water-soluble and fat-soluble vitamins. Provide examples of each type.
3. Explain the role of vitamin C in the human body. What are the consequences of vitamin C deficiency?
4. How does vitamin D contribute to bone health? Discuss sources of vitamin D and factors influencing its synthesis in the body.
5. Describe the functions of B-complex vitamins. Provide examples of foods rich in B vitamins.
6. What are antioxidants, and how do vitamins E and C act as antioxidants in the body?
7. Discuss the potential risks associated with excessive intake of fat-soluble vitamins. Give examples and explain the concept of vitamin toxicity.
8. Explain the importance of vitamin A in vision and overall health. What are dietary sources of vitamin A?
9. Describe the role of vitamin K in blood clotting. Discuss sources of vitamin K and conditions that may lead to vitamin K deficiency.
10. What strategies can individuals use to ensure adequate intake of vitamins through diet? Discuss the role of supplementation in vitamin intake.

REFERENCES

1. Adeyemi, O. M., Vibhakar, S., & Evans, A. T. (2009). Obesity and lymphocyte subsets in virologically suppressed HIV-infected patients. *Metabolism Clinical and Experimental, 58*(9), 1285–1287. https://doi.org/10.1016/j.metabol.2009.04.006.
2. Belitz, H., Grosch, W., & Schieberle, P. (2004). *Food Chemistry*. Springer Verlag.
3. Bender, D. A. (2003). *Nutritional Biochemistry of the Vitamins*. Cambridge University Press.
4. Cancello, R., & Clement, K. (2006). Is obesity an inflammatory illness? Role of low-grade inflammation and macrophage infiltration in human white adipose tissue. *BJOG: An International Journal of Obstetrics & Gynecology, 113*(10), 1141–1147. https://doi.org/10.1111/j.1471-0528.2006.01006.x.
5. Combs, G. F. (2008). Discovery of vitamins. In G. F. Combs (Ed.), *The Vitamins: Fundamental Aspects in Nutrition and Health* (3rd ed., pp. 1–26). Elsevier.
6. Dixon, J. B. (2010). The effect of obesity on health outcomes. *Molecular and Cellular Endocrinology, 316*(2), 104–108. https://doi.org/10.1016/j.mce.2009.07.008.

7. Fain, J. N. (2006). Release of interleukins and other inflammatory cytokines by human adipose tissue is enhanced in obesity and primarily due to the nonfat cells. *Vitamins and Hormones, 74*, 443–477. https://doi.org/10.1016/S0083-6729(06)74018-3.
8. Fortmann, S. P., Burda, B. U., Senger, C. A., Lin, J. S., & Whitlock, E. P. (2013). Vitamin and mineral supplements in the primary prevention of cardiovascular disease and cancer: An updated systematic evidence review for the U.S. Preventive Services Task Force. *Annals of Internal Medicine, 159*(12), 824–834. https://doi.org/10.7326/0003-4819-159-12-201312170-00729.
9. Furukawa, S., Fujita, T., Shimabukuro, M., Iwaki, M., Yamada, Y., Nakajima, Y., & Shimomura, I. (2004). Increased oxidative stress in obesity and its impact on metabolic syndrome. *The Journal of Clinical Investigation, 114*(12), 1752–1761. https://doi.org/10.1172/JCI21625.
10. Habib, S. A., Saad, E. A., Elsharkawy, A. A., & Attia, Z. R. (2015). Pro-inflammatory adipocytokines, oxidative stress, insulin, Zn and Cu: Interrelations with obesity in Egyptian non-diabetic obese children and adolescents. *Advances in Medical Sciences, 60*(2), 179–185. https://doi.org/10.1016/j.advms.2015.01.002.
11. Higdon, J. (2011). Vitamin E recommendations. Linus Pauling Institute's Micronutrient Information Center. Retrieved from: https://lpi.oregonstate.edu/mic/vitamins/vitamin-E (accessed on 3 July 2024).
12. Jenkins, D. J., Spence, J. D., Giovannucci, E. L., Kim, Y. I., Josse, R., Vieth, R., & Sievenpiper, J. L. (2018). Supplemental vitamins and minerals for CVD prevention and treatment. *Journal of the American College of Cardiology, 71*(22), 2570–2584. https://doi.org/10.1016/j.jacc.2018.03.517.
13. Maqbool, A., & Stallings, V. A. (2008). Update on fat-soluble vitamins in cystic fibrosis. *Current Opinion in Pulmonary Medicine, 14*(6), 574–581. https://doi.org/10.1097/MCP.0b013e3283114c6e.
14. Marti, A., Marcos, A., & Martinez, J. A. (2001). Obesity and immune function relationships. *Obesity Reviews, 2*(3), 131–140. https://doi.org/10.1046/j.1467-789x.2001.00031.x.
15. O'Rourke, R. W., Kay, T., Scholz, M. H., Diggs, B., Jobe, B. A., Lewinsohn, D. M., & Bakke, A. C. (2005). Alterations in T-cell subset frequency in peripheral blood in obesity. *Obesity Surgery, 15*(10), 1463–1468. https://doi.org/10.1381/096089205774512429.
16. Price, C. (2015). *Vitamania: Our Obsessive Quest for Nutritional Perfection*. Penguin Press.
17. Rohde, L. E., de Assis, M. C., & Rabelo, E. R. (2007). Dietary vitamin K intake and anticoagulation in elderly patients. *Current Opinion in Clinical Nutrition and Metabolic Care, 10*(1), 120–124. https://doi.org/10.1097/MCO.0b013e328010c6a3.
18. Young, S., & Woodside, J. V. (2001). Antioxidants in health and disease. *Journal of Clinical Pathology, 54*(3), 176–186. https://doi.org/10.1136/jcp.54.3.176.

CHAPTER

7 Minerals

LEARNING OBJECTIVES

After studying this chapter, you will be able to:

- Define chemistry of minerals
- Discus nutritional aspects of minerals
- Explain the mineral composition of foods

KEY TERMS FROM THIS CHAPTER

Oxides	Phosphates	Halides
Nitrates	Phosphoric	Pyrophosphoric
Tripolyphosphoric acids	Triiodothyronine	Aminocarboxylic acids
Cysteamine	Putrescine	Cadaverine

INTRODUCTION

The crust of the earth naturally contains 90 chemical elements. There are about 25 of them that are found in living cells because they are known to be necessary for life (Figure 7.1). These 25 components should be present in foods since living plants or animals are ultimately the source of our food. Since living systems can accumulate unnecessary elements from their surroundings, foods also contain other elements. Furthermore, substances can contaminate food during its production, processing, and storage, or they can be added on purpose.

In the context of food and nutrition, the term "mineral" refers to elements other than C, H, O, and N that are found in foods, though there is no agreed-upon definition of the term. These four nonmineral elements make up 99% of all the atoms in living systems and are mostly found in organic molecules and water. Therefore, food contains relatively small amounts of mineral elements. However, they are essential to the functioning of both foods and living systems.

In the past, minerals were categorized as major or trace based on the amounts they were found in plants and animals. This classification developed during a period when analytical techniques were not highly accurate in measuring small concentrations of elements. As a result, the existence of an element that was impossible to measure precisely was denoted by the term "trace." Nearly every element in the periodic table can now be measured with extreme precision and accuracy thanks to advancements in technology and equipment (Figure 7.1). Nevertheless, mineral elements in biological systems are still referred to by the terms major and trace. Calcium, phosphorus, magnesium, sodium, potassium, and chloride are important minerals. Iron, iodine, zinc, selenium, copper, fluorine, lead, and tin are examples of trace elements.

I-A	II-A	III-B	IV-B	V-B	VI-B	VII-B	VIII	VIII	VIII	I-B	II-B	III-A	IV-A	V-A	VI-A	VIIA	O
H																	He
Li	Be											B	C	N	O	F	Ne
Na	Mg											Al	Si	P	S	Cl	Ar
K	Ca	Sc	Ti	V	Cr	Mn	Fe	Co	Ni	Cu	Zn	Ga	Ge	As	Se	Br	Kr
Rb	Sr	Y	Zr	Nb	Mo	Te	Ru	Rh	Pd	Ag	Cd	In	Sn	Sb	Te	I	Xe
Cs	Ba	Ln	Hf	Ta	W	Re	Os	Ir	Pt	Au	Hg	Tl	Pb	Bi	Po	At	Rn
Fr	Ra	Ac	Th	Pa	U												

Figure 7.1. *Periodic table of the naturally occurring elements. Shaded elements are believed to be essential nutrients for animals and humans.*

Source: Owen R. Fennema; Food Chemistry Third Edition; ISBN 0-8247-9691-8.

7.1. CHEMISTRY OF MINERALS

Food-borne minerals interact with water, with the organic matter that is there, and with each other. The bioavailability of nutrients in the diet is impacted by these interactions.

Food composition, pH, the potential for metal ion hydration, the system's redox potential, the associated potential to alter the element's degree of oxidation, and other factors all influence the element's chemical state in food. The resorption characteristics of the minerals are intimately related to the solubility of the individual components. A variety of substances found in food, including lignin, phytic acid, proteins, carbohydrates, amino acids, peptides, and organic acids, have the ability to bind minerals and therefore affect their biological availability (Barrett, & Ranum, 1985).

7.1.1. Bonding Possibilities of Elements

The way an element interacts with the organic food matrix depends critically on its chemical properties, which vary based on the form of the products. These characteristics are a function of the element's electron configuration within its atom and are dictated by the element's position in the Periodic Table.

In biological systems, non-metals and metalloids with medium electronegativity values (P, As, S, and Se) form covalent compounds, such as sulfur heterocycles, arsenic-analogous amino acids, phosphoric, pyrophosphoric, and tripolyphosphoric acid esters.

High electronegativity elements (halogens like Cl and I) and extremely low electronegativity elements (alkali metals, alkaline earth metals, like Na, Ca, and Mg) are primarily found in biological materials as free ions and are ideally involved in electrostatic interactions.

Even so, these elements can combine to form complex compounds (containing some metal ions as central atoms and chlorides as ligands), less soluble compounds (like calcium oxalate), and covalent compounds (like the hormones thyroxine and triiodothyronine, which are iodinated aromatic amino acids). An atom, ion, or molecule that can function as an acceptor of an electron pair to form a coordinate covalent bond with the central ion is known as a ligand. Moreover, covalent compounds with Cd and Hg are common.

Complex compounds are highly likely to be formed by transition metals and certain post-transition metals (such as Al, Pb, and Zn). This is caused, in part, by their ions' greater electric charge and smaller radius. In biological materials, these elements have a wide range of possible reaction partners, or ligands.

The kind of ligand donor atoms, the metal and its oxidation state, and steric factors all affect how stable a complex is. The most stable complexes are formed by divalent cations of the first transitional series, such as the physiologically significant Cu^{2+} ions. Cu^{2+} ions prefer to bind to ligands that have multiple nitrogen atoms or sulfur and nitrogen as donors. On the other hand, Mn^{2+} ions ligands with sulfur and nitrogen entities that have oxygen donor atoms can also function as ligands; these complexes are formed rather reluctantly (Figure 7.2).

Metal ions in oxidation states III and IV are generally more stable than the corresponding cationic complexes in oxidation state II. Metal cations with higher charge may, however, in the neutral or weakly acidic environments typical of foods, form insoluble hydroxides, for example $Fe(OH)_3$, oxides (MnO_2) or oxocationts (TiO^{2+}, VO^{2+} and BiO^+). Some metal ions in a middle oxidation state, unless they are bound in complexes, are susceptible to disproportionation (e.g., $2\ Mn^{3+} + 2H_2O \rightarrow Mn^{2+} + MnO_2 + 4H^+$). With the exception of Mo, which is commonly in the very stable oxidation state VI, transition metals do not occur in high oxidation states in biological materials (e.g., oxoanions MnO_4^-, MnO_4^{2-} and CrO_4^{2-}). Various forms of minerals in foods are summarized in Table 7.1.

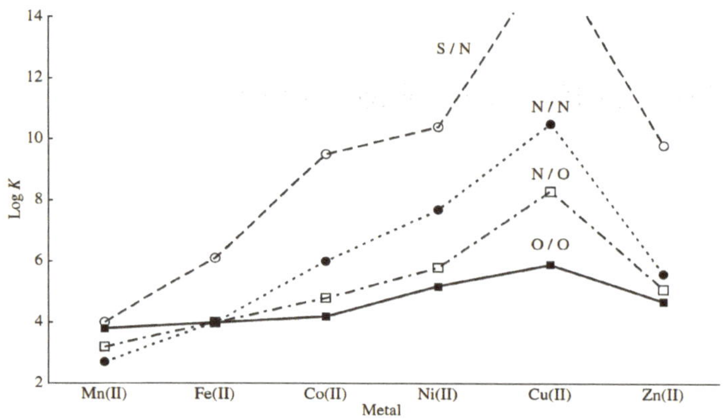

Figure 7.2. *Dependence of the logarithm of stability constants (log K) of divalent metal cations complexes of the first transitional series on the type of ligand according to the type of donor atoms: (oxalic acid–oxygen/oxygen; ·–·–·–· glycine–nitrogen/oxygen; ······· ethylenediamine–nitrogen/nitrogen; ------cysteine–sulfur/nitrogen).*

Source: Jan Velíšek; The Chemistry of Food; ISBN 978-1-118-38381-0.

Table 7.1. Overview of Forms of Minerals in Foods

Form	Example
Elemental form	Fe in some fortified foods
Free and hydrated ions of metals and non-metals in different oxidation states	Cu^{2+}, $Cu(H_2O)_4{}^{2+}$, $Fe(H_2O)_6{}^{2+}$, $Fe(H_2O)_6{}^{3+}$, $AsO_3{}^{3-}$, $AsO_4{}^{3-}$
Complex compounds of metals with inorganic ligands	$CuCl_4{}^{2-}$, $Cu(NH_3)_4{}^{2+}$
Complex compounds of metals with organic ligands	Compounds with amino acids, peptides, proteins, carbohydrates, phytic acid, hydroxycarboxylic acids, plant phenols (flavonoids) and porphyrins
Minerals bound to insoluble biopolymers	Binding to various components of fiber
Slightly soluble compounds	Sulfides, sulfates, phosphates, oxalates, hydroxides, phytates
Covalent compounds of non-metallic and semimetallic elements	Phytic acid, sulfur amino acids and their selenium analogues, arsenic analogues of amino compounds
Organometallic compounds	Methylmercury, dimethylmercury, tetraethyllead

Source: Jan Velıšek; The Chemistry of Food; ISBN 978–1-118–38381–0.

7.1.2. Interaction with Organic Food Components

7.1.2.1. Aminocarboxylic Acids

Aminocarboxylic acids can use their dissociated carboxyl and amino groups to bind metal ions as coordination compounds. Both the nitrogen and the oxygen atoms in the carboxyl and amino groups are donor atoms that can contribute an electron pair to form coordinate covalent bonds with the central metal ion. The order in which the bound metal affects the stability of complexes of bivalent metal cations with amino acids is as follows.

$$Cu^{2+} > Ni^{2+} > Zn^{2+} > Co^{2+} > Cd^{2+} > Fe^{2+} > Mn^{2+}$$

However, the stability constants of complexes containing different amino acids can vary by more than two orders of magnitude for a given metal. The amphoteric nature of amino acids and their derived compounds, such as peptides and proteins, allows them to exist in aqueous media in a variety of forms that vary in how the functional groups dissociate, contingent on the pH. The formation of complexes with metal ions is significantly influenced by the presence of these groups in the ligand molecule as well as the pH level. The participation of various functional groups in metal binding depends on two conditions:

- how the relevant functional group competes with other groups in the vicinity of the metal ions; and
- how metal ions compete with hydrogen ions for the binding sites in the ligand functional groups.

Thus, the formation of complexes is influenced by the dissociation constant values of different functional groups of amino acids. The ligand donor atom's capacity to bind the metal increases with decreasing pK value. More functional groups can function as electron acceptors in a specific pH range, resulting in the formation of cyclic complexes, or chelates, which are more stable thermodynamically than complexes containing monofunctional ligands. If the configuration of the ligand molecule permits the formation of rings with five or seven members, it will be possible to form a chelate with an appropriate coordination geometry. Despite having high pK values, amino acid groups are frequently involved in coordination bonds with metals in addition to the anion of the amino acid. Coordination is made easier by the nitrogen atom's nucleophilic nature.

Zinc-glycine complexes serve as an example of how the acid-base equilibrium affects the structure of amino acid–metal complexes. Zinc is bound by both of glycine's (Gly) functional groups in neutral media, resulting in the chelate $Zn(Gly)_2{\cdot}2H_2O$ (7–1). Zinc salt of glycine, $Zn(Gly)_2.4H_2O$ (7–2), tetrahydrate, is formed when amino groups are protonized in acidic media, leaving only carboxyl groups to bind the metal. The coordination is facilitated by nitrogen's nucleophilic nature.

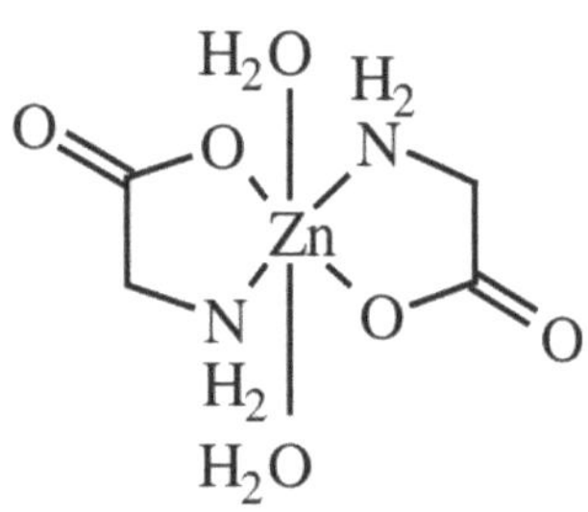

7–1, diaqua-bis(glycinato)zinc complex

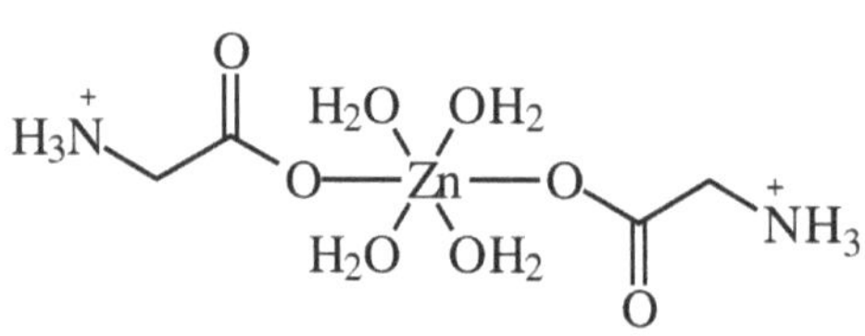

7–2, glycine zinc salt tetrahydrate

Amino acid functional groups with different functions may also help form complexes with metals and stabilize them by adding more bonds. Very stable compounds containing thiols are produced by metals and metalloids (e.g., Hg, Cd, Ag, Cu, Bi, Sb, and As) that have a particular affinity for sulfur. A more stable complex is produced when the cysteine sulfhydryl group forms bonds with metal ions (7–3). Cysteine complexes containing Mn^{2+}, Fe^{2+}, Co^{2+}, Ni^{2+}, Zn^{2+}, and Pb^{2+} have stability constants that are noticeably higher than those of the corresponding glycine and histidine complexes. The hydroxyl groups in serine, threonine, and tyrosine do not play a major role in the metal binding process, in contrast to sulfhydryl groups. Dallman, P. R. (1990).

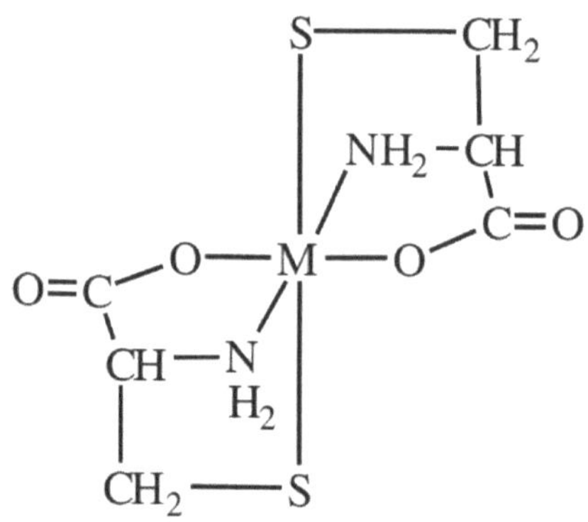

7–3, complex of metal (M) with cysteine (M = Cu, Zn, Cd, Mn, Ni, Fe, Pb)

Often, however, the imidazole nucleus of histidine (free and bound histidine in peptides and proteins) is involved in metal binding. The order of dissociation of functional groups in the molecule of histidine is evident from the following pK values: pK_1 (COOH) = 1.80, pK_2 (imidazolium) = 6.04, pK_3 (NH_3^+) = 9.33, pK_4 (imidazole) = 14. This results in the following stages during the formation of histidine complexes with metals:

- Coordination at the oxygen atom of the carboxyl, even at low ph value
- Metal binding to the nitrogen atom of the imidazole nucleus
- Metal binding to the amino group nitrogen.

Complexes like bis-l-histidinatocobalt or bis-l-histidinatonickel contain all three types of bonds (7–4). Additionally, amines and diamines (byproducts of the decarboxylation of amino acids) like cadaverine, putrescine, histamine, and cysteamine are formed by stable complexes.

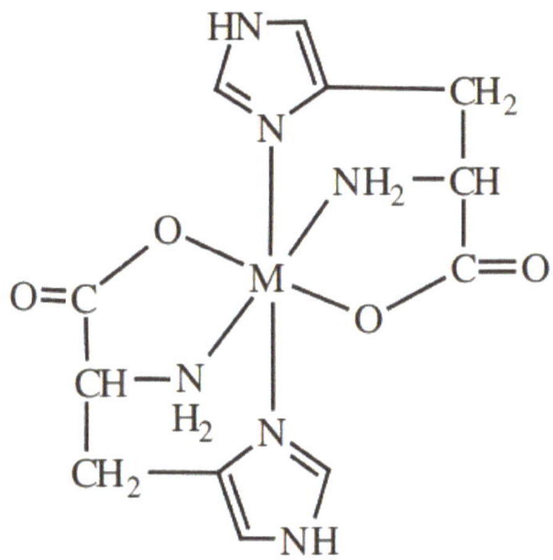

6–4, complex of metal (M) with histidine (M = Co, Ni)

7.1.2.2. Peptides and Proteins

Metals can be bound by peptides and proteins via side chain functional groups, C-terminal carboxyl, and N-terminal amino groups of amino acid residues. Amino acids like lysine (Lys), ornithine (Orn), aspartic acid (Asp), glutamic acid (Glu), cysteine (Cys), and histidine (His) are especially examples of this. An additional choice is to attach the metal to the peptide bond's nitrogen or oxygen atom.

Weak metal binding results from the oxygen atom of the peptide bond's carbonyl group having only a small nucleophilic character. Chelate stabilization may be provided by the nearby amino group. When metal ions bind to the peptide bond's nitrogen atom, it is assumed that the NH group's ability to dissociate as an electron pair on nitrogen is unavailable.

Certain oligopeptides, like Gly-Gly, GlyGly-Gly-Gly, or Gly-Gly-His, have metal complexes where a peptide chain encircles the central metal ion and forms bonds with the help of all sterically accessible donor groups. Water molecules, on the other hand, occupy free coordination sites (7–5 to 7–7). Certain simple peptides, like Gly-Gly, can have complex metal complex structures, leading to the formation of multinucleated complexes like $Zn_2(Gly\text{-}Gly)_4.2H_2O$.

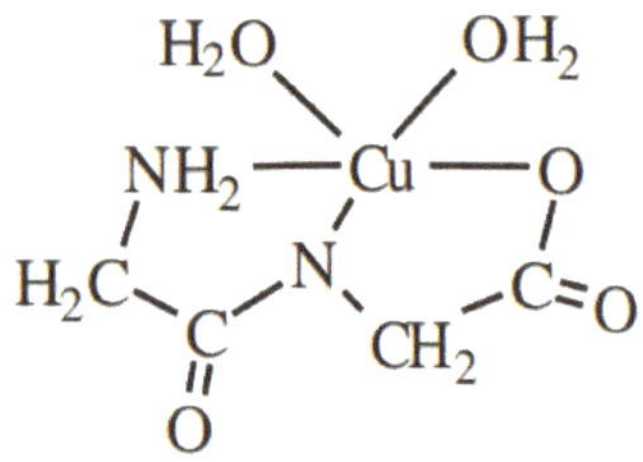

7–5, complex of copper with dipeptide Gly-Gly

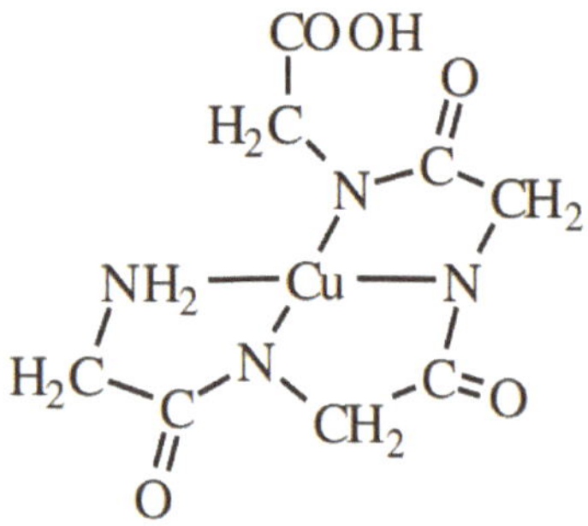

7–6, complex of copper with tetrapeptide Gly-Gly-Gly-Gly

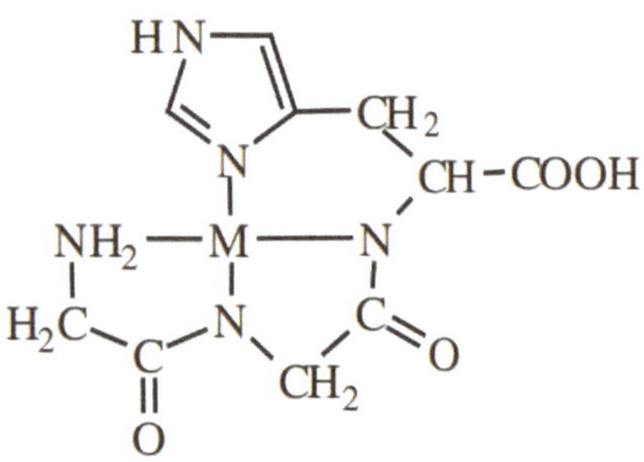

7–7, complex of metal (M) with tripeptide Gly-Gly-His

Among the naturally occurring substances that function as metal ligands, the metallothionein family of cysteine-rich metal-binding peptides and low molecular weight proteins (molecular weight of 3–14 kDa) plays a unique role. Numerous microorganisms, plants, fungi, invertebrates, vertebrates, and humans all contain metallothioneins in their bodies. They were first discovered in the internal organs of mammals, including the pancreas, liver, kidneys, intestines, and spleen of cattle, horses, and other animals. Metallothioneins produce stable complexes that guarantee the detoxification of harmful metals (like Cd). Consistent combinations with necessary metals (e.g., Zinc and copper) are used to store metals in tissues temporarily until they are needed for the synthesis of other substances or metalloproteins. Additionally, they offer defense against oxidative stress. Three classes were established by the initial classification of metallothioneins (MTs): class I comprises metallothioneins that share homology with horse metallothionein; class II comprises the remaining metallothioneins that do not share homology with horse metallothionein; and class III comprises phytochelatins. At least the four primary isoforms of class I metallothioneins—MT1, MT2, MT3, and MT4—are expressed by all mammals. While MT3 and MT4 are tissue-specific, MT1 and MT2 are expressed in nearly all tissues.

Metallothioneins belonging to class I are peptides that are thermostable and resistant to acid. They have a molecular weight of 6–7 kDa and contain 60–68 amino acid residues, of which 20 are cysteine residues (which account for approximately 30% of the amino acid content of metallothioneins). Additionally, they can bind seven equivalents of bivalent metal ions. Typically, aromatic amino acids are not present. Within peptide chains, the positions of the cysteine and lysine units are nearly indistinguishable across different mammalian metallothionein types. N-acetylmethionine is the amino acid that comes at position N (Figure 7.3).

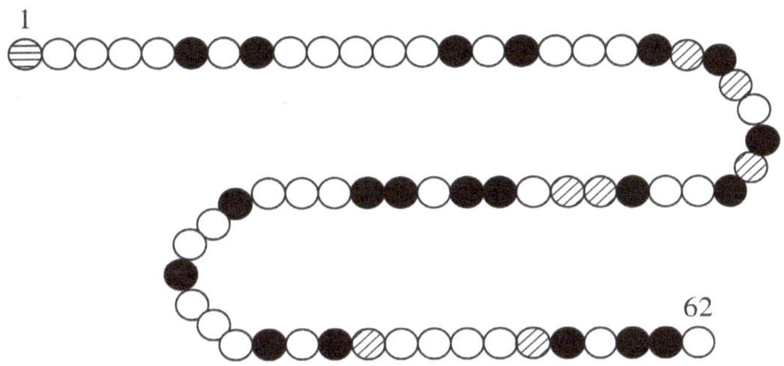

Figure 7.3. *Cysteine (full circles), lysine (obliquely shaded circles) and N-acetylmethionine (horizontally hatched circle) positions in the sequence of amino acids in mammalian metallothioneine.*

Source: Jan Velíšek; The Chemistry of Food; ISBN 978-1-118-38381-0.

Through mercaptide bonds, all cysteine residues are found in the reduced form and are coordinated to the metal ions. Organism-specific species, metallothioneins share common structural characteristics. Higher dietary intake of metals (Cd, Cu, Zn, Ni, Pb, Co, Bi, Hg, Au, and Ag) or exposure in the environment where an organism lives cause their increased synthesis in the body. The most potent inducer of metallothionein synthesis is Cd. Divalent metal atoms can be bound by a single peptide molecule up to seven times. Complexes that are separated from animal tissues nearly invariably include Cd and Zn. Nonetheless, these materials retain their ability to bind, enabling the other metal ions to be fixed.

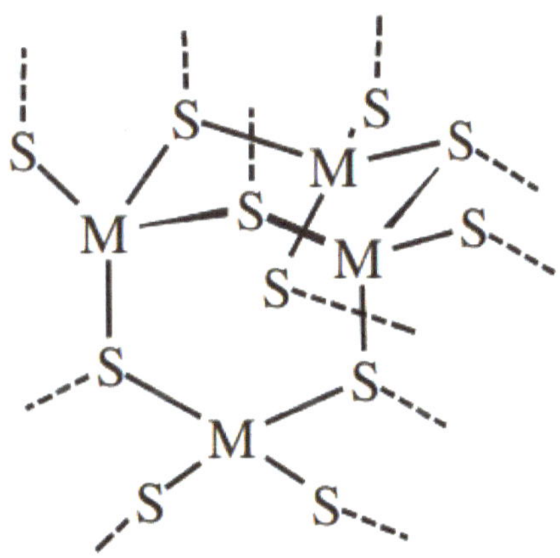

7–8, α-domain of metallothioneine (M = metal)

The metallothionein structure has been thoroughly studied, and it has been discovered that the metal atoms are bound in two domains, α and β. Domain α (7–8) is situated in close proximity to the C-terminus of the peptide chain, wherein the four metal atoms are fixed by the participation of 11 sulfur atoms. Domain β (7–9) is situated in close proximity to the peptide chain's N-terminus, where three metal atoms (M) are fixed by nine sulfur atoms. Even greater amounts of Cu in the monovalent form can be bound in complexes with metallothionein (12 Cu^{I} atoms in one molecule). When exposed to UV light, cuthioneins exhibit a characteristic orange to red fluorescence.

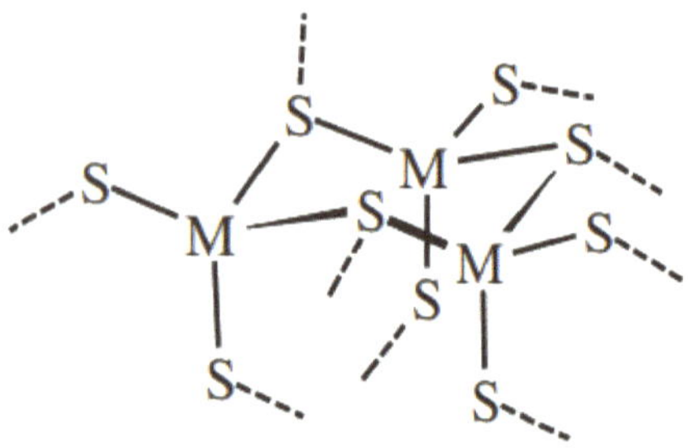

7–9, β-domain of metallothioneine (M = metal)

Brain tissue contains a different form of class I metallothionein called MT3, which has bound copper and zinc. With 68 amino acids, a typical representative resembles the metallothioneins found in crustaceans, which have 18 cysteine residues. Comparable binding peptides were also discovered in fish and invertebrates, among other animals exposed to harmful substances.

Class II metallothioneins also have a high percentage of cysteine units, though they are arranged very differently in the group's polypeptides. The distinctive repeating units Cys-Cys, Cys-X-Cys, or CysX-X-Cys, where X is an amino acid other than cysteine, are present in a molecule of these peptides. These polypeptides were discovered in a number of economically significant crops, including cotton, wheat, maize, rice, buckwheat, and maize, but frequently only in particular sections of the plant and during particular growth stages.

In biological materials, proteins play a major role in mediating the binding of metals. Similar mechanisms to those for amino acids and peptides are used by the protein molecule to bind the metal ions. It is not always the case that adjacent amino acid residues in polypeptide chains correspond to the functional groups of proteins that are involved in metal binding. The portions of the macromolecule that are relatively far apart in the amino acid sequence can also be in close contact with the metal ion because of the tertiary structure of the proteins. Another possibility is that metals can bind by interacting with phosphoproteins' phosphate groups (e.g., Phosvitin, a glycophosphoprotein found in egg yolks, combines with Fe^{3+}, Ca^{2+}, or Mg^{2+} through its serine residues. Temperature, pH, the kind of metal ion, the presence of additional components in the primary structure, and protein conformation all affect the possibility of interactions between proteins and metal ions as well as the stability of the complexes that result. Food-borne metal complexes of proteins can be classified into two categories. The initial set of complexes appears at random and is comparatively unstable. Under the right circumstances (pH value), almost all proteins interact with metal ions due to their abundance of reactive functional groups. Such a complex's composition is irregular and shows the statistical distribution of the constituent particles, which represents the extent to which the protein's functional groups have dissociated. It is necessary to conduct experiments in each specific case because the stability and properties of such complexes can only be partially estimated based on theory. Denaturation of proteins can occur from higher concentrations of certain metal ions.

Metalloproteins, which have a regular structure, metal binding sites for the metal ion, and a distinctive binding mechanism, make up the second class of metal complexes with proteins. As an example, prosthetic groups can bind the metal to the protein macromolecule in addition to amino acids (Fe bound in the porphyrin structure of haem in hemoglobin, myoglobin, and haem enzymes). Certain metalloproteins perform vital biological tasks, such as serving as storage, transport, and catalysts (metalloenzymes) (Table 7.2). The pertinent sections that address each of the individual elements mention a number of significant metalloenzymes.

Table 7.2. Some Important Metalloproteins

Element	Metalloprotein	Occurrence	Element	Metalloprotein	Occurrence
Ca	Calmodulin	Muscle	Cu	Plantacyanin	Spinach
	Parvalbumin	Muscle		Ceruloplasmin	Blood plasma
	Troponin C	Muscle		Cerebrocuprein	Brain
Fe	Myoglobin	Muscle		Haemocuprein	Blood
	Haemoglobin	Erythrocytes	Mo	Xanthinoxidase	Liver
	Cytochromes, catalases, peroxidases	Widespread	Fe (Cu, Zn, Mn)	Conalbumin	Egg white
	Transferrin	Blood, liver	Ni	Nickelplasmin	Krevní plasma
	Ferritin	Spleen		Urease	Soy, rice
	Lactoferrin	Milk	Mn	Pyruvate decarboxylase	Widespread
	Ferredoxins	Spinach		Arginase	Liver

Source: Jan Velıšek; The Chemistry of Food; ISBN 978–1-118–38381–0.

7.1.2.3. Saccharides and their Derivatives

Since the chelating ligands have an affinity for metal ions, polyhydroxy compounds like saccharides and sugar alcohols theoretically also have chelating properties. But these compounds (apart from the unique steric arrangement) only appear in extraordinary chemical environments, which are not typically found in foods. Compared to atoms of nitrogen or sulfur, the oxygen atom of the hydroxyl group is a far weaker donor of an electron pair for the majority of metals. In an alkaline medium, metal ions combine with sugar alcohols, disaccharides, and monosaccharides to form complexes. The stoichiometric ratio of iron to sugar is 1:1, and complexes containing ferric ions range in color from yellow to dark brown.

The discovery that adding some sugars to the diet—particularly d-fructose and d-glucitol—increases the bioavailability of iron is what prompted the investigation of these compounds. Aldoses and ketoses can combine with metals to form cyclic (furanoses and pyranoses) or acyclic (hydrates) complexes.

For instance, d-altrose (also known as d-arabinose and d-idose) primarily forms complexes in its acyclic forms with hexavalent Mo ions in acidic media. The secondary hydroxyl groups at C-2, C-3, and C-4 as well as the hydroxyl group of the hydrated carbonyl group function as donors in the binding of ions.

D-allose and d-gulose, two aldopyranoses, combine to form complexes with hydroxyls at C-2, C-3, C-4, and C-6. Pentose d-ribose forms comparable complexes. Carbohydrate derivatives with an even higher propensity to form complexes include aldonic and alduronic acids, 2-amino-2-deoxysugars, and sugar phosphates.

Food containing metal ions and carbohydrates may naturally form complexes when the hydroxyl groups on the three neighboring carbon atoms in the chair conformation's six-membered rings are arranged in an alternating pattern of axial, equatorial, and axial directions. For the epi-inositol–Ca^{2+} complex, this is the case (7–10).

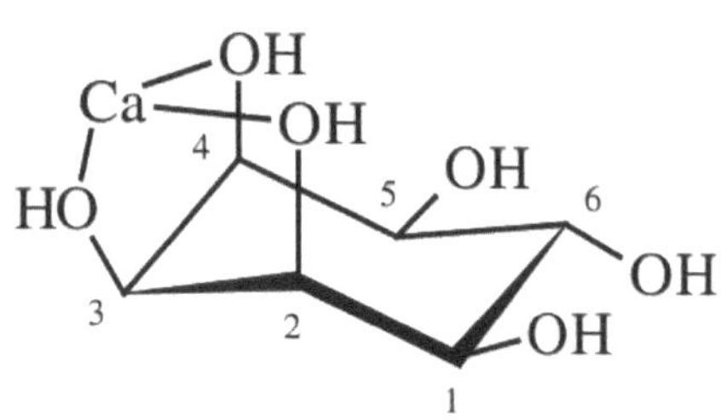

7–10, complex of calcium with epi-inositol

Substrates for organic phosphorylation (e.g., Sugar phosphates are weak acids that combine with different metals to form salts. Phytic acid, also known as myo-inositol hexakisdihydrogen phosphate, is a significant binding agent for metallic elements.

The formula in 7–11 represents the conformation of phytic acid in solutions with pH values between 0 and 9 (one phosphate group in the axial position and five phosphate groups in the equatorial positions). Five phosphate groups are bound in the axial positions and one in the equatorial positions in solutions with a pH greater than 9.05. Both chair conformations are present in equimolar concentrations at pH 9.5.

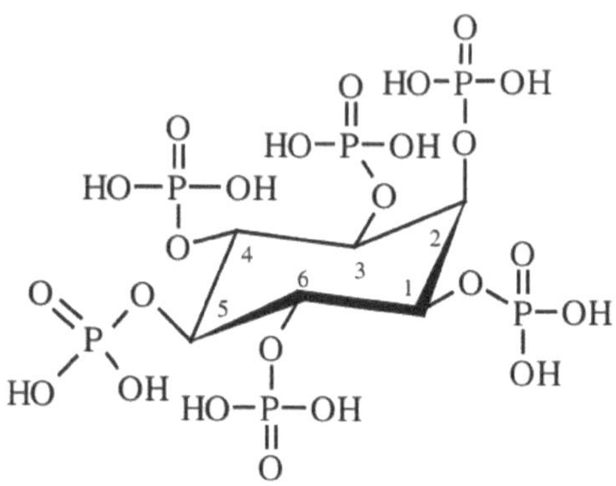

7–11, phytic acid

In ratios of 1:1 to 1:6, phytic acid forms stable complexes, or phytates, with calcium, magnesium, iron, zinc, and other metal ions. Phytin, also called calcium-magnesium phytate, is the calcium-magnesium complex of phytic acid. The sequence of metals' binding to phytic acid at pH = 7.4 is as follows: Cu > Zn > Ni > Co > Mn > Fe > Ca. However, in diets with higher amounts of phytic acid and phytin, the bioavailability of the bound elements is decreased due to the strong binding of these compounds' elements and their low solubility (iron phytate). This may prevent colon cancer by reducing oxidative stress in the intestinal tract lumen. Foods derived from plants, particularly cereals, legumes, oil seeds, and nut hulls, contain phytotic acid and its salts containing metals.

Owing to the dissociated phosphate groups' negative charges, phytic acid—even when bound to metal—can react with certain phospholipids or ionized proteins (Figure 7.4), forming the complexes phytic acid–metal–phospholipid, phytic acid–protein, or metal–phytic acid–protein. The positively charged amino groups of the basic amino acids (Lys, Arg, and His) and the terminal amino groups bind the proteins to phytic acid in an acidic medium. Weak interactions only occur at pH values up to the protein isoelectric point pH. Phytate complexes with metals bind to proteins through ionized carboxyl groups in neutral and alkaline media.

For instance, 50–340 mg/kg of phytate are present in crude soybean oil. This phytate is mostly eliminated during oil degumming and is combined with the phospholipid fraction to form the lecithin fraction. Phytate levels in degummed oil range from 4 to 50 mg/kg. During the sugar juice clarification using calcium hydroxide, saccharose dissociates in an alkaline medium in several steps (at 25°C: pK_1 = 12.60; pK_2 = 13.52; pK_3 = 13.72; pK_4 = 13.77). The most reactive atoms are hydrogens of the primary hydroxyl at C-6 of the glucose moiety, and C-1 and C-6 of fructose moiety. It is reported that in solutions of pH 11, about 10% of sucrose is present in the form of monovalent anions, at pH 12.2 about 50% of the molecules, and at pH 12.5 bivalent anions are also present.

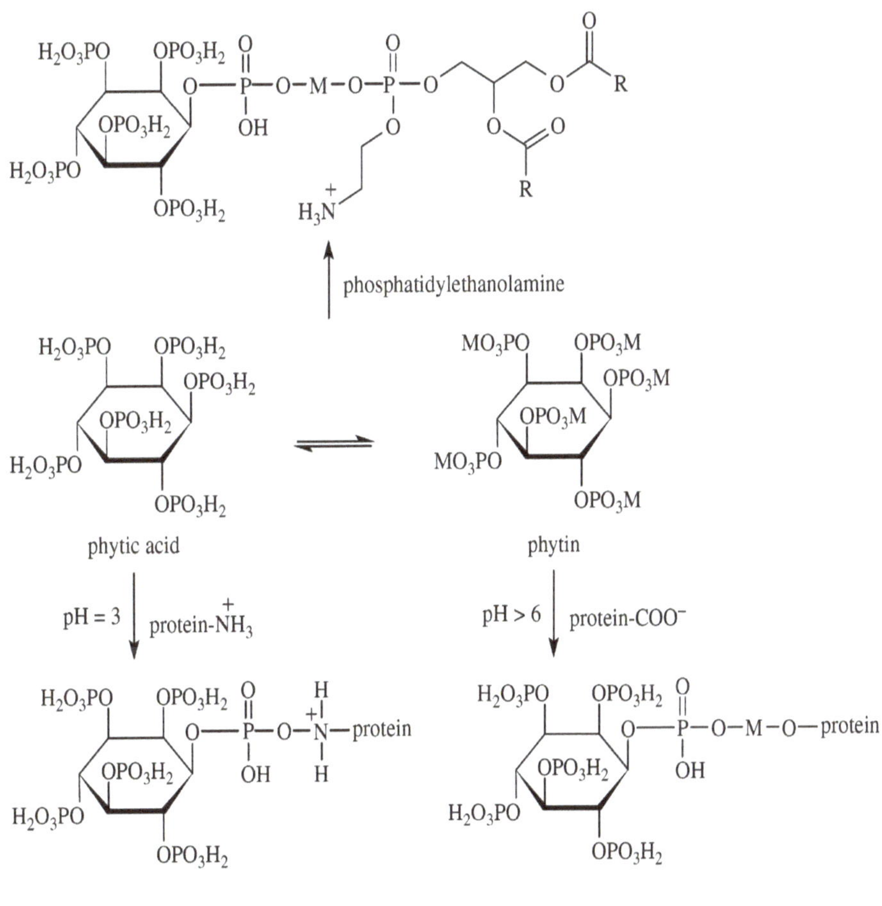

Figure 7.4. *Reactions of phytic acid and phytin with metal ions (M = Ca, Mg, Zn, Fe, Cu, Mn), proteins and phospholipids (R = fatty acid residue).*

Source: Jan Velíšek; The Chemistry of Food; ISBN 978–1-118–38381–0.

Depending on the amount of added lime milk (pH of the juice) and the temperature, initially soluble monocalcium saccharate (with the molecular formula $C_{12}H_{22}O_{11}·CaO$) and slightly soluble dicalcium saccharate ($C_{12}H_{22}O_{11}·2CaO$) form, followed at higher pH and higher temperatures by insoluble tricalcium saccharate ($C_{12}H_{22}O_{11}·3CaO$). In solutions, the ions $(C_{12}H_{21}O_{11}Ca)^+$ or $(C_{12}H_{21}O_{11}Ca_2)^{3+}$ are apparently present. After saturation of juice by carbon dioxide, which decreases the pH to 8.9–9.2, saccharates decompose. Formation of insoluble tricalcium saccharate is used to obtain saccharose from molasses. Similar complexes, as with calcium ions, are similarly formed with Sr^{2+} and Ba^{2+}ions.

Disaccharide lactose occupies a special place in terms of complex formation potential. It was discovered that it forms complexes in the pH range of 2 to 6.5 and in a ratio of 1 to metals (Ca, Ba, Sr, Mg, Mn, Zn, Na, and Li). Increased calcium resorption from milk is thought to be the outcome of lactose binding to calcium.

Gels are created when strong complexes between polysaccharides derived from glycuronic acids (like pectins and alginates) and Ca^{2+} ions are formed. Insoluble polysaccharides that are not useful (e.g., Minerals can also be bound by lignin, which is an insoluble fiber, and cellulose and certain hemicelluloses.

The gastrointestinal tract's ability to resorb bound elements can be significantly impacted by these interactions. The strength of interaction decreases in the following order when investigating metal binding to cellulose and hemicelluloses (such as arabinoxylans): Cu > Zn > Ca. Cupric ions have a strong affinity for cellulose in slightly acidic media, but their binding to cellulose molecules is significantly decreased when glycine is present. Lignin contains two different kinds of metal-binding sites. The decreasing order of interaction strength for binding sites with high affinity is Fe > Cu > Zn, whereas the decreasing order for binding sites with low affinity is Cu > Fe > Zn.

Ion-exchange capacities of different dietary fiber components also differ significantly. Long-term overconsumption of dietary fiber can result in deficiencies in calcium, iron, and zinc. This decreased mineral resorption is particularly noticeable when high dosages of phytic acid and fiber are consumed.

7.1.2.4. Carboxylic Acids

Benzoic acid is one of the many aromatic acids, hydroxycarboxylic acids (citric, tartaric, malic, and lactic acids), oxocarboxylic acids (pyruvic acid), and aliphatic mono- and dicarboxylic acids (acetic and oxalic acid, respectively) are common ingredients in plant and animal diets.

They combine with metal ions to form salts, and their anions have ligand-like characteristics as well. The main metal binding partners in fruits and certain vegetables are carboxylic acids. Additionally, they play a major role in the humic substances' binding of minerals (metal ions) in soils and sediments.

Together with oxalic acid, calcium forms a stable insoluble salt. Higher oxalic acid concentrations in plant cells can cause calcium oxalate to crystallize. It has been demonstrated that some plants can bind metals in mixed complexes. For instance, nickel and zinc can combine to form a citrate–malate complex, and chromium can be bound in an oxalate–malate complex. It has been demonstrated that citric acid binds calcium in casein micelles and functions as a low molecular weight zinc ligand in human milk. As it is also used as a food additive (acidulant, antioxidant synergist, and sequestrant), the way in which its complexes with metal ions form has received a lot of attention. Because the naturally occurring iron is released from phytic acid salts (phytates), the addition to cereal products increases the solubility of the iron.

Ascorbic acid and other acidic substances can also form metal complexes. The ternary complex of ascorbic acid with oxygen and Fe^{3+} or Cu^{2+} ions, which is created when vitamin C is autoxidized, is of special interest. Metal ions can also be decreased by ascorbic acid. As a consequence of the reduction of Fe^{3+} ions to Fe^{2+} ions, hydroxyl radicals are produced. Ascorbic acid serves as a prooxidant in this reaction. Citrate succinate ≈ ascorbate ≈ malate > lactate is the decreasing order in which the stability of ferric complexes of aliphatic organic acids occurs.

Examples of aliphatic and aromatic carboxylic acid complexes with metal ions are shown in Figure 7.5 and in formulae (7–12) to (7–16). In a molar ratio of 1:1, aromatic 1,2-dicarboxylic acids and M^{2+} ions can form metal complexes (7–13). Alternatively, the metals can be coordinated with a ligand that is attached to two or more acid molecules (7–14).

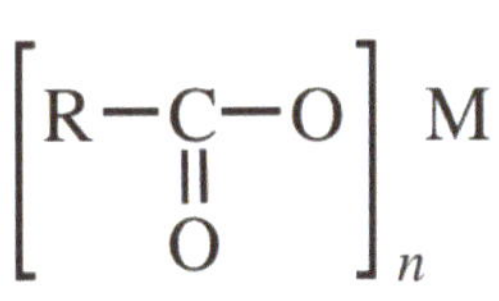

7–12, complex of metal (M) ion with aliphatic carboxylic acid

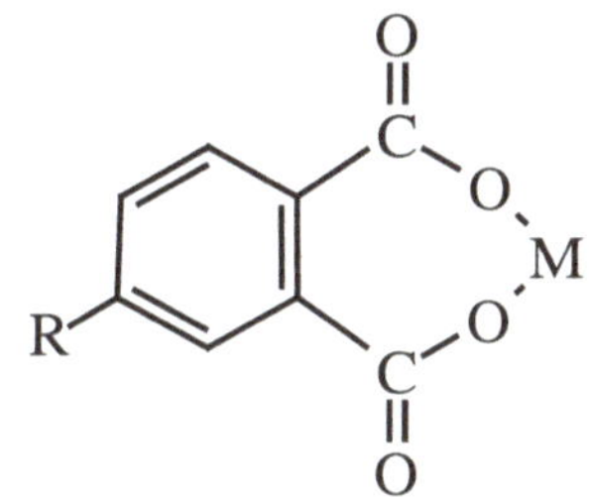

7–13, complex of metal (M) ion with aromatic dicarboxylic acid

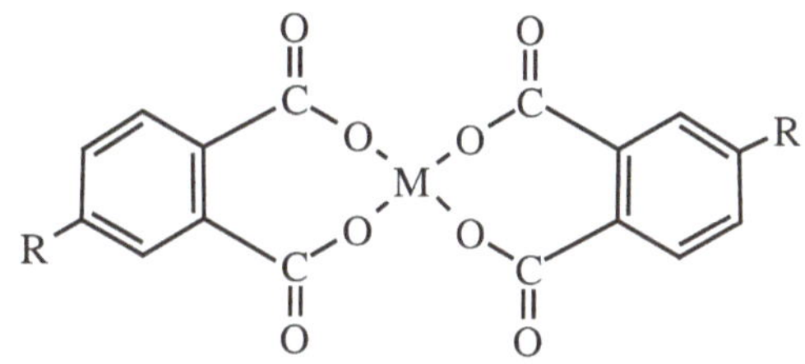

7–14, complex of metal (M) ion with two molecules of aromatic dicarboxylic acid

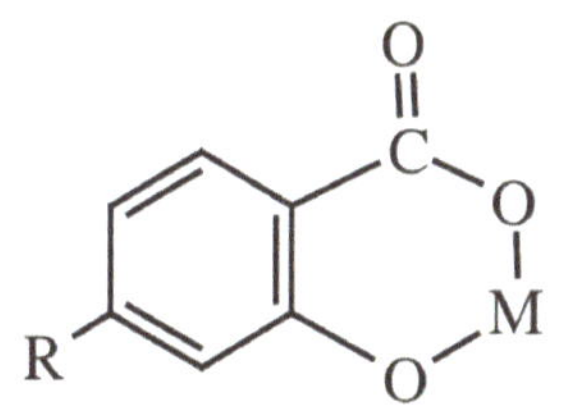

7–15, complex of metal (M) ion with an aromatic 2-hydroxycarboxylic acid (salicylic acid derivative)

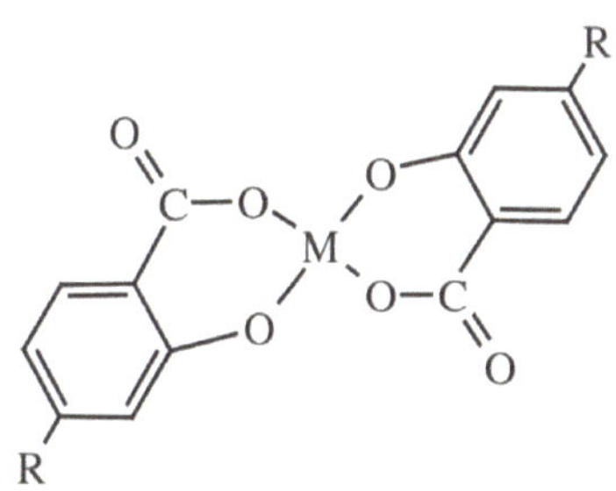

7–16, complex of metal (M) ion with two molecules of an aromatic

2-hydroxycarboxylic acid (salicylic acid derivative).

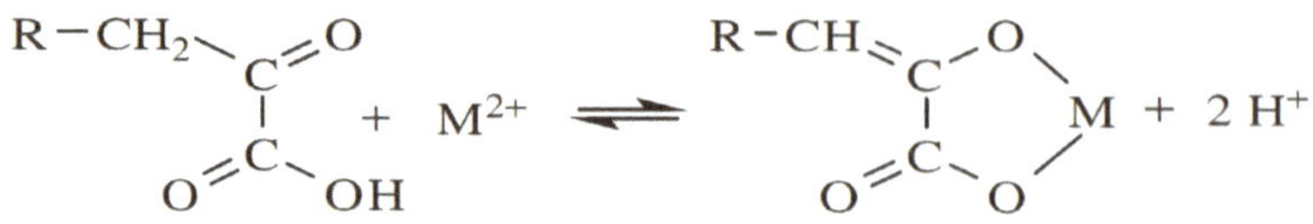

Figure 7.5. *Formation of metal (M) ion complex of 2-oxocarboxylic acid.*

Source: Jan Velıšek; The Chemistry of Food; ISBN 978–1-118–38381–0.

7.1.2.4. Flavonoids and Other Plant Phenols

Many aromatic compounds that have two adjacent carbon atoms in the aromatic ring bearing hydroxyl groups or hydroxyl and carbonyl groups can form complexes with the transition element ions. Many plant-based phenolic compounds share this structural motif. The antioxidant activity of phenols is connected to their reaction with metal ions. The highest antioxidant activity, for instance, is found in flavonoids containing two hydroxyl groups in positions C-3, and C-4,, a carbonyl group in position C-4, a free hydroxyl at position C-3, or two hydroxyls in positions C-3 and C-5. As flavonols are a prominent yellow pigment found in plants and are highly effective antioxidants, Figure 7.6 illustrates the reaction of metal ions with flavonols. For instance, it has been determined that the primary zinc ligands in tea infusions are flavonol glycosides. Catechins bind aluminum in tea leaves more strongly than chlorogenic acid and other phenolic compounds.

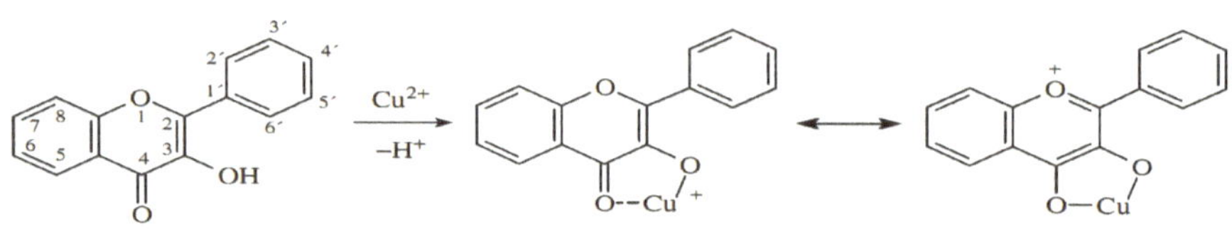

Figure 7.6. *Complex of flavonols with Cu ions.*

Source: Jan Velıšek; The Chemistry of Food; ISBN 978–1-118–38381–0.

7.1.2.5. Other Complexing Agents

Phosphorylated organic substrates can form salts with different metals and have acidic properties (e.g., phosphoric acid esters of sugars). Nucleotides and other compounds with heterocyclic nitrogen have even stronger binding properties than sugar phosphates. Adenosine 5'-triphosphate (ATP), for instance, combines with magnesium ions to form complexes (7–17).

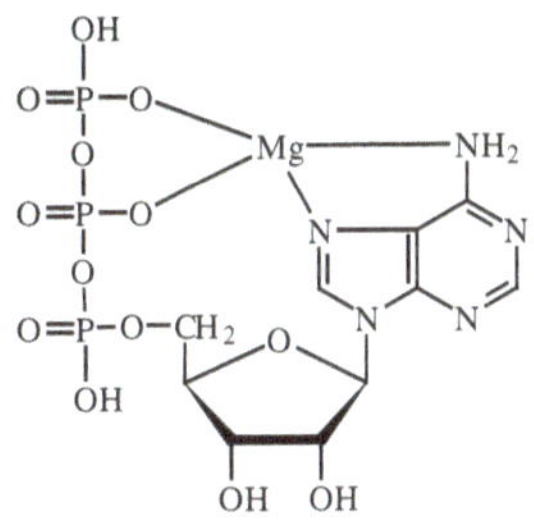

7–17, complex of magnesium with ATP

Similar to this, certain vitamins (like folic acid and riboflavin) and their byproducts (like riboflavin 5' -phosphate and flavoprotein FAD) have the ability to chelate because they contain donor heteroatoms in the right places. Compounds derived from isoalloxazine, like riboflavin, yield the chelates indicated in formula 7–18. Comparable pyranopterins with two thiol groups were found to be ligands of tungsten and molybdenum in cofactors of some metalloenzymes of bacteria and mammals, including xanthinoxidase found in cow's milk. The phosphorylated pyranopterin known as molybdopterin has an ene-dithiol group that coordinates molybdenum. Formula 7–19 shows the molybdopterin complex's structure (R stands for hydrogen or adenosine, and free valences of Mo are available for binding other pyranopterin molecules or other ligands).

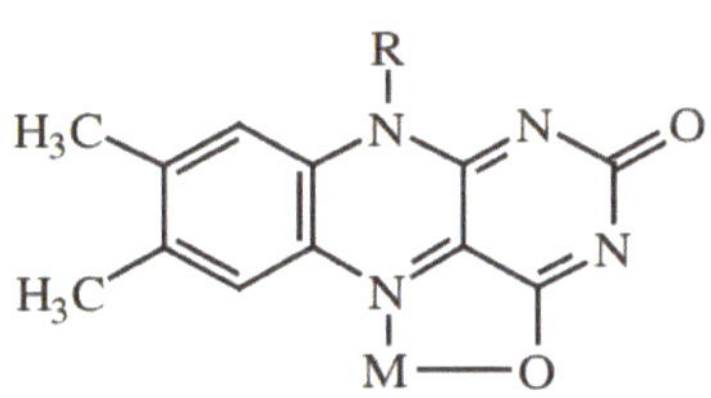

7–18, metal (M) complex of riboflavin and of
other isoalloxazine derivatives

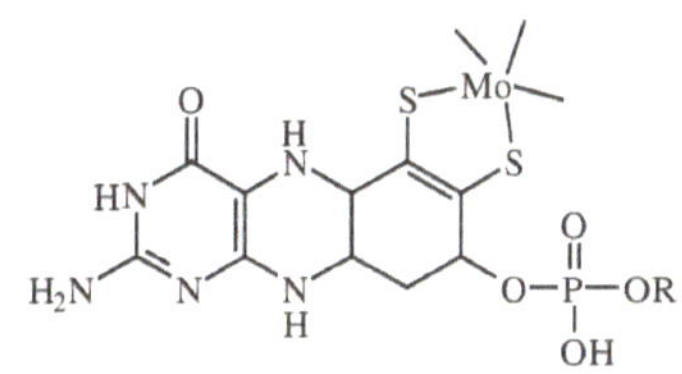

7–19, complex of molybdopterin with molybdene (R = H or adenosine)

Additionally, chelates can be created by reacting amino acids with nicotinic acid or amino acids with imines (such as the products of the pyridoxal reaction with amino acids) (7–20). Additionally, β-dicarbonyl compounds and metal ions form very stable chelates (Figure 7.7). Analytical chemistry has made extensive use of this characteristic.

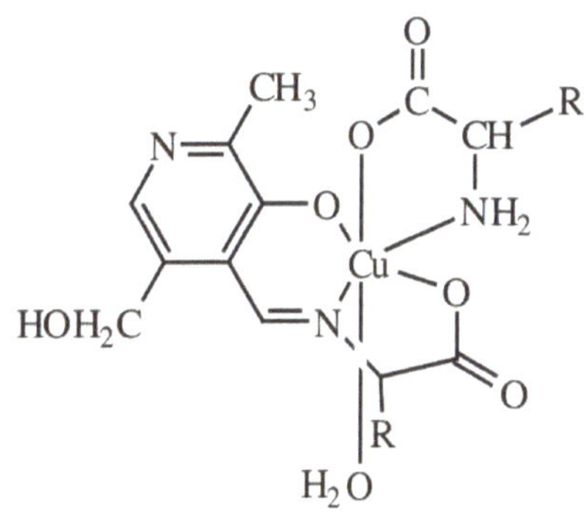

7–20, copper complex of pyridoxal imine with an amino acid
(R = amino acid residue)

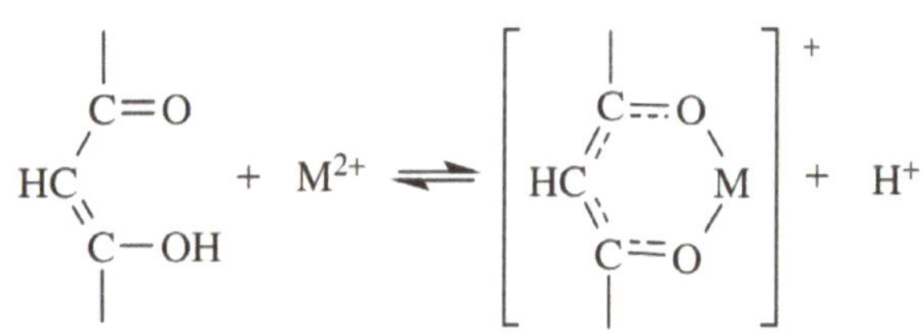

Figure 7.7. *Complex of metal ions (M) with β-diketones (enolized).*

Source: Jan Velišek; The Chemistry of Food; ISBN 978-1-118-38381-0.

7.2. NUTRITIONAL ASPECTS OF MINERALS

7.2.1. Essential Mineral Elements

There are various definitions of what it means to say that an element is "essential for life." A commonly recognized definition states that an element is considered necessary for life if it "results in a consistent and reproducible impairment of a physiological function" when removed from the diet or through another exposure pathway. Therefore, essentiality can be proven by giving humans or experimental animals diets deficient in a certain element and observing for symptoms of impaired function. The amount of essential minerals that humans need varies from a few micrograms to approximately one gram per day. Deficiency symptoms will appear if intakes are low for a while. On the other hand, toxicity may arise from overly high intakes. Thankfully, there is a broad range of acceptable and safe intake for most minerals, so if a varied diet is followed, deficiencies or toxicity are not common. Figure 7.8 provides an illustration of this idea. Because higher organisms have homeostatic mechanisms to deal with both low and high exposures to essential nutrients, a broad range of safe and adequate intakes are possible (Dallman, 1990).

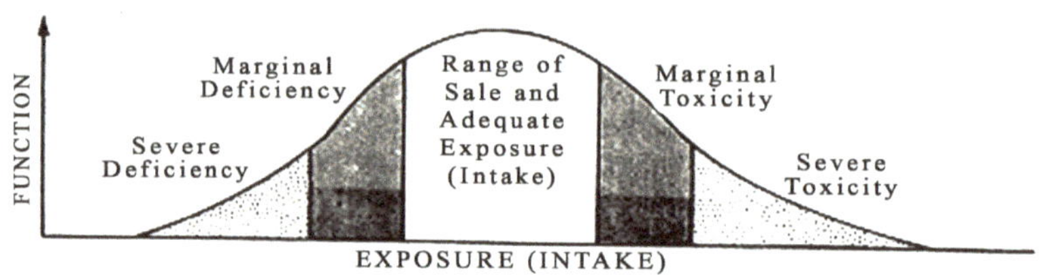

Figure 7.8. *Dose-response relationship between mineral intake (dose) and an associated physiological function. Function is impaired at both excessively low and excessively high intakes of most mineral nutrients.*

Source: Owen R. Fennema; Food Chemistry Third Edition; ISBN 0-8247-9691-8.

The mechanisms by which an organism keeps the amounts of nutrients in its tissues within a specific, continuous range are referred to as homeostasis. Higher organisms' homeostasis is maintained by a highly intricate web of mechanisms that control nutrient absorption, excretion, metabolism, and storage. Nutrient intakes would need to be strictly regulated in the absence of homeostatic mechanisms in order to avoid toxicity or deficiency.

When dietary levels are abnormally low or high for prolonged periods of time, homeostasis may be disrupted. It is not unusual for populations living in poverty, where access to a wide range of foods is frequently restricted, to have persistently low intakes of mineral nutrients. Although there has been a lot of discussion recently regarding potential connections between high dietary levels of iron and chronic diseases, toxicities resulting from excessive mineral intakes are less common.

7.2.2. Recommended Dietary Allowances for Mineral Nutrients (U.S.)

"The levels of intake of essential nutrients that, on the basis of scientific knowledge, are judged by the Food and Nutrition Board to be adequate to meet the known nutrient needs of practically all healthy persons" is the definition of recommended dietary allowances, or RDAs.

Estimating the requirement for the absorbed nutrient, accounting for incomplete utilization of the ingested nutrient, and adding a safety factor to account for individual variability are the steps involved in establishing RDAs. RDA values, therefore, are typically two standard deviations above the mean, which is greater than the required value.

This implies that people who consume less than the recommended daily allowance of certain nutrients do not always do so insufficiently. However, the likelihood that some people will experience a deficiency rises when a sizable portion of the population consumes significantly less than the RDA over an extended period of time. Only 7 of the 20 known essential minerals have RDAs published by the RDA committee (Table 7.3). An RDA is only created when there is enough information to calculate a trustworthy number. When there is insufficient data to support an RDA, the RDA Committee has nevertheless published "estimated minimum requirements" for a number of minerals. According to Table 7.4, these are potassium, sodium, and chloride. For copper, manganese, fluoride, chromium, and molybdenum, "Estimated Safe and Adequate Intakes" have been published (Table 7.5).

Deficiencies of the minerals listed in Table 7.4 and 7.5 are rare in the United States. However, for many people, intakes of sodium tend to be higher than desirable for optimal health. Epidemiological data show that blood pressure is positively associated with level of salt intake. Populations with salt intakes greater than 6 g/day have progressively increasing blood pressure with age, while blood pressure rise with age is not seen in populations with salt intakes below 4.5 g/day. Thus, dietary guidelines for the prevention of chronic disease usually include the recommendation that intakes of salt be less than 6 g/day. Sodium chloride is 40% sodium by weight, so 6 g salt provides 2,400 mg sodium.

Table 7.3. Recommended Dietary Allowances (RDAs) for the Mineral Nutrients

Category	Age (years) or condition	Calcium (mg)	Phosphorus (mg)	Magnesium (mg)	Iron (mg)	Zinc (mg)	Iodine (µg)	Seleniu (mg)
Infants	0.0–0.5	400	300	40	6	5	40	10
	0.5–1.0	600	500	60	10	5	50	15
Children	1–3	800	800	80	10	10	70	20
	4–6	800	800	120	10	10	90	20
	7–10	800	800	170	10	10	120	30
Males	11–14	1200	1200	270	12	15	150	40
	15–18	1200	1200	400	12	15	150	50
	19–24	1200	1200	350	10	15	150	70
	25–50	800	800	350	10	15	150	70
	51+	800	800	350	10	15	150	70
Females	11–14	1200	1200	280	15	12	150	45
	15–18	1200	1200	300	15	12	150	50
	19–24	1200	1200	280	15	12	150	55
	25–50	800	800	280	15	12	150	55
	51+	800	800	280	10	12	150	55
Pregnant		1200	1200	320	30	15	175	65
Lactating	1st 6 months	1200	1200	355	15	19	200	75
	2nd 6 months	1200	1200	340	15	16	200	75

Source: Owen R. Fennema; Food Chemistry Third Edition; ISBN 0-8247-9691-8.

Table 7.4. Estimated Minimum Requirements for Electrolytes

Age	Weight (kg)	Sodium (mg)	Chloride (mg)	Potassium (mg)
Months				
0.–5	4.5	120	180	500
6–11	8.9	200	300	700
Years				
1	11.0	225	350	1000
2–5	16.0	300	500	1400
6–9	25.0	400	600	1600
10–18	50.0	500	750	2000
>18	70.0	500	50	2000

Source: Owen R. Fennema; Food Chemistry Third Edition; ISBN 0-8247-9691-8.

Table 7.5. Estimated Safe and Adequate Daily Dietary Intakes of Selected Minerals

Category	Age (years)	Cu (mg)	Mn (mg)	F (mg)	Cr (µg)	Mo (µg)
Infants	0.–0.5	0.4–0.6	0.3–0.6	0.1–0.5	10–40	15–30
	0.5–1	0.6–0.7	0.6–1.0	0.2–1.0	20–60	20–40
Children and adolescents	1–3	0.7–1.0	1.0–1.5	0.5–1.5	20–80	25–50
	4–6	1.0–1.5	1.5–2.0	1.0–2.5	30–120	30–75
	7–10	1.0–2.0	2.0–3.0	1.5–2.5	50–200	50–150
	11+	1.5–2.5	2.0–5.0	1.5–2.5	50–200	5–250
Adults		1.5–3.0	2.0–5.0	1.5–4.0	50–200	75–250

Source: Owen R. Fennema; Food Chemistry Third Edition; ISBN 0-8247-9691-8.

7.2.3. Bioavailability

For the past century or more, it has been understood that the amount of a nutrient present in a food does not always indicate how valuable the food is as a source of that nutrient. This prompted the development of the idea of nutrient bioavailability by nutritionists. The amount of a nutrient in food that can be used in metabolic processes after ingestion is known as its bioavailability. When it comes to mineral nutrients, the effectiveness of absorption from the intestinal lumen into the blood is what essentially determines bioavailability. However, sometimes nutrients that are absorbed come in a form that is not useful. As an illustration, certain chelates contain iron so tightly that, even if the chelate is absorbed, the iron will not be released to cells for incorporation into iron proteins; instead, the intact chelate will be eliminated through the urine.

The bioavailabilities of mineral nutrients range from more than 90% for sodium and potassium to less than 1% for certain forms of iron. Given that a number of variables interact to determine a nutrient's final bioavailability, there are numerous and intricate explanations for this broad range.

7.2.3.1. Nutritional Utilization of Minerals

Food is ground in the mouth, and salivary amylase starts to break down carbohydrates. Just minor alterations in mineral species take place at this point. After that, the meal is swallowed and goes into the stomach, where gastric acid gradually lowers the pH to roughly 2. At this point, mineral species undergo significant changes. Protein denaturation and hydrolysis, as well as the changed pH, can alter the stabilization of complexes. Minerals have the ability to form complexes with various ligands when they are released into solution. Furthermore, when the pH is lowered, transition metals like iron could experience a valance change. Iron's redox behavior is highly pH dependent. Ferric iron will not be reduced at neutral pH, even in the presence of excess reducing agents like ascorbic acid. But as the pH drops, ascorbic acid quickly converts Fe^{3+} to Fe^{2+}. Since Fe^{2+} is less ligand-affinity than Fe^{3+}, this reduction will encourage the release of iron from food complexes. The partially digested food in the stomach is emptied into the proximal small intestine during the next stage of digestion. Here, pancreatic secretions containing digestive enzymes and sodium bicarbonate raise the pH and carry out the digestion of proteins and carbohydrates. Furthermore, lipases start breaking down triacylglycerols. More ligands are created during digestion, and those that already exist undergo changes that surely impact how affinitous they are for metal ions. As a result, the small intestine's lumen undergoes additional mineral species reorganization that produces a complex mixture of soluble and insoluble, high and low molecular weight species. It is possible for soluble species to spread to the intestinal mucosa's brush border surface, where they can either be absorbed by the mucosal cell or move between cells (the paracellular route). A membrane carrier or an ion channel may help with uptake, which can also be saturable, an active process requiring energy, and subject to physiological regulation.

7.2.3.2. Iron Bioavailability

The effectiveness of iron absorption in the gut determines iron bioavailability nearly entirely. The amount of iron absorbed depends on the total amount of iron consumed, the makeup of the diet, and the iron status of the person consuming the diet. Industrialized nations such as the United States regularly provide roughly 6 mg of iron per 1000 kcal (4186 kJ) in their diets. Food-borne iron species can be broadly classified as heme or nonheme. Heme iron is tightly bound to a porphyrin ring in the middle, and it doesn't separate from this ligand until intestinal mucosal cells have absorbed it. It is mainly found in meat, poultry, and fish because it occurs as hemoglobin or myoglobin. Nonheme iron makes up almost all of the iron in plant-based diets and between 40 and 60% of the iron in animal tissues. It is mainly bound to proteins, but other ligands such as citrate, phytate, oxalate, or polyphenolics can also form complexes with it.

The diet's composition generally has little effect on the bioavailability of heme iron, which is typically much higher than that of nonheme iron. The nutritional makeup of an individual has a significant impact on the bioavailability of nonheme iron. It is commonly believed that during digestion, nonheme iron from all sources—foods and iron from fortification—enters a common pool, and that the amount of ligands present in the small intestine at the absorption site determines how much iron is absorbed from this pool. There are a number of known enhancers and inhibitors of nonheme iron absorption. Meat, poultry, fish, ascorbic acid, and EDTA (in diets with low bioavailabilities) are examples of enhancers. Polyphenolics (tannins found in tea, legumes, and sorghum), phytates (found in whole-grain cereals and legumes), some plant proteins (particularly those found in legumes), calcium, and phosphates are examples of inhibitors. The intricate interactions between the enhancers and inhibitors in a diet determine the total bioavailability of iron. Even in individuals with low iron status, the amount of iron absorbed from diets high in roots, tubers, legumes, and cereals and low in meat and ascorbic acid may only be approximately 5%. Only roughly 0.7 mg of absorbable iron would be provided by such a diet each day, which is insufficient to meet the needs of many people. Diets rich in roots, cereals, legumes, along with some meat, poultry, or fish, and foods high in ascorbic acid, may absorb about 10% of the iron in the diet. Around 1.4 mg of absorbable iron per day is provided by these diets, which is sufficient for the majority of men and postmenopausal women but insufficient for up to 50% of women who are childbearing age. Over 2 mg of absorbable iron per day can be obtained from diets rich in meat, poultry, fish, and foods high in ascorbic acid. This is more than enough to meet the needs of almost all healthy individuals.

7.2.4. Minerals of Particular Nutritional Concern

Deficits in certain mineral elements are uncommon or nonexistent while they are common for others due to a variety of factors. Furthermore, the prevalence of particular deficiencies varies greatly between socioeconomic and geographic categories. There have been reports of dietary deficiencies in humans for calcium, iron, zinc, chromium, cobalt (also known as vitamin B_{12}), and iodine. Foods contain bound forms of calcium, chromium, iron, and zinc, and depending on the meal's composition, the bioavailabilities may be low. Thus, low intake and poor bioavailability combine to cause deficiencies of these minerals. Foods and water contain iodine mostly in its ionic, unbound form, which has a high bioavailability. Low intake is the main cause of iodine deficiency. Foods contain selenium as the chelate selenomethionine, but because it is effectively used, low intakes of the mineral lead to deficiencies. Only those who follow strict vegetarian diets, which are low in this vitamin, are at risk for vitamin B_{12} deficiency. These findings highlight the intricacies of the mineral bioavailability narrative. While some bound forms are highly bioavailable, others have low bioavailability. In general, unbound forms exhibit high bioavailability. Figure 7.9 provides an overview of current perspectives on mineral deficiencies and bioavailability.

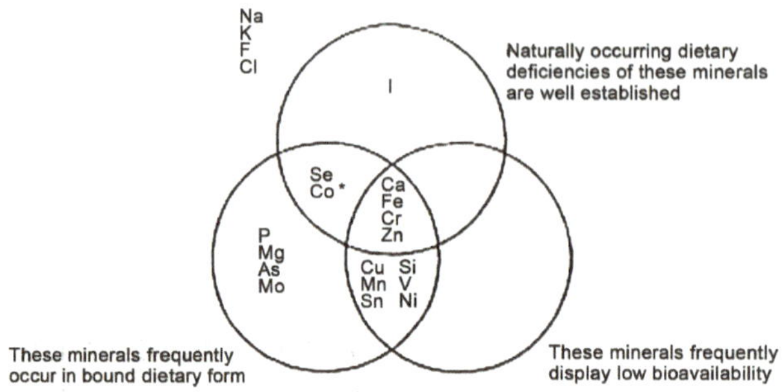

Figure 7.9. *Essential minerals grouped by chemical form (free ions in solution or bound to food ligands), bioavailability, and occurrence of nutritional deficiency. As vitamin B*[12].

Source: Owen R. Fennema; Food Chemistry Third Edition; ISBN 0–8247–9691–8.

7.3. MINERAL COMPOSITION OF FOODS

7.3.1. Ash

Nutrient databases list "ash" as one of the proximate ingredients in food items. It offers a rough estimate of the total amount of minerals found in food. Publications from government agencies provide methods for calculating the amount of ash in particular foods and food groups. Metal oxides, sulfates, phosphates, nitrates, chlorides, and other halides are among the minerals found in ash. Because oxygen is present in many of the anions, ash content thus significantly overestimates total mineral content. Nonetheless, it gives a rough indication of the mineral content and is necessary to determine the total amount of carbohydrates in the proximate analysis scheme (Harbers, 1994).

7.3.2. Individual Minerals

Foods can be ash-treated, the ash dissolved (often in acid), and the concentrations of the individual minerals in the resulting solution can be measured. Mineral concentrations can be measured using both chemical and instrumental techniques, although instrumental techniques are typically faster, more accurate, and more precise. Since it became available in the 1960s, atomic absorption spectroscopy is still frequently used. Although it measures only one mineral at a time, the method is dependable. Since they can measure multiple mineral elements at once from a single sample, inductively coupled plasma spectrometers have become more and more common. The amounts of specific minerals in a variety of foods are listed in Table 7.6. A few sources are the U. S. Journal articles, manufacturer data, and the Department of Agriculture database. Values are averages; reported data may differ significantly from data obtained from specific foods.

Table 7.6. Mineral Composition of Selected Foods

Quantity	Food	Weight (g)	kcal[2]	Ca	Mg	P	Na	K	Fe	Zn	Cu	Se
1 egg	Scrambled	100	157	57	13	269	290	138	2.1	2.0	0.06	8
1 slice	White bread	28	75	35	6	30	144	31	0.8	0.2	0.04	8
1 slice	Whole wheat bread	28	70	20	26	74	180	50	1.5	1.0	0.10	16
0.5 cup	Spaghetti, cooked without salt	70	99	5	13	38	1	22	1.0	0.4	0.07	19.0
0.5 cup	Brown rice, cooked	98	108	10	42	81	5	42	0.4	0.6	0.01	13.0
0.5 cup	White rice, parboiled, cooked	88	100	17	11	37	3	32	1.0	0.3	0.08	8.30
0.5 cup	Black beans, cooked	86	113	24	61	120	1	305	2.0	1.0	0.18	6.9
0.5 cup	Red kidney beans	89	112	25	40	126	2	356	3.0	0.9	0.21	1.9
1 cup	Whole milk	244	150	291	33	228	120	370	0.1	0.9	0.05	3.0
1 cup	Skim milk/nonfat milk	245	86	302	28	247	126	406	0.1	0.9	0.05	6.6
1.5 oz.	American cheese, processed	43	159	261	10	316	608	69	0.2	1.3	0.01	3.8
1.5 cup	Cheddar cheese	43	171	305	12	219	264	42	0.3	1.3	0.01	6.0
0.5 cup	Cottage cheese, creamed, small curd	105	108	63	6	139	425	89	0.1	0.4	0.03	6.3
1 cup	Yogurt, low-fat, plain	227	144	415	10	326	159	531	0.2	2.0	0.10	5.5
0.5 cup	Ice cream, regular vanilla	67	134	88	9	67	58	128	0.1	0.7	0.01	4.7
1 each	Baked potato with skin	202	220	20	55	115	16	844	2.8	0.7	0.62	1.8
1 each	Peeled potato, boiled	135	116	10	26	54	7	443	0.4	0.4	0.23	1.2
3 each	Broccoli, raw spears	453	126	216	114	297	123	1470	4.0	2.0	0.40	0.9
3 each	Broccoli spear, cooked from fresh	540	151	249	130	318	141	1575	4.5	2.1	0.23	1.1
0.5 cup	Raw carrot, grated	55	24	15	8	24	19	178	0.3	0.1	0.03	0.8
0.5 cup 1	Cooked carrots, from frozen Tomato,	73	26	21	7	19	43	115	0.4	0.2	0.05	0.9
each 0.75	fresh, whole, average Tomato juice,	123	26	6	14	30	11	273	0.6	0.1	0.09	0.6
cup 0.75	canned	183	31	17	20	35	661	403	1.0	0.3	0.18	0.4
cup	Orange juice prepared from frozen	187	83	17	18	30	2	356	0.2	0.1	0.08	0.4
1 each	Orange, average, 2 5/8 in diameter	131	60	52	13	18	0	237	0.1	0.1	0.06	1.2
1 each	Apple with peel, 2.75 in diameter	138	80	10	6	10	1	159	0.3	0.1	0.06	0.6
1 each	Banana (peeled weight)	114	85	7	32	22	1	451	0.4	0.2	0.12	1.1
3 oz.	Beef round, roasted	85	205	5	21	176	50	305	1.6	3.7	0.08	–
3 oz.	Veal, round, roasted	85	160	6	28	234	68	389	0.9	3.0	0.13	–
3 oz.	Chicken, white meat, roasted	85	140	13	25	194	63	218	0.9	0.8	0.04	–
3 oz.	Chicken, leg meat, roasted Salmon,	85	162	10	20	156	77	206	1.1	2.4	0.07	–
3 oz.	cooked	85	183	6	26	234	56	319	0.5	0.4	0.06	–
3 oz.	Salmon, canned, with bones	85	130	203	25	277	458	231	0.9	0.9	0.07	–

[2]kcal × 4.186 = kJ.

Source: Owen R. Fennema; Food Chemistry Third Edition; ISBN 0-8247-9691-8.

7.3.3. Factors Affecting the Mineral Composition of Foods

Many factors interact to affect the mineral composition of foods, so compositions can vary greatly.

7.3.3.1. Factors Affecting the Mineral Composition of Plant Foods

Plants require water and vital mineral nutrients from the soil in order to grow. Nutrients are transferred to different sections of the plant by the roots. The final makeup of a plant's edible portions is determined and impacted by the soil's fertility, the plant's genetic makeup, and its growing environment. Wheat grain serves as an example of how mineral content can differ significantly even within a plant species. Zinc concentrations for grain grown in Australia, North America, and the United Kingdom range from 4 to 37 mg/kg, while iron concentrations range from 23 to 74 mg/kg.

7.3.3.2. Adequacy of Plant Foods for Supplying the Nutrient Needs of Animals and Humans

There are several relevant questions. The list of essential minerals for humans and plants is similar, but not identical. Do humans and plants require the same mineral nutrients? Can plants' mineral concentrations meet human needs? Can plants' mineral concentrations be changed through genetics or agriculture to improve their nutritional quality? Do plants grown on depleted soils have lower nutritional value than plants grown on more fertile soils? While not necessary for most plants, F, Se, and I are for humans. Therefore, in populations that rely on locally grown plants where soil concentrations of these elements are low, we might anticipate human deficiencies of these elements. Indeed, there are severe iodine and selenium shortages in humans in a number of places across the globe. Given that plant foods will inevitably contain the nutrients needed by both plants and animals, we might anticipate that human deficiencies in these nutrients would be less of an issue. Regretfully, human needs may not always be satisfied by the concentrations of minerals in plants, or the minerals may exist in forms that are inefficient for human use. These circumstances relate to iron and calcium, respectively. There are certain plants that have incredibly little calcium. For instance, rice only has roughly 10 mg of calcium per 100 kcal. As a result, people who eat a diet high in rice must rely on other foods to meet their calcium needs. Iron is found in plant foods more evenly than calcium is, but because of its extremely low bioavailability, diets high in cereals and legumes are frequently deficient in iron. Although agronomic techniques and plant breeding can sometimes improve the nutritional value of crops, the transfer of mineral nutrients from the soil to the plant and from the plant to the animal or human is a very intricate process. Soils have a wide range of mineral compositions. Furthermore, because the chemical form of an element and the pH of the soil have a significant impact on the mineral bioavailability to plants, the concentration of an element in the soil may not be a reliable indicator of the amount that can be absorbed by plant roots. For instance, adding lime to the soil will raise the pH and increase the availability of molybdenum and selenium to plants while decreasing their availability of iron, zinc, manganese, and nickel. Furthermore, plants typically have physiological systems in place to control how much nutrient uptake occurs from the soil. Consequently, it stands to reason that attempts to change the mineral makeup of food crops would yield contradictory outcomes. For instance, adding fertilizer does not considerably raise the amount of calcium, iron, or manganese in food crops. However, it has been demonstrated that fertilizing with zinc at levels higher than what the plant needs will raise the zinc content of pea seeds.

7.3.3.3. Factors Affecting the Mineral Composition of Animal Foods

Compared to plant foods, animal foods have less variation in mineral concentrations. Mineral concentrations in meat, milk, and eggs are generally not significantly affected by changes in the animal's diet. This is because homeostatic mechanisms operating in the animal regulate tissue concentrations of essential nutrients.

7.3.3.4. Adequacy of Animal Foods for Supplying the Nutrient Needs of Humans

Animal foods should be a good source of nutrients because animal tissues and human tissues are similar in composition. Iron, zinc, phosphate, and cobalt (as vitamin B_{12}) are all found in meat, poultry, and fish. These products don't provide enough calcium unless you eat bones, which isn't the case in most cases. Furthermore, animal foods might not contain much iodine. Products made from dairy are a great way to get calcium. To ensure adequate intake of all essential minerals, it is therefore best to consume a variety of plant foods in addition to a variety of animal foods.

7.3.4. Effects of Processing

Mineral elements are not oxidized by heat, light, oxidizing agents, pH extremes, or other conditions that affect organic nutrients, unlike vitamins and amino acids. Minerals are essentially unbreakable. It is possible to extract minerals from food through physical separation or leaching, though. Additionally, the previously listed factors may change the bioavailabilities of minerals. Cereal milling is the main factor responsible for food-borne mineral loss. Grain kernels typically contain concentrated amounts of mineral elements in the bran layers and the germ.

Thus, after bran and germ are removed, only pure endosperm—which is low in minerals—remains. Table 7.7 displays the mineral contents of whole wheat, white flour, wheat bran, and wheat germ. Rice and other cereals are milled with comparable losses. These are significant losses. Iron is the only mineral that is frequently added to milled products in the United States for fortification.

The conditions under which cheese is manufactured can have a significant impact on its calcium retention. When whey is drained from cheeses with low pH, significant calcium losses take place. The calcium content of different cheeses is displayed in Table 7.8. Both a Ca:protein ratio and milligrams per 100 grams of cheese are used to express compositions.

Table 7.7. Concentrations of Selected Trace Minerals in Wheat and Milled Wheat Products

Mineral	Whole heat	White flour	Wheat germ	Millfeeds (bran)	Loss from wheat to flour (%)
Iron	43	10.5	67	47–78	76
Zinc	35	8	101	54–130	78
Manganese	46	6.5	137	64–119	86
Copper	5	2	7	7–17	68
Selenium	0.6	0.5	1.1	0.5–0.8	16

Source: Owen R. Fennema; Food Chemistry Third Edition; ISBN 0-8247-9691-8.

Since different varieties of cheeses have different water contents, the latter expression provides a more accurate comparison of calcium losses. Because cottage cheese usually has a pH of less than 5 when the whey is removed, it has the lowest concentration of calcium.

Table 7.8. Protein, Calcium, and Phosphate Contents of Selected Cheeses

Cheese variety	Protein (%)	Ca (mg/100 g)	Ca:protein (mg:g)	PO_4 (mg/100 g)	PO_4:protein (mg:g)
Cottage	15.2	80	5.4	90	16.7
Cheddar	25.4	800	31.5	860	27.3
Emmenthal	27.9	920	33.1	980	29.6

Source: Owen R. Fennema; Food Chemistry Third Edition; ISBN 0-8247-9691-8.

In cheddar and Emmenthal cheeses, the whey is normally drained at pH 6.1 and 6.5, respectively, and this causes the calcium content of Emmenthal cheese to be greater than that of cheddar. It makes sense to assume that cooking in water would cause some mineral losses because many minerals do dissolve significantly in water. Regretfully, not much controlled research has been carried out. Vegetables lose more minerals when they are boiled in water than when they are steamed. When pasta is cooked, very little iron is lost, but potassium is lost in excess of 50%. This makes sense because iron is bound to proteins and other high-molecular-weight ligands in food, whereas potassium is present in food as a free ion.

CASE STUDY

THE IMPACT OF BIOSTIMULANTS ON THE GROWTH OF SELENIUM-ENRICHED WHEAT

The essential trace mineral, selenium, is of fundamental importance to human health. It plays an important role in many biological processes such as guarding against oxidants, the formation of thyroid hormones, DNA synthesis, fertility and reproduction. At present, insufficient levels of Se in the diet affects up to 14% of the population worldwide, bringing with it the associated risk of developing chronic degenerative diseases. Plants can transform naturally occurring inorganic Se species (present in soil) into organic Se species – the desired form of selenium in the human diet. Effectively, the Se level in soil usually has a direct influence on the concentration of Se present in food and subsequently in the human body. However, high Se concentration in soil induces stress in the plant and may hamper its normal development.

1. **The Challenge**

As with other nourishing minerals, food bio-fortified with selenium is becoming increasingly common; however, introducing additional Se into the plant can induce toxicity. To solve this issue, plant biostimulants have been used to enhance nutritional efficiency, and address abiotic stress tolerance and crop quality. Using a plant biostimulant we can significantly reduce the stress applied to the plants. It was therefore critical to understand the effect that the biostimulant has on Se-induced toxicity at different growing stages, to determine the optimal conditions when Se should be added as a supplement.

2. **The Solution**

Researchers from the Universitat Autònoma de Barcelona used I18 at Diamond to study the spatial distribution of Se, as well as Zn, Cu, Fe, K and Mn used in plant metabolism and to determine the chemical state of Se at different growing stages. Combined microfocus X-ray fluroescence (μ-XRF) and microfocus X-ray absorption near edge structure (μ-XANES) measurements on grain sections showed that during both growing stages (tillering and heading) the organic Se species found in wheat grain were co-located with minerals in the outer layer and embryo parts of the grain. These parts of the grain contain a higher concentration of proteins. The proportion of organic Se species was at its highest at the tillering stage (>90%), whereas during the heading stage, the Se content was reduced to around 80%.

3. **The Benefits**

This research showed that the chosen biostimulant played a key role in the enhancement of wheat grains and their dry biomass without hindering the Se enrichment process. This knowledge is crucial to reduce both plant toxicity and the cost of producing plants with a healthy amount of selenium.

CLASS ACTIVITY

1. **Objective:** Students will investigate and identify various minerals based on their physical properties and uses.
2. **Materials:**

- Mineral samples (e.g., quartz, feldspar, mica, calcite, gypsum, halite, etc.)
- Streak plates
- Magnifying glasses
- Mohs hardness scale kits
- Vinegar (for acid test)
- Magnet
- Data sheets for recording observations
- Identification keys

3. **Hands-on Identification:**

Have students perform the following tests on each mineral sample:

- **Color:** Observe and record the color of the mineral.
- **Luster:** Determine if the mineral has a metallic or non-metallic luster.
- **Streak:** Rub the mineral on a streak plate and observe the color of the streak.
- **Hardness:** Use the Mohs hardness scale to determine the mineral's hardness by scratching it with objects of known hardness (e.g., fingernail, copper penny, glass).
- **Cleavage and Fracture:** Observe how the mineral breaks – does it have flat, even surfaces (cleavage) or irregular surfaces (fracture)?
- **Acid Test:** Drop a small amount of vinegar on the mineral to see if it fizzes (indicating the presence of carbonate).
- **Magnetism:** Test if the mineral is magnetic by using a magnet.
- **Other Properties**: Note any other distinguishing features (e.g., specific gravity, taste for halite).

Divide students into small groups and assign each group the task of designing a segment of a highway or road. They should consider factors such as horizontal and vertical alignment, sight distance, superelevation, and lane width.

Objective: To understand the principles of road design and the factors that influence it.

Materials: Graph paper, rulers, compasses, calculators, and design guidelines.

SUMMARY

- Minerals in food (dietary minerals, also known as mineral nutrients) are the chemical elements required by living organisms (other than carbon, hydrogen, nitrogen and oxygen present in common organic molecules) that are usually defined as the elements contained in the ash of a food, or more accurately as the elements that remain after the complete oxidation of the organic fraction of food to carbon dioxide, water and other oxidation products. The mineral fraction for most foods accounts for 0.5–3% by weight.
- Mineral substances found in food interact with water, with the organic matter present and with one another. These interactions affect the bioavailability of elements in the diet. The chemical state of the element in the food is determined by food composition, pH, the possibility of hydration of the metal ions, redox potential of the system and the related possibility of changing the degree of oxidation of the element and other factors.
- Majority mineral elements Or quantity elements, formerly referred to as macroelements, which occur in food in larger amounts, usually in hundredths to units of weight percent (hundreds to 10,000 thousands mg/kg) and include alkali metals sodium (Na) and potassium (K), alkaline earth metals magnesium (Mg) and calcium (Ca), halogen chlorine (Cl) and non-metals phosphorus (P) and sulfur (S).
- Minority mineral elements are found in foods in smaller amounts, representing the tens to hundreds mg/kg; a transition between the majority elements and trace elements; this category usually includes transition metals iron (Fe) and zinc (Zn).
- Binding of metals in biological materials is largely mediated by proteins. This binding of the metal ions by the protein molecule proceeds by a similar mechanisms to that for the amino acids and peptides. The functional groups of

proteins that are involved in metal binding are not necessarily the amino acid residues that are adjacent in the polypeptide chains.

- Aliphatic mono- and dicarboxylic acids (such as acetic acid and oxalic acid, respectively), oxocarboxylic acids (such as pyruvic acid), hydroxycarboxylic acids (citric, tartaric, malic and lactic acids) and numerous aromatic acids (such as benzoic acid) are normal constituents of plant and animal foods.
- Recommended dietary allowances (RDAs) are defined as "the levels of intake of essential nutrients that, on the basis of scientific knowledge, are judged by the Food and Nutrition Board to be adequate to meet the known nutrient needs of practically all healthy persons."

REVIEW QUESTIONS

1. What are the key chemical properties that determine the bioavailability of minerals in the body?
2. How do minerals interact with organic molecules such as proteins and carbohydrates during digestion and absorption?
3. Discuss the role of minerals in stabilizing protein structures. Provide examples of minerals involved and their mechanisms.
4. List and briefly describe the functions of five essential mineral elements crucial for human health.
5. Compare and contrast the absorption mechanisms of iron and calcium in the digestive system.
6. Analyze the mineral composition of a common food item (e.g., spinach, milk, salmon). Highlight the major minerals present and their nutritional significance.

REFERENCES

1. Allen, R. R. (1987). Theory of hydrogenation and isomerization. In R. Hastert (Ed.), *Hydrogenation: Proceedings of an AOCS Colloquium* (pp. 1–10). American Oil Chemists' Society, Champaign, IL.
2. Barrett, F., & Ranum, P. (1985). Wheat and blended cereal foods. In F. M. Clydesdale & K. L. Wiemer (Eds.), *Iron Fortification of Foods* (pp. 75–109). Academic Press, Orlando, FL.
3. Batten, G. D. (1994). Concentrations of elements in wheat grains grown in Australia, North America, and the United Kingdom. *Australian Journal of Experimental Agriculture, 34*, 51–56.
4. Chrichton, R. R. (1991). Inorganic biochemistry of iron metabolism. In J. Burgess (Ed.), *Ellis Horwood Series in Inorganic Chemistry*. Ellis Horwood.
5. Dallman, P. R. (1990). Iron. In M. L. Brown (Ed.), *Present Knowledge in Nutrition* (pp. 241–250). International Life Sciences Institute, Washington, DC.

6. Guinee, T. P., Pudja, P. D., & Farkye, N. Y. (1993). Fresh acid-curd cheese varieties. In P. F. Fox (Ed.), *Cheese: Chemistry, Physics and Microbiology, vol. 2, 2nd ed.* (pp. 369–371). Chapman & Hall, London.
7. Harbers, L. H. (1994). Ash analysis. In S. S. Nielson (Ed.), *Introduction to the Chemical Analysis of Foods* (pp. 113–121). Jones and Bartlett, Boston.
8. Hazell, T. (1985). Minerals in foods: Dietary sources, chemical forms, interactions, bioavailability. *World Review of Nutrition and Dietetics, 46*, 1–123.
9. Heaney, R. P., Weaver, C. M., & Fitzsimmons, M. L. (1990). Influence of calcium load on absorption fraction. *Journal of Bone and Mineral Research, 5*, 1135–1138.
10. Hendricks, D. G. (1994). Mineral analysis by traditional methods. In S. S. Nielson (Ed.), *Introduction to the Chemical Analysis of Foods* (pp. 123–135). Jones and Bartlett, Boston.
11. Hurrell, R. F. (1985). Nonelemental sources. In F. M. Clydesdale & K. L. Wiemer (Eds.), *Iron Fortification of Foods* (pp. 39–53). Academic Press, Orlando, FL.
12. Lachance, P. A., & Fisher, M. C. (1988). Effects of food preparation procedures on nutrient retention with emphasis on food service practices. In E. Karmas & R. S. Harris (Eds.), *Nutritional Evaluation of Food Processing* (pp. 505–556). Van Nostrand Reinhold, New York.
13. Lucey, J. A., & Fox, P. F. (1993). Importance of calcium and phosphate in cheese manufacture: A review. *Journal of Dairy Science, 76*(6), 1714–1724.
14. McGee, H. J., Long, S. R., & Briggs, W. R. (1984). Why whip egg whites in copper bowls? *Nature, 308*, 667–668.
15. Miller, D. D. (1994). Atomic absorption and emission spectroscopy. In S. S. Nielson (Ed.), *Introduction to the Chemical Analysis of Foods* (pp. 353–370). Jones and Bartlett, Boston.
16. Patrick, J. (1985). Elemental sources. In F. M. Clyesdale & K. L. Wiemer (Eds.), *Iron Fortification of Foods* (pp. 31–38). Academic Press, Orlando, FL.
17. Phillips, L. G., Haque, Z., & Kinsella, J. E. (1987). A method for measurement of foam formation and stability. *Journal of Food Science, 52*, 1047–1049.

CHAPTER

8 Colorants

LEARNING OBJECTIVES

After studying this chapter, you will be able to:

- Explain food additives, food industries, and their demands
- Define food colorants
- Describe food preservatives
- Discuss potential compounds with dual activity
- Illustration of food colorants analysis in foods

KEY TERMS FROM THIS CHAPTER

Food additives	Codex Alimentarius	Toxicity
Food colorant	Color additive	Artificial colors
Synthetic food colors	Genotoxic effect	Turmeric pigment
Phycocyanin pigments	Antibrowning agents	Color retention
Color formation	Antimicrobial activity	Anthocyanins

INTRODUCTION

Enhancing the safety and quality of food is crucial for maintaining human health. From the beginning of time, humans have been enhancing their diets and hunting techniques, domesticating animals and plants, physically preserving food, and ultimately adding molecules to food to improve its flavors or prolong its shelf life (Branen et al., 2001).

Food additives and technological advancements have played critical roles in ensuring that the food supply is cheap, wholesome, tasty, colorful, and safe. Over time, a number of ingredients have served beneficial purposes in a variety of foods. They play a crucial role in the food industry by reducing loss, improving quality, extending shelf life, developing new formulations, and standardizing processes—all of which help to meet the ever more demanding demands of the consumer. Additives are found in all kinds of food and are becoming more and more common and significant in human nutrition. Despite the controversy surrounding them, which stems from differing government interpretations and contradicting findings from numerous studies involving these compounds, they are strictly regulated. Hundreds of additives are added to food today worldwide, and many more have been outlawed over time (Branen et al., 2001).

While maintaining food safety and quality, it is crucial to lower production costs and monetize current products in an increasingly competitive market. Several studies have been focused on developing new solutions and/or enhancing current alternatives because coloring and preservation additives are among the most significant ones in the food industry, improving the appearance and preservation of foodstuffs. A viable approach may be to identify substances that possess both colorant and preservative properties while also acting as bioactive agents. However, the concentration of the compounds commonly used as dyes is typically high to obtain benefits like antioxidant or antimicrobial activity, not meeting the admissible daily intake (ADI) requirement. The scientific community may choose to go around the challenges and make money off of the use of these additive molecules in the food industry by conducting research and development of new molecules using new chemical approaches, along with the modification of natural molecules already known, in order to develop a better and double performance (colorant plus preservative).

8.1. FOOD ADDITIVES, FOOD INDUSTRIES, AND THEIR DEMANDS

Customers are very concerned about the use of pesticides, radiation, food additives, and even trace substances in the production and preservation of food because of the potential short-, medium-, and long-term health effects (Martins et al., 2017).

With the rise in highly processed and modified food products over the past few decades and their higher additive load, these problems and health concerns become even more apparent. In order to provide food products that are both safe and healthy and that are tailored to the modern lifestyle (i.e., aesthetically pleasing and flawless foods with specific organoleptic characteristics and a long shelf life), there is currently a general push for innovation in the food and biotechnology industries.

A variety of procedures and materials, including food additives (colors, sweeteners, preservatives, emulsifiers, acidifiers, etc.), are used to meet the majority of these requirements.) have been employed, which were created to enhance the flavor and appearance of food as well as stop it from spoiling. Studies on their long-term impacts are still necessary, though. Food additives should, in general, only be used when their technological impact is justified and cannot be satisfied by other workable solutions. Furthermore, at the recommended level of use, these compounds cannot endanger the health of users. The European Food Safety Authority (EFSA) in the European Union (EU) and the Food and Drug Administration (FDA) in the United States (US) are the two most significant regulatory bodies in the world that establish standards, enact laws, and supervise the approval and standardization of food additives. Moreover, JECFA, the Codex Alimentarius, and the Joint Expert Committee on Food Additives of the Food and Agriculture Organization (FAO)/World Health Organization (WHO) are important organizations in charge of conducting research, evaluating safety risks, and making declarations. When it comes to meeting local, national, and international quality standards for processed food exports, Codex Alimentarius serves as the standard for other official food authorities worldwide. Food additives are categorized in the EU based on their function, which is broken down into 26 functional classes. However, in the US, additives are categorized into two or more classes; the FDA permits the use of 3,000 of these classes, which include nutritional additives, texturizers, flavorings, colorants, and preservatives (Aparicio et al., 2017).

To facilitate consumer comprehension of food labels from various European countries, all food additives in Europe are assigned an E-number, which is a code consisting of three or four digits after the letter E. Carefully considered calculations are made to ensure that the amount of additive added does not exceed the acceptable daily intake (ADI), which is the total amount of a particular additive that, when consumed on a daily basis, does not pose a health risk. The Codex Alimentarius also uses the same code number (Gallen et al., 2013).

Notwithstanding their apparent advantages, it is important to remember that a significant amount of permitted food additives are synthetic in nature and can be carcinogenic when misused, leading to a variety of negative side effects, toxicity, and other adverse reactions. Global regulatory bodies have implemented rigorous assessment processes in this regard, and as a result, a number of artificial additives have been outlawed owing to adverse reactions related to the nervous system, gastrointestinal tract, respiratory system, and skin.

Nonetheless, a worldwide agreement regarding food additive laws has not yet been reached. In the U.S., certain substances can be added to food. S. and prohibited in the EU, including the colors FD and C green no., as well as the antimicrobials calcium sorbate (E203) and sodium sorbate (E201). Citric Red No. 3 (E143), Fast Green 3. 2 (E121). However, the coloring agents carmoisine (E122), amaranth (E123), and patent blue (E131) as well as the antimicrobial sodium methyl p-hydroxybenzoate (E219) are prohibited in the United States but permitted in the European Union. S. All of this results in legislative contradictions, which impede international trade and cause issues with food safety. Preservatives, antioxidants, colorants, and natural sweeteners also lack a defined definition; the only substances covered by identical laws in the EU and the US are natural flavorings. S. has resulted in incorrect application across the board for all additive classes, misinterpreting what is natural and what is synthetic. Due to their widespread use, particularly in developed nations, it is obvious that there is a need for consistent legislation on natural additives (Arvanitoyannis et al., 2005). Concurrently, there is an increasing need for food items that have the ability to prevent and promote health. Foods that "have been satisfactorily shown to affect one or more target functions in the body, in addition to appropriate nutritional effects, in a way that is relevant to an improved state of health and well-being and/or reduced risk of disease" are referred to as functional foods, and this definition is presently accepted worldwide. Because they have a high biological quality and the potential to promote health, natural additives may also be included in this concept.

8.2. FOOD COLORANTS

Food acceptance, choice, and consumption are primarily influenced by organoleptic properties. One of the most striking and alluring aspects of food is its color. While naturally occurring food products naturally have their own color, various processes and environmental elements like oxygen, metals, light, pH, and water activity can alter their color unintentionally. The food industry heavily uses chemical compounds that affect color to get around this issue, and as these developments happen, more stringent regulations are put in place to guarantee safe manufacturing processes and complete consumer protection.

"Any colorant, pigment, or substance that, when added or applied to a food, drug, or cosmetic, or to the human body, is capable (alone or through reactions with other substances) of imparting color" is what the FDA defines as a color additive, also known as a food colorant (Moral Turiel, 2017).

Colorants can be categorized based on a number of factors, including origin (natural, natural-identical, or synthetic; organic and inorganic), solubility (soluble and insoluble), and hiding power (transparent and opaque). Variations in chemical structures, sources, and purposes of use can make classifications of colorants complex. The most common form is based on whether they are man-made or natural. Plant tissue can be used to make natural dyes (e.g., betalains, anthocyanins, curcumin, carotenoids, and chlorophylls), animal cells (kermesic and carminic acid), microorganism metabolism (carotenoids and chlorophylls), or mineral sources (calcium carbonate or titanic dioxide). Chemical synthesis is used to create artificial colorants, which are not present in nature (Amchova et al., 2015).

When added at a specific concentration, none of these dyes leave an aftertaste, and the food industry primarily uses them in meat, dairy products, confections, and bakeries (Abo-EL-Sooud et al., 2018).

8.2.1. Synthetic Colorants

The industry prefers synthetic colorants, which are products that have been chemically or physically modified and have desirable properties for manufacturing, including low cost when compared to natural colorants, high purity, coloring capacity, stability, brightness, and a wide range of shades.

The food industry is using artificial colors more frequently as a result of the advancement of science, which has led to the development of more stable and versatile artificial colors. Nevertheless, despite ongoing research into synthetic compounds—many of which are intended for human use—natural alternatives are growing in popularity among consumers worldwide as a result of recent findings about the toxicity and adverse effects of some synthetic dyes. Furthermore, extremely specialized techniques and procedures have been developed to identify the presence of synthetic food coloring in food. These techniques also enable the determination of the likelihood of medium- and long-term toxicity and side effects (Bhatt D., et al., 2018).

Consuming artificial colorings, primarily nitro derivatives of the azo type (E102, E110, E122, E123, E124, and E129), has been linked to a number of health issues, according to studies. Foods containing one or more of these color additives had to be labeled with the name or E-number information along with a warning that the food "may have an adverse effect on activity and attention in children," according to a 2008 European Parliament decree. One of the most divisive color additives in terms of safety is tartrazine (E102), a lemon yellow that is used in candy, ice cream, cereals, soups, jellies, cakes, soft drinks, and other foods. Due to its interaction with human serum proteins, it has been linked to mutagenic potential as well as obsessive-compulsive disorders and hyperactivity in children.

Recent research using mice administered tartrazine at ADI levels revealed elevated DNA abnormalities, altered biochemical markers in brain tissue hematotoxin, immunotoxin effects, and increased lipid oxidation. A higher pro-inflammatory activity has been linked to petroleum-derived aromatic compounds that yield Sunset Yellow (E110), as well as carmoisine (E122), allura red (E129), and ponceau 4R (E124), all of which bind to human and bovine serum albumin. Amarath (E123), a red food coloring that is permitted in the EU but prohibited in the US, gives candy, ice cream, and beverages their characteristic red color. S. has demonstrated a strong genotoxic effect on cultured human lymphocytes because of its carcinogenicity. It is significant to remember that while certain colors are still acceptable in the European Union, they are no longer in the US. Research on sources of natural origin is becoming more and more important due to these contradictions and the negative health effects of artificial colorings. This is because natural sources are safe and have health-promoting qualities.

Given the importance of natural colors to the food industry, there is significant pressure to find them even in the face of controversy surrounding food colors. In the United States. s. and the EU, some excellent substitutes have been developed; molecules like carotenoids, anthocyanins, annatto, and paprika stand out as having potential qualities that could replace their synthetic counterparts. Nevertheless, these substances have drawbacks like temperature and pH instability, oxidation-related color loss, increased production costs, and higher product yields compared to synthetic colorants (MacDougall, 2002).

8.2.2. Natural Colorants

Food dyes used until the middle of the 20th century came from natural sources, such as saffron, which is made from the stigma and styles of the Crocus sativa L. flower.), alizarin (obtained from tropical woods), caramel (sugar paste turned into syrup), red beet (aqueous extract of the red beet root), cochineal (obtained from certain insects of the family Coccidae, a parasite of some cacti), and indigo (from the indigo plant or glasto, a European shrub). There are now techniques for gathering, extracting, purifying, stabilizing, and standardizing different natural food colors, such as anthocyanins, betalains, chlorophylls, and carotenoids. These comprise various groups of chemical compounds that, depending on the compound or its stabilized form, can be used directly as dyes or in chemically modified forms to produce a variety of hues, from green to yellow, orange, red, blue, and violet.

The pure turmeric pigment known as curcumin (E100) is taken from the desiccated rhizomes of Curcuma longa L. is a popular food coloring that gives foods like mustard, yogurt, ice cream, dairy products, and salad dressings their orange hue. The primary pigment found in Dactylopius coccus Costa (cochineal) is carminic acid (E120), which complexes with aluminum to produce a vivid red color. Due to its stability, this dye is thought to be significant technologically even though it is more costly than other natural reds like anthocyanins. It takes the place of potentially harmful synthetic dyes, but because it contains animal products, the food item cannot be considered kosher or vegan/vegetarian. It is a component of bakery goods, dairy products, jams, jellies, and noncarbonated drinks. Plants naturally contain chlorophylls and chlorophyllins (E140); only two of the many chlorophylls are utilized as colorants in food manufacturing, primarily because of their challenging stabilization; more stable copper complexes of these chlorophylls are also permitted (E141).

The commercial chlorophyll dyes are made from alfalfa and are used in dairy products, soups, drinks, and candies made with sugar. Extracted from the seeds of the fruits of Bixa orellana L. is anatto (E160b). tree, which is slightly redder than β-carotene, is primarily composed of the yellow-to orange carotenoids bixin and norbixin. Cakes, cookies, rice, dairy, flour, fish, soft drinks, snacks, and meat products all contain it. Carotenoids, which range in color from orange to red, are also present in capsanthin, capsorubin (E160c), and paprika extract. A variety of additional carotenoids found in food, such as lutein, violaxanthin, neoxanthin, β-cryptoxanthin, fucoxanthin, lycopene, and astaxanthin, are primarily used in marinades, sauces, toppings, drinks, and other preparations. With over 700 anthocyanins identified, the primary anthocyanins (E163) found in nature are glycosides of the aglycones cyanidin, delphinidin, malvidin, pelargonidin, peonidin, and petunidin. Their colors range from red to purple, violet, and blue.

Their primary applications are in fruit preparations, confections, and soft drinks. Similar in hue to anthocyanins, betalamins and betataxanthins are pigments that range from reddish-violet to yellow-orange. But the only betalain that is approved for use by law is betanin (E162), which is made from beets and is the most widely used natural pigment in industry to take the place of Allura red AC (E129; FDandC Red 40). It can be used to give candies, ice cream, meat alternatives, and drinks a red color. Talaains have a range of colors and are more water soluble than other natural dyes. They also have strong dyeability and stability in neutral and acidic pHs, which makes them a great option for food coloring applications even though their sensitivity to light and heat limits their use.

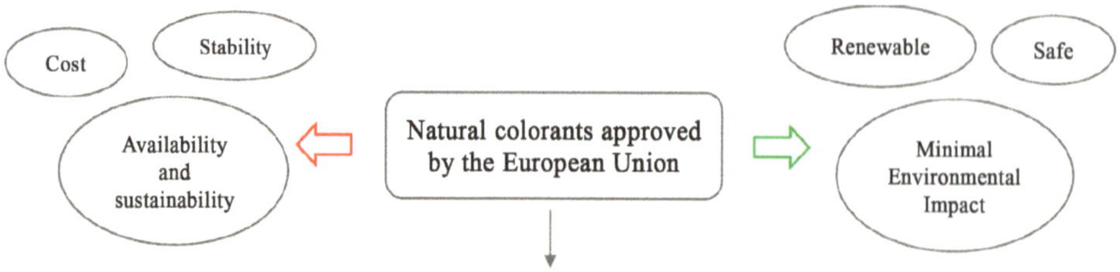

E Code	Denomination	Tonality
E 100	Curcumin	Yellow orange
E 101	Riboflavin, Lactoflavin, Rivoflavin-5'-phosphate	Yellow orange
E 120	Cochineal, Carminic acid, carmines	Red
E 140	Chlorophylls and chlorophyllins	Green
E 141	Copper complexes of chlorophylls and chlorophyllins	Green
E 150 a-d	Caramel	Brown
E 153	Vegetable carbon	Black
E 160 a	Carotenes	Yellow orange
E 160 b	Anatto, Bixin, Norbixin	Yellow to Orange
E 160 c	Paprika extract, Capsanthin, Capsorubin	Red
E 160 d	Lycopene	Red
E 160 e	Apocarotenal	Yellow to red
E 161 b	Lutein	Orange yellow
E 161 g	Canthaxanthin	Violet
E 162	Beetroot Red, Betanin	Violet
E 163	Anthocyanins	Red, blue, or violet
E 170	Calcium carbonate	White
E 171	Titanium dioxide	White
E 172	Iron oxides and Hydroxides	Brown
E 173	Aluminum	Silver to dark gray
E 174	Silver	Silver
E 175	Gold	Golden

Figure 8.1. *Natural colorants approved by the European Union, their shade, and functionality.*

Source. https://www.ncbi.nlm.nih.gov/pmc/articles/PMC9776543/bin/jf1c07533_0001.jpg.

The phycocyanin pigments from Spirulina (A) could be used as an alternative to natural blue dyes. platensis). As anthocyanins, which can have a blue color between pH 5 and 7, tend to change their hue depending on the pH value and water content, these would be a good substitute for the elusive bright and stable source of natural blue dyes that have been sought after. A phycocyanin protein complex can be obtained from Spirulina, a Gram-negative cyanobacterium that yields isolable picocyanobacteria. The only naturally occurring blue dye that is currently recognized in Asia and the United States is picocyanobacteria from Spirulina, though it is not yet permitted in the European Union. Originally, the only products colored with phycocyanins were candies and chewing gum, but advancements in food processing have made it possible to color dairy products, soft drinks, and cosmetics as well. Although their primary application has been in beverage products, low pH stability problems still prevent wider use. Microorganisms like fungi and bacteria could be another way to create natural dyes. These are less limited by seasonal variation problems, produce a wider variety of natural pigments than botanical sources, and are easily extracted and scaled up for use in various industries. Because they have a shorter life cycle and are generally easier to genetically modify, bacteria present a promising avenue for further research into the still-poorly understood process of pigment production. A lab at Troy, New York's Rensselaer Polytechnic Institute has created a method for using E. coli and glucose to generate anthocyanins at g/L levels that are industrially relevant. As far as fungi go, harmless Monascus sp. has not yet acquired U.S. approval. It makes six substantial polyethylene pigments, ranging from yellow to red. S. or EU approval despite the fact that it has been used successfully in Asia for hundreds of years due to worries about the production of citrinin, a strong fungal toxin. But there are issues with consumers accepting food coloring made from microorganisms.

Some natural food colors have been found to be just as effective as those derived from chemical synthesis, despite the instability associated with these dyes. These natural food colors also have the added benefit of being safer and offering health benefits in addition to imparting organoleptic characteristics, meaning they can serve multiple purposes as food ingredients. In actuality, a number of food additives that have coloring effects also serve as preservatives and antioxidants, giving food items additional functional qualities. Functional foods are becoming more and more popular because of the health benefits they offer when consumed in moderation as part of a regular diet.

Given that artificial colorants have both health benefits and the ability to pigment, many businesses have been replacing natural colorants with artificial ones in order to meet consumer demand.

It should be noted that stability is generally regarded as one of the most crucial aspects of using natural food colorants, as the application of unstable food colors can result in unpleasant organoleptic features and consumer rejection of the food. Thus, a number of strategies have been developed to address this issue, and while some research has concentrated on identifying various sources of promising food colorants, an increasing number of studies is looking into new sources of natural colorants; in fact, there are already a number of reports available that explore the potential of novel alternatives derived from plants, insects, fungi, bacteria, and algae. The objective of this assessment is to enable the application of extraction and stabilization techniques in the food industry, enabling them to rival artificial solutions in terms of coloring efficiency and color stability throughout the product's storage period.

8.3. FOOD PRESERVATIVES

Microbiological, enzymatic, physical, or chemical processes can cause food products to deteriorate, lowering their nutritional value, safety, and quality while also causing unfavorable alterations to their physicochemical and sensory characteristics. In light of this, numerous biological, physical, and chemical techniques for food preservation have been created in order to increase food shelf life and ensure consumer safety while maintaining sensory qualities. One of the modern methods for food preservation that helps prevent contaminations that can cause foodborne illnesses in addition to preserving food quality is the use of preservatives in food processing. These materials fall into three categories: antimicrobials, antioxidants, and anti-browning agents. They can be synthetic or natural. Antimicrobials are used to suppress and stop microorganisms from naturally spoiling food; antioxidants are used as preservatives to stop or slow down biological and chemical deterioration of food by stopping pigments, flavors, lipids, and vitamins from auto-oxidizing; and lastly, anti-browning agents are used to stop food from browning, which can happen at any point during handling, processing, and storage (Davidson et al., 2003).

8.3.1. Antimicrobials

To ensure food safety and quality, antimicrobial compounds can be added to food or exist naturally. They can delay and/or prevent the growth and proliferation of natural food spoilage microorganisms like molds, yeasts, and bacteria, as well as prevent or control the contamination of food by pathogenic microorganisms (Kalpana et al., 2019).

The choice of antimicrobial additives depends on the antimicrobial properties and spectrum of activity, the physicochemical composition and properties of the food matrix, and the type of storage and preservation techniques. While the use of antimicrobial additives extends the shelf life of food, high concentrations of these compounds can cause an unpleasant taste, strong odor, altered viscosity, and color retention. They may be synthetic or natural.

8.3.1.1. Synthetic Antimicrobials

The regulatory community has formally acknowledged the use of a number of artificial preservatives as food antimicrobials. These are chemical compounds that stop or limit the growth and multiplication of molds, yeasts, and bacteria. Weak organic acids like sodium propionate, benzoic acid, and sodium benzoate, as well as inorganic acids and their sodium salts like nitrite and sulfate, are among those that have been approved for use in food applications. The use of chemical antimicrobials in food is still up for debate and controversy despite the compounds' significant and beneficial roles in food product preservation. This is because the results of published research have not consistently shown a clear pattern of action. In fact, it can be challenging to draw firm conclusions about the safety and effectiveness of synthetic antimicrobial food additives like sorbates, nitrates, and sulfites, as they have been shown to have a variety of negative effects on human health. The only way to lessen their use in the food industry would be to find a nontoxic alternative that has the same preserving effect, despite the unsettling reports. Utilizing these substances lowers the risk of food poisoning and spoilage. Although it hasn't been demonstrated, long-term exposure to sodium benzoate (E211) is thought to be safe; conversely, some researchers believe that sodium sulfate (E221) poses health risks. Studies have indicated that, depending on the dose used, sodium sorbate may be genotoxic to human lymphocytes *in vitro*, which is one of the many questions and concerns surrounding this substance. Many investigations into sorbates (E200–E203) and their potential health effects have been carried out over the years. These studies have suggested that sorbates may be genotoxic and mutagenic, although other studies find no connection between these claims. Furthermore, because they have no taste or odor, parabens (E214–E218), the generic name for the class of antimicrobials that are alkyl esters of p-hydroxybenzoic acid, are widely used. However, studies have shown that using parabens *in vitro* can cause invasive and migrating behavior in human breast cancer cells. Other antimicrobials found in food include nitrites (E249 and E250) and nitrates (E251 and E252). The second group can only be added to meat for slow curing and was recently restricted in the EU. The only food additives that can inhibit botulinum toxin, nitrites are used in meat for color formation, flavor enhancement, and antimicrobial activity. This justifies their use in the food industry despite the fact that they have been shown to have carcinogenic effects among other negative effects on humans. Only the lowest possible dosage is allowed for their use in the European Union. Naturally occurring in untreated fruits and vegetables, nitrates and nitrites may also play a role in the synthesis of nitrosamines. Foods containing sulfites (E220–228) include wine, dried fruit, dehydrated cookies, and fish. They can function alone or in combination with organic acids. They have, however, been shown to have cytotoxic and carcinogenic effects on rats and humans, as well as to cause adverse reactions in sensitive individuals and to cause food-borne thiamine (vitamin B1) depletion. Propionic acid (E280) and its potassium (E283), calcium (E282), and sodium (E281) salts are mostly used in bakery goods to guard against the growth of mold and other fungi. Few studies have been conducted on their toxicity, but they have been shown to cause abnormalities in the tips of onion roots, which can lead to chromotoxic effects and chromosomal abnormalities. They have also been connected to irritability, restlessness, inattention, and sleep disturbances in certain children. Benzoic, lactic, and acetic acids are further potent antimicrobial agents. Despite having little antimicrobial activity, malic (E296), fumaric (E297), citric (E330), and other organic acids are also utilized; in fact, they are more frequently used as flavoring agents. The inhibition of nutrients is typically the basis for the mechanisms of action of weak organic acids. The primary ingredient in vinegar, acetic acid (E260), is frequently added to condiments like mayonnaise, mustard, and salad dressings in order to prevent the growth of bacteria and fungus. Compared to molds, it is more effective against bacteria and yeasts. In foods with low pH levels, like fruits, mayonnaises, pickled vegetables, and beverages, benzoic acid and sodium benzoate (E210 and E211) are frequently used to exert their antimicrobial activity against yeasts and fungi, including the inhibition of aflatoxin production. To add extra stability and safety to food products that have been acidified, lactic acid (E270) is added. It works well to lower microbial communities on meat surfaces and reduces E even more effectively than acetic acid. Coli O157:H7 (Sdot Mamur et al., 2012).

8.3.1.2. Natural Antimicrobials

Consumers and businesses are attempting to replace synthetic preservatives with natural preservatives, which can be obtained from sources like plants, bacteria, fungi, animals, and algae and are thought to be safer for both humans and the environment, because synthetic preservatives have the potential to cause health issues (Quinto et al., 2019).

Figure 8.2. *Natural compounds with preservative capacity, in use or under research.*

Source: https://www.ncbi.nlm.nih.gov/pmc/articles/PMC9776543/bin/jf1c07533_0002.jpg.

Note: *Additive allowed as a colorant but not as an antioxidant.

***Additive still under study, not yet permitted in the EU.*

8.3.1.3. Plant-Based Antimicrobials

Plant extracts vary in type, nature, and concentration. Herbs and spices possess antimicrobial properties that act against a range of microorganisms, making them effective preservatives that extend the shelf life of food. As several researchers have already noted, these can be alternate sources of novel antimicrobial compounds because of compounds like phenols, alcohols, aldehydes, and ketones, among others. Alkaloids, polypeptides, phenolic compounds, essential oils, and lectins are some of the antimicrobial substances found in plants. Further groups of compounds have also demonstrated potential as natural antimicrobials, including glycosides, polyamines, organic acids, and glucosinolates (Mamur et al., 2012).

Essential oils (EO) are complex, bioactive liquid volatile compounds with strong flavor and aroma that are mostly produced by aromatic plants. They serve primarily as flavoring agents in the food industry, but they can also function as natural antioxidants and antimicrobials to preserve food. Lemongrass, peppermint, lavender, geranium, thyme, oregano, sage, rosemary, basil, vanilla, clove, fennel, cumin, cinnamon, and anise are the main plants used to extract essential oils. Terpenes, aldehydes, ketones, aliphatic alcohols, organic acids, phenolic compounds, saponins, glucosinolates, and thiosulfinates are the principal substances found in essential oil (EO) that have antimicrobial activity. Rich in terpenes, including limonene and its precursors, citral, eugenol (from clove and cinnamon), thymol (from thyme and oregano), carvone, carvacrol (from oregano), and many other compounds, essential oil (EO) has antimicrobial properties against a wide range of food-borne microorganisms, including pathogenic bacteria and fungi and spoilage. For instance, carvacrol, which is present in large amounts in oregano, possesses antibacterial and antifungal properties even at low concentrations. If cinnamaldehyde and nisin are also present, these properties will be further amplified. An isomer of carvacrol called thymol exhibits similar antibacterial and fungicidal properties, including the ability to inhibit the growth of Aspergillus species *in vitro*. Ozogul and associates. demonstrated that the ability of an EO-based nanoemulsion (thyme and rosemary) to enhance the microbiological and organoleptic characteristics of rainbow trout filets makes it suitable for use as a preservative ingredient. Since EO is a lipophilic compound that can penetrate Gram-positive bacteria, it generally has a greater inhibitory effect against Gram-positive bacteria than Gram-negative bacteria. Depending on the application, their potential can be further increased through synergy with bacteriocins or even food ingredients. Although direct mixing, immersion, and spraying are the usual ways that essential oils are added to food, they have also recently been used in bioactive packaging, encapsulation, and nanotechnology. Edible films and coatings containing bioactive substances are also being considered for use. These materials have been specifically focused on food preservation, serving as carriers of essential oils (EO) and thereby regulating microbial growth. In this particular context, it has been demonstrated that the EO of Origanum Majorana encapsulated in terpinen-4-ol enriched chitosan nanoemulsion is a significant preservative of food products against lipid peroxidation, aflatoxin B1 (AFB1), and fungal contamination. As a result, these bioactively encapsulated substances may be authorized as an appropriate antifungal agent to increase the longevity of food items. Furthermore, because these herbs are rich in phenolic compounds, using their essential oils alone or in combination has shown the ability to limit the oxidative process in food matrixes like meats and dairy products. Regarding this, a number of EO have demonstrated exceptional antioxidant activity; they are used as preservatives in fish supplements and are thought to be useful in delaying the oxidation of lipids in fish and meat products. The only disadvantage of EOs is that they can be toxic to humans even at low concentrations. Therefore, more research is needed to understand the true effects of EOs on the body and to establish an acceptable daily intake. Conversely, weak organic acids (such as lactic, tartaric, malic, acetic, and citric.), which are naturally occurring flavoring and acidifying agents in acidified foods and can inhibit a variety of contaminating microbes, are found in oranges, lemons, apples, grapes, and other fruits and juices (Mamur et al., 2012).

8.3.1.4. Animal and Microbial-Based Antimicrobials

Animal secretions or defense systems can provide additional bioactive compounds for food preservation. For instance, oligopeptides known as antimicrobial peptides (AMPs) have a wide range of activity against bacteria, fungi, protozoa, and certain viruses. They come from natural sources like plants, insects, crustaceans, amphibians, and marine life, and some that are already used as food additives have been shown to have beneficial health effects while having unknown toxicity. Defensins, chitosan, pleurocidin, lactoferrin, ovotransferrin, and lysozyme, among others. are additional potential antimicrobials derived from animals. Different sources of lysozyme exhibit antimicrobial activity, especially against Gram-positive bacteria. Although it only partially inhibits bacteria and fungi, this enzyme is widely used as a preservative in fruits, vegetables, meat, fish, milk, and dairy products. Its effectiveness can be enhanced by combining it with other preservatives, such as nisin. Lactoferrin is a substance that can bind iron and is present in secretions such as milk. Pleurocidin and protamine, two AMPs that have been isolated from fish, are effective against food-borne microorganisms like Listeria monocytogenes. Mammals contain cathledicin, which exhibits antibacterial, antifungal, and antiviral properties. It has been discovered that salmin, a cationic AMP made from salmon milt, inhibits the growth of Listeria monocytogenes in smoked salmon. The food and pharmaceutical industries primarily utilize chitosan and its derivatives, which are derived from the exoskeletons of crustaceans and arthropods, for their preservative properties. They are used in combination with other molecules like glucose and xylan, as well as in the creation of edible films and coatings for the food industry. In actuality, chitosan fibers and flavonoids have demonstrated outstanding antioxidant activity. By lowering the thiobarbituric acid value and delaying lipid oxidation during storage, the application of chitosan to meat and seafood provided notable oxidative stability. Because it is used as a natural antimicrobial agent in food packaging, lactoperoxidase (LPO) is one enzyme that has garnered attention for its bacteriostatic or bactericidal qualities. The quality of the trout filets was maintained by coating them with chitosan and LPO. Likewise, Listeria monocytogenes and E were inhibited by the addition of LPO to alginate coatings. rainbow trout filets' coli bacteria (Samaranayaka et al., 2011).

In addition to the proven effectiveness of AMPs against a range of food-borne pathogens, numerous studies have also shown that protein hydrolysates and peptides derived from goat milk, blue mussels, Pacific hake filet, tuna backbone, or beef can exhibit a greater antioxidant activity than α-tocopherol, BHT, and BHA (butylated hydroxytoluene and butylated hydroxyanisole, respectively) additives. As a result, these substances can be valuable substitute preservatives and useful functional ingredients in food that can reduce oxidative changes during storage.

One option to chemical treatments is food biopreservation, which increases food safety and extends shelf life by utilizing microbes and microbial metabolites. These days, fermented foods like cheeses, yogurts, and sausages are made mostly from lactic acid bacteria (LAB), of which several species have been cultured. LAB produces antibacterial substances like bacteriocins, organic acids, reuterin, diacetyl, ethanol, CO_2, H_2O_2, and derivatives of lactic acid (Mamur et al., 2012).

Because of their bactericidal or bacteriostatic properties, bacteria produce ribosomally synthesized antimicrobial peptides called bacteriocins, which are used to preserve meat and vegetable products. The main bacteriocins found in food are nisin, pediocin, diplococcin, plantaricin, acetylphilin, helveticin, bulgarican, and lactacin. Bacteriocins are produced by strains of Bacillus, Leuconostoc, Lactobacillus, Pediococcus, and Propionibacterium. By preventing the growth of microorganisms on the food surface through contact with the packaging material, they can also be used in bioactive packaging applications to control food pathogens. The only bacteriocin with outstanding antibacterial qualities and GRAS (generally recognized as safe) status that is advised for use as an additive in the food industry worldwide is nisin, an antimicrobial peptide that is naturally produced by Lactococcus lactis. It works against S and other Gram-positive bacteria. bacterium aureus and L. monocytogenes inhibits the growth of various Bacillus and Clostridium species spores. Molds, yeasts, and Gram-negative bacteria are not significantly affected by it, but the addition of chelating agents (e.g., EDTA) might strengthen its impact. Since it can withstand both high and low temperatures as well as acidity, it has been used with success in the wine and beer industries, as well as in the meat, cheese, and dairy products, seafood, and wine industries. In addition, it has proven to be able to stop human squamous carcinoma cell lines from progressing, although it also caused toxic effects in mice. Nisin's ability to work in concert with carvacrol and lysozyme, as well as its potential for encapsulation, have all been investigated with encouraging results.

Conversely, reuterin, which is only generated by the enzyme glycerol dehydratase from L, is a broad-spectrum antimicrobial agent. repertori. This food preservative is effective due to its high water solubility, heat resistance, stability across a broad pH range, and lipolytic and proteolytic enzymes. In milk, dairy, and meat products, it regulates the growth of Aeromonas hydrophila, L. and harmful bacteria, both Gram-positive and Gram-negative. monocytogenes and E. H7, C. jejuni, Y. coli O157. S. enterocolitica, as well. *Oryza* (Ray et al., 2015).

When it comes to ensuring food safety, organic acids are crucial to the LAB starter cultures' capacity to suppress unwanted microorganisms. They lower the pH, which prevents spoiling and harmful bacteria. The primary substance produced during food fermentation is lactic acid, which is followed by propionic and acetic acids. Some of the strains that produce lactic acid as their primary product are Enterococcus, Lactococcus, Lactobacillus, Streptococcus, Pediococcus, and Leuconostoc; these acids are what keep yogurts, olives, sauerkrauts, and fermented sausages alive. Propionibacterium in cheese produces propionic acid. When it comes to pathogenic microorganisms like pseudomonas spp. and spoiling, the hydrogen peroxide generated by LAB is highly effective. along with S. aureus. Additionally, produced during the fermentation of vegetables, like sauerkraut, carbon dioxide aids in the establishment of anaerobic conditions, is toxic to some aerobic food microorganisms, and can lower intracellular pH and impede enzymatic reactions. It usually inhibits the growth of facultative microorganisms like Enterobacteriaceae and inhibits the growth of obligate aerobes like fungus. Gram-negative bacteria, which are more sensitive to diacetyl than Gram-positive bacteria, are among the microorganisms against which the diacetyl produced by strains from various LAB exhibits antibacterial activity. Even though ethanol has little effect on biopreservation, it can make low pH and lactic acid in E more deadly. coli, albeit in greater quantities than those that are probably going to happen in lactic fermentations (Liu, 2006).

There are many different types of natural antimicrobial compounds, as mentioned. Some, though, have not yet received approval as food antimicrobials, partly because their safety necessitates lengthy toxicological testing. Therefore, more investigation is required to determine the ideal concentrations of antioxidants and antimicrobial agents that can be safely added to food systems without significantly changing sensory or physicochemical properties (Ayala et al., 2014).

8.3.2. Antioxidants

Foods undergo two types of oxidations during storage: rancidification and lipid peroxidation. The main process that causes food to degrade is called molecular oxygen oxidation, which also gives food an unfavorable taste and smell and encourages a loss of nutritional value. Vitamins can undergo auto-oxidation in food, but fat (unsaturated fatty acids, cholesterol, and phospholipids) is the main food component where it is most commonly seen. In addition to changing the color, texture, and imparting a rancid taste, it also causes the formation of toxic compounds that have an impact on the organoleptic properties. The food industry uses antioxidants, which scavenge oxygen and free radicals and stop peroxidation at the initiation or propagation stages, to prevent or delay these reactions. This prolongs food shelf life and prevents food decomposition without altering the food's flavor or appearance (Berdahl et al., 2010). The growing interest in using natural food additives makes it increasingly important to distinguish between synthetic/artificial and natural additives. The official tables should also include the quantities and authorizations for the use of artificial and natural antioxidant sources in each category of food.

BHA (E320) and BHT (E221) are the primary synthetic antioxidants that are used, primarily to prevent the oxidation of fatty acids. BHA functions better in animal fats than in vegetable fats and is used in foods and coatings. The recommended daily intake (ADI) for children and adults is 1 mg/kg body weight (bw), and exposure levels should not be exceeded. One of the least expensive antioxidants to produce, BHT, is typically found in animal fats and dry breakfast cereals, with an ADI of no more than 0.5 mg/kg bw. Given their chemical structures, these antioxidants behave synergistically when added to food, and it is not advised to use them in processes that involve heat, such as cooking or frying. Because of their detrimental effects on human health, their use has been restricted or outlawed in some countries. Tert-butylhydroquinone, or TBHQ (E319), is another widely used ingredient that finds use in meats, oils, cereals, and margarines, but it is ineffective in the bread and pastry sector. BHT and BHA can be used alone or in conjunction with gallate compounds (E310–E312). They can be used in dried milk, fats, oils, nut butters, potato products, chewing gum, cereals, meats, nuts, and food supplements, but their uses are extremely limited and case-specific. Additional chemical antioxidants approved by the EFSA are ascorbic acid and its derivatives (E300–E304), erythorbic acid (E315), sodium erythorbate (E316), EDTA (E385), citrates (E330–E380), lactates (E325–E327), and tartrates (E334–E354) (Błaszczyk et al., 2014).

Strong antioxidants like ascorbic acid (E300) are practically useful everywhere. Because of its superior ability to regenerate other antioxidants, it is frequently used in conjunction with synthetic antioxidants like BHT and BHA. It is also closely linked to tocopherols. To stop proteins and fats from oxidizing, sodium ascorbate (E301), which is used in a variety of foods, was tested in fermented dry sausages. However, because of its pro-oxidant nature, its use as a food additive in place of nitrites is extremely restricted. Calcium ascorbate (E302) is added to dairy products, cooked or cured meat products, fruit dips to keep it from browning, cured meats to enhance their color, and antioxidants that work in concert with one another. Ascorbic acid fatty acid esters (E304) can be found in dairy products, chewing gum, meats, cereals, desserts, salad dressings, sauces, and more. Mostly found in cured meats, frozen fruits and vegetables, oils, fats, seafood, and fish, erythrorbates (E315–E316) have the effect of preventing the formation of nitrosamines during the curing or cooking process. Sodium lactate (E325) and potassium lactate (E326) are utilized as antimicrobials in fish, sausages, meat burgers, and processed and unprocessed meats. Strong synergy exists between citric acid (E330) and ascorbic acid and chitosan in particular. It also functions as a chelating agent, extending the shelf life of food and preventing browning. Sodium citrate (E331), which is mostly used in meats and skim milk, is another potent synergist with organic acids, particularly with citric, tartaric, malic, and lactic acids. It also has chelating, pH buffering, retarding, antimicrobial, flavor-enhancing, and antioxidant effects. In jellies, marmalades, and cheeses, potassium citrate (E332) and calcium citrate (E333) are employed as emulsifying agents, pH buffers, sequestrants, and antioxidants; their primary purpose is to balance out strong flavors. Cheeses, fats, oils, meats, sausages, chocolates, marmalades, gelatins, canned food, and fresh pasta all contain tartaric acid (E334) and its derivatives, sodium (E335), potassium (E336), and calcium (E336) salts. Phosphoric acid (E338) and its salts are used in soft drinks, fruit jellies, cheese, and yeast powders. They work in tandem with citric acid to stop fat oxidation. Sodium phosphate (E339) is used in pasta, meat, powdered milk, fruit, cheese, snacks, and ready-made desserts. Its chelating and antimicrobial properties are boosted by its synergy with nisin. Calcium phosphate (E341) is used in baking, fruit preserves, powdered juices, flour, cheese, and porridge. Ammonium phosphate (E342) and magnesium phosphate (E343) are less common but can be used in breads, pasta, cookies, and pancakes. Potassium phosphate (E340) is used in meats, breads, pasta, eggs, pasta, and sausages. They are all regarded as GRAS molecules. Adepic acid and its salts work in concert with sodium metabisulfite (E223) to produce cheeses, jellies, and canned fruit. Dairy products, cooked foods, and chicken meat are all treated with succinic acid. Because it is an antioxidant and a metal chelating compound, calcium disodium EDTA (E385) is found in foods like poultry meats, processed meats, vegetables, fruits, beer, and fruit juices. Its antimicrobial potential is further increased when it interacts with lysozyme, nisin, monolaurin, and monocapin. In addition, it is frequently added to fresh and processed meats, fish, sauces, cereals, seafood, lecithins, BHA, BHT, and PG (propyl gallate) for the purpose of anointing fats.

Regarding natural antioxidant additives, the food industry uses plant extracts, such as rosemary extract (E392), to preserve food. This makes it possible to find and potentially use additional natural antioxidant sources. The phenolic compounds found in rosemary extract, which include carnosic acid, rosmarinic acid, carnosol, and rosmaridiphenol, among others, have a strong antioxidant effect and are used in a variety of products throughout the European Union. These products include dehydrated milk, fats, processed potatoes, fine pastry, processed and heat-treated meat, fish, and eggs, mustard, soups, sauces, and food supplements. Because of its safety, it has not been given an admissible daily intake (AID) value and is regarded as a quantum satisfier (it can be added to food in the amount necessary to achieve the desired technological effect). To increase its antioxidant effect, it can also be used in synergy with other antioxidants like nisin, polyphenols, BHA, and BHT.

8.3.3. Natural Antioxidant Preservatives

Natural antioxidants have become incredibly popular, and there is no denying their significance in food and *in vivo*. In order to stabilize and fortify food, natural and even functional additives that can be found in various fruits, vegetables, algae, and mushrooms have been proposed as an alternative to synthetic additives. Certain substances, like vitamins, polyphenols, and carotenoids, are considered to be among the most significant natural antioxidant compounds because they can even exhibit protective effects that are comparable to those of BHT and BHA. Because of their strong antioxidant capacity, potential health benefits, and consumer acceptance (since they are natural molecules), polyphenols are a popular choice for the food industry, particularly when it comes to replacing artificial additives. They can also preserve food. Phenolic acids (hydroxybenzoic and cinnamic acids), flavonoids (anthocyanidins, flavones, flavonols, isoflavones, flavanols, and flavanones), lignans, and stilbenes are the primary groups of polyphenols. Certain phenolic compounds are more noticeable than others and can be added to food on their own, as plant extracts to capitalize on their synergistic effects, or after the molecules have been purified. Antioxidants have been added to food through the use of polyphenolic plant extracts, such as those from rosemary that were previously mentioned and recognized as food additives. Industries frequently give preference to particular molecules despite the benefits of compound synergy. This is why it's thought that carnosic acid, which comes from the hydroxybenzoic acid component of rosemary extract and is used in oils, animal fats, sauces, roasts, meat, and fish burgers, among other things, has the strongest antioxidant potential. Ferulic acid is a hydroxycinnamic acid that has been used in food gels and edible films, as well as being used as an antioxidant and precursor to other preservatives. The effects of catechin, also known as flavan-3-ol, can be amplified and potentiated by adding it straight to food, combining it with other natural ingredients, or encapsulating it. E300, or ascorbic acid, i.e., Through its role in the regeneration of oxidized tocopherols and phenolic oxidants, vitamin C has the ability to stabilize lipids and oils. Its consumption is safe, according to EFSA, which did not establish an ADI.

Carotenoids are another class of naturally occurring food additives that have the potential to be antioxidants, but their application is restricted due to their high sensitivity to light-induced oxidation. The most prevalent ones in food are lutein, β-carotene, and lycopene. They can be found in a variety of foods, including dairy products, meat, fish, fruit, cereal products, and pastries. Lycopene (E160d), which is mostly found in tomatoes, is the most commonly used carotenoid in food, although it is not frequently used as an antioxidant. Conversely, β-carotene (E160a) serves as an oxygen singlet in a variety of foods, including eggs, dairy, and baked goods. The carotenoid pigment astaxanthin, which is not permitted by the EU, is being researched as a potential food preservative due to its strong antioxidant capacity. The stability of these molecules during storage must be ensured by the use of carotenoids in the food industry. Synergies are facilitated by the combination of carotenoids with ascorbic acid or vitamin E. This combination, which can be produced artificially or extracted from plants and fruits, has been examined by the EFSA Scientific Panel, which has ruled out any potential toxicity from ingestion. Tocopherols are another substance that ascorbic acid can interact with and function as extremely potent antioxidants. They function primarily as an inhibitor of lipid peroxidation and rancidity in oils, meats, vegetable oils, and dairy animal fats, among other things. They are utilized in films, coatings, and additives (E306–E309).

Additional types of carotenoids are also added to food. For instance, anatto is a concoction of various carotenoids that are extracted from the Bixa Orellana shrub, the most notable of which is bixin. While annatto is primarily used as a food coloring, it also has anti-oxidant properties and has been shown to prevent the auto-oxidation of triacylglycerols in seeds. According to other reports, norbixin, which is known to be a stronger antioxidant than tocopherols, and tocopherols may work synergistically to increase antioxidant potential. With an ADI of 0.065 mg/kg bw and the code E160b, its use is approved in the EU. Because lutein has a variety of biological purposes, including preventing lipid autoxidation in skin cells, its use in the cosmetics industry is warranted by its antioxidant properties. Its activity surpasses that of β-carotene by far. According to EFSA, its application in the food industry is restricted to soft drinks, baked goods, bread-based foods, sauces, and confections; it is not permitted to be used as a food antioxidant.

8.3.4. Anti-Browning Agents

Foods can undergo a number of changes throughout the production process, including browning processes. In order to preserve the stability and preservation of food, additives that can reverse this reaction are necessary. Sulfates are the most widely used of these additives. Contrary to popular belief, there are natural substitutes for these artificial additives. These include amino acids like l-cysteine (E920), erythorbic acid (E315), which is used in beverages to preserve flavor, and 4-hexylresorcinol (E586), an organic compound that is only permitted in shrimp. Their mechanism of action is based on the inhibition of further compound formation through interactions with quinolone intermediates and the reconversion of quinone intermediates to the phenolic form. Calcium lactate is also used to prevent food from browning by preserving its structure, particularly in fruits. Additionally, it is used in concert with phosphates to increase meat's antioxidant capacity.

8.4. POTENTIAL COMPOUNDS WITH DUAL ACTIVITY

The food industry is constantly seeking new technological advancements, as well as methods, applications, and solutions to create innovative products that meet consumer demands. The search for new solutions and/or the enhancement of current alternatives has been prompted by the significance of color additives and preservatives for the marketing and preservation of food products. Certain compounds found in nature have the ability to extend a product's shelf life in addition to stabilizing or improving color. The food industry has shown a great deal of interest in evaluating various natural matrixes that may be able to serve as preservatives and sources of color. These characteristics may be provided by anthocyanins, carotenoids, betalains, and their derivatives because of their pigmentation, antioxidant potential, and/or antimicrobial activity. As an illustration, Ab Rashid et al. They recently created an antibacterial food colorant based on the anthocyanins found in *Clitoria ternatea* flowers. It exhibits a broad range of antibacterial activity and can keep its color for 21 days at -20 to 4°C, which makes it a viable substitute for biopreservatives and colorants in food products. Another study looked into the antibacterial qualities of eight plant-derived food colorings, including *Acacia catechu* L. *Bixa Orellana* L. *Auriculata Cassia* L., *Pinus granatum* L., and *Emillica officinalis* Gaertn. *Lubiscum tinctorum* L., *Tagetes erecta* L. Furthermore, *Terminalia chebula* Retz.), it was determined that the red pigments exhibited superior antibacterial activity while the yellow pigments demonstrated superior antifungal activity. The antioxidant activity was also examined in this study, and the red pigments once again demonstrated higher activity than the yellow ones; these characteristics would permit the use of these dyes in food, giving them preservative qualities.

In order to assess the antioxidant and antimicrobial activity of the Murta (*Ugni molinae* Turkcz) fruit, the phenolic composition of the methanolic and ethanolic extracts was examined; the ethanolic extracts demonstrated a high antioxidant capacity, whereas the methanol extracts produced inhibitory activity against *Escherichia coli* and *S. typhi* bacteria that are comparable to those treated with conventional antibiotics, providing opportunities to investigate these extracts as possible biopreservatives that have the ability to color due to their high anthocyanin content. In a different investigation, a curcumin microcapsule's antibacterial and antifungal properties were assessed against a variety of foodborne pathogens and spoilage microbes, such as *Escherichia coli, Yersinia enterocolitica*, *Staphylococcus aureus, Bacillus subtilis, Bacillus cereus, Aspergillus Niger*, *Penicllium notatum*, and *Saccharomyces cerevisiae*.

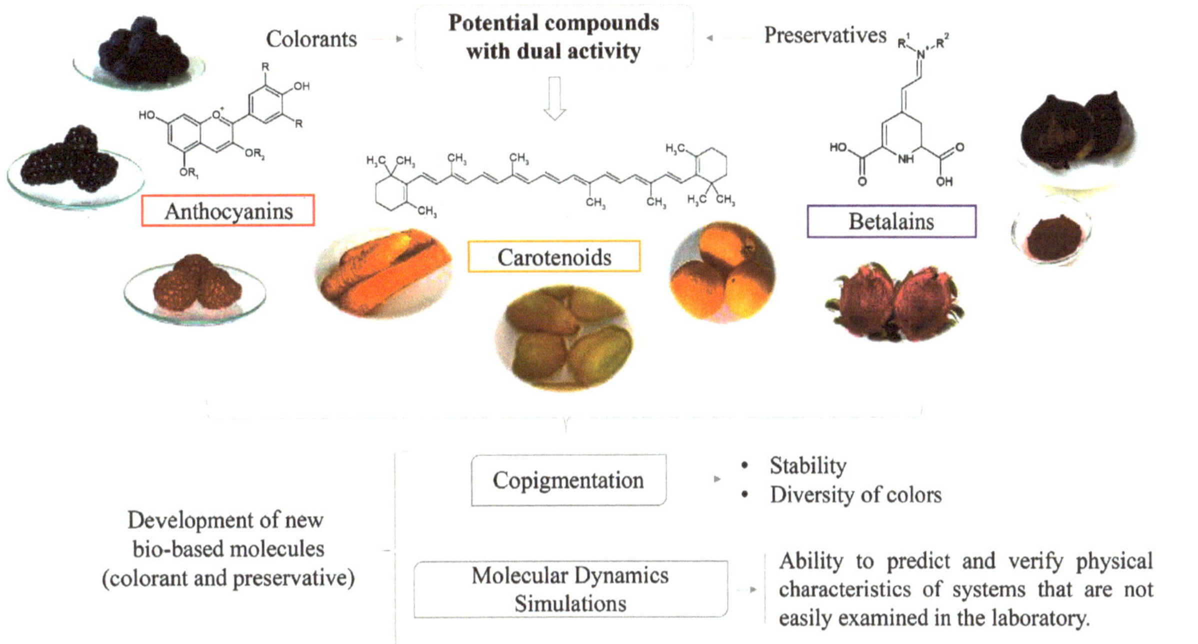

Figure 8.3. *Schematic illustration of the development of new molecules, through new chemical approaches, with modification of already known natural molecules to develop a better and double performance (dye plus preservative).*

Source. https://www.ncbi.nlm.nih.gov/pmc/articles/PMC9776543/bin/jf1c07533_0003.jpg.

In addition to their notable coloring qualities, betalains are a good example of natural food colorants because they have a variety of biological activities, including antioxidant and antiradical activity, which provides protection against oxidative damage, antimicrobial properties, antiproliferative, cytotoxic, and neuroprotective properties, and so on. These biological activities come with health benefits in addition to their coloring and preservative properties.

Carotenoids are an additional significant class of compounds with dual actions. α-carotene, β-carotene, and lycopene are among the most significant; they are already authorized for use as food colorings. In addition to adding color and/or preserving food, their strong antioxidant content makes them excellent natural food additives that pose no toxicological risk when used in food in the required amounts. Precursor of vitamin A β-carotene has been successfully added to a variety of foods, particularly nutraceuticals and functional beverages, and its use in both new and existing products is anticipated to rise in the future. When added to minced meat, lycopene, one of the primary phytochemicals found in tomatoes, has been shown to improve color, increase stability, and offer health benefits.

But it's crucial to stress that in order to achieve functional benefits like antioxidant and/or antimicrobial activity, the concentrations of compounds commonly used as dyes differ significantly (and are typically higher) from those used for other purposes, and they don't meet the regulatory agencies' acceptable daily intake (ADI) requirements, which prevent health risks. This demonstrates the necessity of investigating novel natural matrixes with the ability to color and preserve food. These matrixes may also offer advantageous qualities to the consumer. This presents a difficult task for the scientific community to address through the creation of novel biobased molecules, the next generation of food additives, and novel chemical techniques that alter existing natural molecules to improve and double their functionality (preservative and coloring).

A naturally occurring process called noncovalent complexation is primarily in charge of stabilizing and enhancing blue, violet, and red hues (e.g., anthocyanins) found in fruits, vegetables, and flowers as well as in foods made from them, such as syrups, jams, and wines. Over time, the notion of copigmentation has undergone refinement, and the atomistic understanding of it has proven indispensable in enhancing it by clarifying the diversity of colors found in nature and their stabilization mechanism. It's amazing how much more interest the food industry has shown in copigmentation as a way to enhance color palettes. It is imperative to have precise (computer-aided) control over supramolecular assemblies of noncovalent supramolecular copigments due to its application in the addition of copigments to food products. In this sense, it is possible to investigate copigmentation with antioxidant/antimicrobial molecules, and the development of novel hybrid compounds with dual functions (staining and preservation) can be aided by the use of cheminformatic models and new tools. These models are based on the understanding that a variety of biomolecules give rise to new biobased molecules that will become the next generation of food additives.

8.4.1. Molecular Dynamics Simulations: Challenges and Opportunities

There is growing interest in both fundamental and technological aspects of the delivery of technology and biofunctionality in processed foods that are all-natural. The topic of physical properties and stability has long piqued the interest of food scientists because they are essential to providing consumers with safe and healthful food. Food is a complex system that changes significantly both chemically and physically during processing and storage. The final product's molecular motions control these changes, which are indicative of the product's overall stability and perceived quality.

For several decades, research has been conducted using computer simulations, and in the past few years, the number of biomolecular system simulations has increased significantly. Molecular dynamics (MD) simulation has been hailed as the most promising tool for the characterization of multicomponent systems. It is distinguished by providing detailed information about the atomic/molecular structure and is not susceptible to physical fallacies. The methodology showed to be helpful in examining intricate geometries and molecules as well as in researching their structural and dynamical properties. Over the past 10 years, MD simulation has made it possible to study a wide range of biological systems, from tiny molecules like small peptides or anesthetics to extremely large protein complexes like the ribosome or virus capsids.

The intricacy of both biological and nonbiological systems has led to a great deal of interest in MD simulations because of their capacity to forecast and validate experimental outcomes, offering a chance to investigate the physical properties of systems that are difficult to study in the lab. One area of study for these features is the development of MD algorithms to simulate the folding and unfolding of proteins. MD simulations have been used to investigate the physical properties of nonbiological nanoparticles in addition to biological applications. Simulated and experimental data comparisons are used to verify the accuracy of the calculated results and offer guidelines for enhancing the methodology. Another important feature of the simulations is that, despite the potentials being approximate, the user has complete control over them. This allows the user to investigate the role of specific contributions in determining a given property by altering or removing them. To use MD simulations to simulate a complex system, a lot of computational power is needed. It is only feasible to simulate a dynamic system in time intervals of the order of femtoseconds with the tools at hand. Nevertheless, it is anticipated that MD simulation will become faster, more accurate, and run for longer simulation times as a result of improved MD simulation software and noticeably faster computer hardware.

Research into natural compounds that serve two purposes is emerging in response to the current demands for cleaner food, the preference for natural additives, and the toxicity and safety concerns with synthetic additives. Regulators have noted that these new molecules require extensive toxicity and safety testing in addition to research before being approved for use as additives. As a result, it may take some time before these compounds are included in the list of additives that they can use. A viable substitute would be to take advantage of currently available natural additive biomolecules that are widely used and, by modifying these molecules and applying novel chemical techniques, create distinctive hybrid compounds that serve two purposes (coloring and preservation). This is due to the fact that their individual use is typically limited to a single purpose, and that their concentrations for other activities (antioxidant and/or antimicrobial) are typically higher than those for dyeing, which could be harmful to one's health. In this way, and in spite of certain drawbacks, the use of molecular dynamics simulations in the development and study of these hybrid molecules can help and expedite research and studies on a topic that is currently fundamental. It can also predict and verify experimental results, allowing one to study certain physical characteristics that are difficult to examine in the laboratory.

8.5. FOOD COLORANTS ANALYSIS IN FOODS

Many food products use a lot of color additives to improve both their nutritional value and appearance. They are defined as "any substance whose intentional addition to a food results, or may be reasonably expected to result, in it or its by-products becoming a component of the food or otherwise affecting the characteristics of such foods" when used in the manufacture, processing, preparation, treatment, packing, packaging, transport, or holding of such food for technological (including organoleptic) purposes. In order to determine and quantify them in food matrices, all recent developments pertaining to sample preparation techniques and analytical methods that are currently in use are thus presented here. Furthermore, in order to support the significance and necessity of appropriate analytical methodologies, all of these advancements are linked to general information about the currently available natural or synthetic food colorants, as well as legislative information and toxicological aspects.

"Any substance whose intentional addition to a food aiming for a technological (including organoleptic) purpose in the manufacture, processing, preparation treatment, packing, packaging, transport or holding of such food results, or may be reasonably expected to result, in it or its by-products becoming a component of the food or otherwise affecting the characteristics of such foods," is the definition of a food additive according to Codex Alimentarius. Carocho and associates. emphasized that the terms "contaminants" and "substances added to food for maintaining or improving nutritional qualities" are not included in the Codex Alimentarius definition (Carocho et al., 2015).

Food colorants, which come in various forms, are chemical substances added to food matrices in food technology to maintain or improve the sensory qualities of the food product—aspects that can be impacted or lost during processing or storage—as well as to preserve the intended color appearance. These are categorized according to multiple factors: firstly, whether they are synthetic or natural in origin, and if not, what kind of material they are—organic or inorganic. They could also be categorized according to how soluble they are (e.g., soluble or insoluble) or capacity to cover (e.g., transparent or opaque), however there may be overlaps in one or more of these categories. The distinction between soluble and insoluble color additives (also known as colorants or pigments), which can then be further divided into natural and synthetic categories, forms the basis of the most popular and extensively used classification (Aberamound et al., 2011).

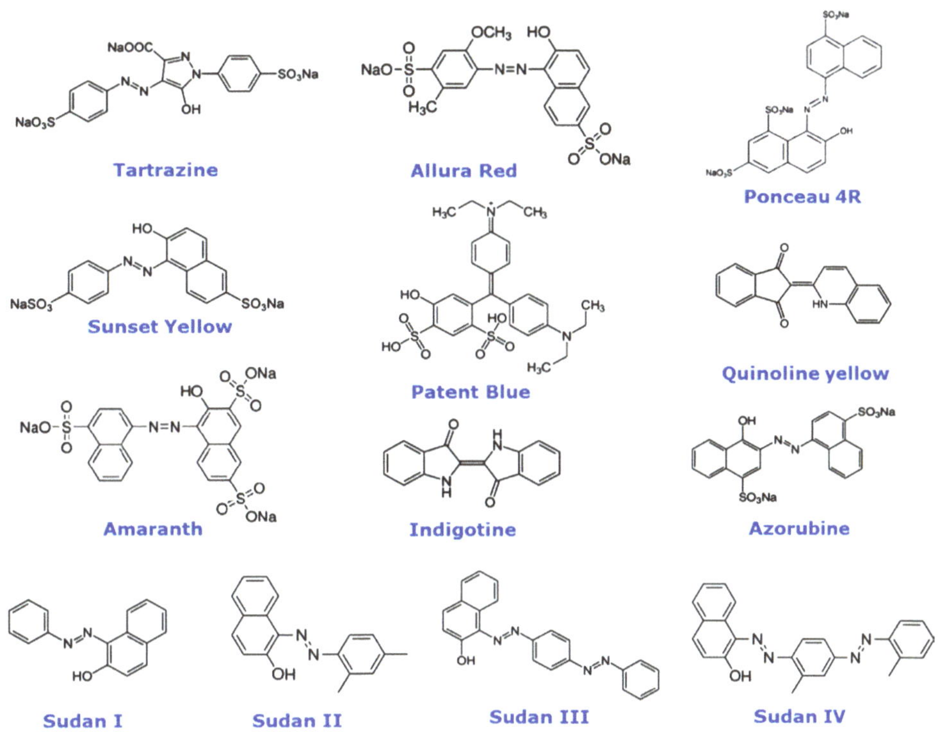

Figure 8.4. *Chemical structures of selected regulated food colorants.*

Source: https://encyclopedia.mdpi.cn/FniRN6GcrFuBq4z9k3OHeHOdfQ8L.

8.5.1. Natural Food Colorants

Since ancient times, natural additives have been utilized. They were utilized in some situations to preserve food. The majority of consumers today appear to prefer the use of natural ingredients over synthetic ones, which the food industry believes to be more effective. Meanwhile, the general elimination of food coloring from food products is a topic of great interest. The vast range of inherent qualities of the coloring compounds causes the classification of naturally derived colorants to become extremely complicated. Natural colorants have a wide range of chemical compositions that affect their properties, solubilities, and stabilities differently because they can be derived from a variety of sources in nature. They can also come from different sources, such as plants or animals (Sigurdson et al., 2017). Naturally occurring food colorants are commonly divided into several subcategories, including phenolic compounds, carotenoids, beet colorants, and anthocyanins. Furthermore, research has been done on annatto, carminic acid, and a few curcuminoids, especially curcumin. Lastly, in order to receive an "E" code authorization, additional colorants still need to be evaluated and assessed.

Anthocyanins, a class of naturally occurring food coloring that ranges in color from red to purple and blue, are primarily found in fruits, flowers, leaves, and even entire plants. They are an extensively researched food coloring group. Commercial anthocyanins have been utilized with success, including cyanidin 3-glucoside, pelargonidin 3-glucoside, and peonidin 3-glucoside.

Another group of naturally occurring pigments with a well-known technological impact are called carotenoids. They are widely used in the food industry as natural preservatives in addition to food colorants because they have coloring qualities as well as certain bioactive and antioxidant qualities. Extracts from algae, yeasts, and aquatic animals, as well as from the roots, flowers, and leaves of plants, are their primary source. The most commonly used carotenoids in this category are lutein, astaxanthin, and lycopene; other carotenoids, like crocin and crocetin, are still being studied.

Beetroot and beets are the sources of reddish-purple colorants (*Beta vulgaris* L.) root is the main source of these natural colorants, but other sources include the fruit of Opuntia *Ficus-indica* [L.], Hylocereus polyrhizus (Weber) Britton, and rose.] *Opuntia stricta* Miller (Haw). Haw and humilis L. Rivina. are also abundant in these colorant compounds, specifically the most researched and approved betacyanins and betalains (E162). They are used in a wide range of food items, including drinks, dairy products, yogurts, burgers, desserts, ice creams, jams, and jellies.

The phenolic compounds, which include flavanones, flavones (4′,5,7-trihydroxyflavones), and flavonols (fisetin, myricetin, myricitin, quercetin, and rutin) are other naturally occurring food coloring agents that have been studied. According to Martins et al. only the goods that are currently on the market are being utilized, i.e., from *Myrica cerifera* L., myricetin and myricitrin. roots). Since the safety, stability, and range of activity of phenolic compounds are still unknown, many of them are currently being researched and evaluated. Phenolic compounds do not currently have an approved "E" code or an ADI value. Curcumin (E100), which is typically extracted from *Curcuma longa* L., is the most well-known and utilized colorant in the group of curcuminoids, another class of natural food coloring. Rhizomes. The annatto (E160b) group and the extracts of *Bixa orellana* L., bixin and norbixin, are other naturally occurring colorants that are used. seeds. Furthermore, a lot of carminic acid (E120), which has an ADI of 5 mg/kg b and a yellow to red-orange food color, is already in use. It can be synthetic or naturally occurring. or crocin. However, additional food colorants are being studied, such as c-phycoerythrin, a reddish-orange pigment derived from blue-green algae, and c-phycocyanin, a blue pigment isolated from *Arthrospira* platensis. Other naturally occurring commercially available pigments under investigation include purple corn color, monascorubrin, and geniposide.

8.5.2. Synthetic Food Colorants

Many artificial food colorants have been developed for use in food production to increase specific quality and organoleptic characteristics in response to growing consumer demand, primarily for visually appealing products. But according to reports, most artificial food colorings were eventually eliminated because of recurring adverse effects, toxicity over the short- and long-terms, and even the possibility of cancer. Thus, a change in consumer expectations has been reported, which is largely in favor of the natural colorants. In addition, from a regulatory perspective, the risk assessment of these food product colorants is gaining more and more attention and interest. e. azo-dyes). The potential carcinogenicity of azo-dyes, which arises from their reduction to carcinogenic metabolites in the intestine, limits the applications of these dyes. The human body produces these metabolites; however, the level of the target substance or colorant ingested determines how harmful they are. Regulatory agencies are reportedly required to routinely review and assess the possible toxicity of food colorings.

8.5.3. Toxicological Aspects and Regulatory Framework

Numerous toxicity effects, such as behavioral effects on children, effects on the respiratory system, connections to allergies, development of attention deficit hyperactivity disorder (ADHD) in children, or neuro-developmental effects at the No-Adverse Effect Limit levels, have been reported based on various scientific findings. In any event, more research is required to evaluate the possible risks that may be connected to these compounds. In a nutshell, Regulation (EU) No contains the authorization procedure within the European regulatory framework. 1331/2008, Regulations (EU) No. 1333/2008 and 1129/2011, which list authorized color additives along with usage guidelines; Regulation (EU) No. 231/2012, which specifies requirements for food additives; and, lastly, Regulations (EU) No. 1331/2008, which deal with labeling. 1333/2008 and 1169/2011. Accordingly, Title 21 CFR Part 70 in the US lists color additives; Title 21 CFR Part 73 lists food additives (exempt from certification, including specifications and conditions of use); and Title 21 CFR Part 80 certifies donor additives. In order to determine and assess acute, sub-chronic, and chronic toxicity, as well as carcinogenicity, mutagenicity, teratotoxicity, genotoxicity, reproductive toxicity, accumulation in the body, bioenergy effects, and immunotoxicity, the use of food colorants in the production of food products is subject to a number of toxicity tests.

8.5.4. Analytical Methodologies for the Determination of Food Colorants

The most common method for analyzing natural food colorings is HPLC coupled with either Diode Array (DAD) or spectrophotometric (UV-Vis) detectors, and then HPLC with MS/MS. Researchers in this field also appear to favor spectrophotometric UV-Vis methods because they are low-cost instruments that don't require specialized knowledge. It should be remembered, though, that the specific characteristics of the spectra acquired for individual colors greatly depend on the solution's or the mobile phase's pH being adjusted with the appropriate acid or alkali. The maximum absorption wavelength, where shifts and intensities based on the different pH can be observed, is undoubtedly affected by the pH adjustment. While the preparation of the samples is far less involved than with LC methods, there is one major drawback to these methods: they cannot simultaneously analyze a greater number of food colorants.

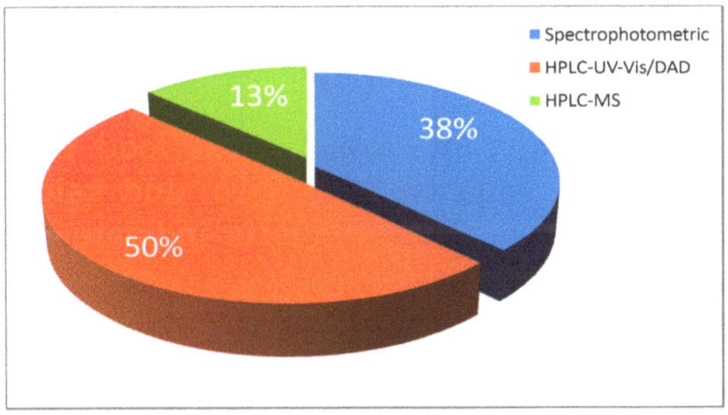

Figure 8.5. *Distribution of techniques used for the analysis of natural food colorants.*

Source: https://encyclopedia.mdpi.cn/FoP8NAmMar46yT60E-LEYIQUSziy.

A greater amount of scientific research has been done in the past 10 years to analyze synthetic colorants in food matrices. This framework includes a large number of LC-MS, LC-MS/MS, and LC-UV/Vis methods that are specifically designed to detect a limited number or a significant number of these analytes simultaneously. The most common technique is clearly LC/HPLC/U(H)PLC, followed by capillary electrophoresis (CE) and enzyme-linked immunosorbent assay (ELISA). With regard to the latter method, it should be noted that monoclonal antibodies have only been developed for single substances; they cannot be used for groups of substances or food colorants.

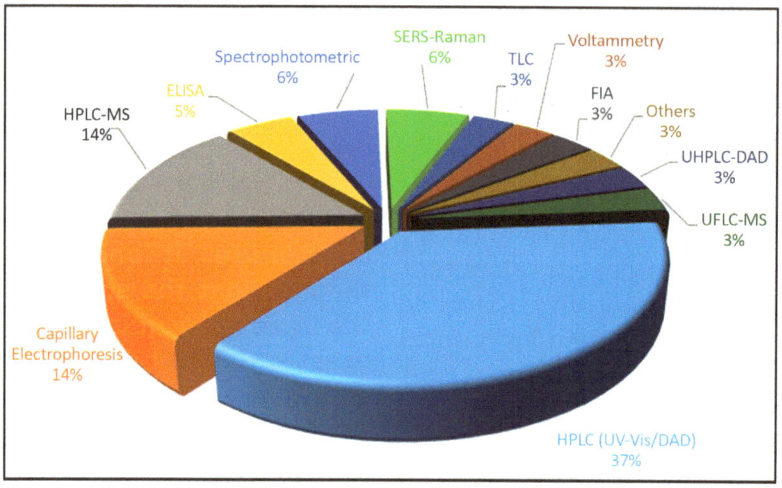

Figure 8.6. *Distribution of techniques for the analysis of synthetic food colorants.*

Source: https://encyclopedia.mdpi.cn/FphAeEHmlJ-ZmAvLMO2BoqMsj8Hr.

Sample preparation is highly sample-dependent for both natural and synthetic colorants. It may entail applying various extraction techniques, such as membrane filtration, liquid-liquid, and solid phase extraction techniques, to clean up the extremely complex matrix of food products. In order to avoid or remove current matrix interferences and develop straightforward, accurate, and selective extraction techniques, sample preparation is crucial and needs to be carefully thought out.

Food colorants are used in food production, which necessitates the development of precise, sensitive, accurate, and selective analytical methods for food analysis and quantification. Global reports indicate a certain level of interest in the effects of food coloring. The analytical problem of the analysis and quantification of either natural or synthetic food colorants is addressed in a multitude of analytical research works.

There is still much to be discovered about analytical methodologies, but based on synthetic colorants, it can be inferred that capillary electrophoresis and HPLC are the most often utilized techniques. The most common detection method is the straightforward UV-Vis/DAD, which is followed by tandem MS. The sample preparation and analytical techniques discussed here cover the methods that have been in use for the past 10 years.

CASE STUDY

CONSUMER COLOR PREFERENCE: FOOD & DRINKS

- **The First Moment of Truth**

Coined by Procter & Gamble, the "First Moment of Truth" is the first 3–7 seconds after a shopper encounters a product on a store shelf. These first seconds are precious to marketers and their best chance of converting a browser into a buyer by appealing to their senses, values and emotions. In the food industry, if a product's look doesn't resonate with its positioned attributes like flavor or wellness, consumers will most often ignore it. Color is an incredibly important part of a consumer's first moment of truth in food and drink products, as it not only delights viewers with an immediate visual experience but also induces flavor perception. Color actually helps bring many products to life.

- **Looking at Food Selection With a Scientific Lens**

It turns out science has a very good explanation for why consumers depend on colors. Scientific Reports recently released the study "The Role of Human Trichromatic Vision in Food Evaluation," which focuses on the visual function of trichromacy. An advanced trait in primates and humans to help them adapt to and survive in changing landscapes, trichromacy refers to our three independent photoreceptors within the retina capable of responding to multiple wavelengths in the visible spectrum of light at once. In simpler terms, the human eye is capable of seeing a color range of violet to red.

- **A Taste for Colorfully Natural Products**

Our attraction to color in natural environments has continued today. Most frequently elicited in the produce sections of retailers, our innate excitement about colors of fruits and vegetables energizes our shopping experience and affects our food selection in other aisles. Our enjoyment of colorful produce sections is probably a blend of our human condition and our more rational, modern desire for less processed foods. The food industry's current push to "naturalness" shouldn't come as a surprise with united demands for the removal of artificial ingredients around the world.

Food color is at the forefront of this global "natural" phenomenon. According to Mintel, natural food colors have taken the lead in all regions and continue to be favored in global food and drink products. In addition, as many as two-thirds of global consumers (68%) said they were willing to pay more for foods without undesirable ingredients (Nielsen). Today's consumers want to see the color from fruit and vegetable juices on product labels.Given this information, we were curious about today's preferred and expected colors from nature in food and drink products. Do we still gravitate towards shades of red? Does color preference change across categories? Etc.

- **Consumer Color Preference Study**

To satisfy our curiosity, Sensient conducted national research to identify retail offerings believed to be the best representation of consumers' expectations in our "age of naturalness." It is important to note our study was completed by North American consumers, but the data applies to global food manufacturers as colors are a universal expression of the human condition. The following question was posed for each survey

If a new food or beverage product was being introduced into the market that was only made from all-natural ingredients and offered a superior nutritional profile, which of the following visuals best represents your expectations of the product's look? Nature not only provides us with boundless color opportunity but also a bounty of natural color sources. Luckily today, expertise in extraction and standardization provides food manufacturers with application-ready color solutions from fruit and vegetable juices to cater to these consumer expectations.

- **Don't Let Nature Lose Its Luster**

Advanced technologies in today's market enable color suppliers to provide application-ready solutions to brands reflective of nature's brilliance. With the abundance of natural color sources, we are able to achieve more shades today than ever before.

Global natural color innovation allows color application specialists to achieve shades that have never been possible from FD&C colors. The number of shade offerings from natural color sources compared to artificial sources is immensely different. However, in the spirit of naturalness and simplicity, some brands choose dull products that are making flavor or wellness claims associated with bright fruits, vegetables, florals, etc.

Why are brands letting nature lose its luster in this age of "naturalness?" It's unclear why brands fear color for natural products when color capabilities from natural sources are better than ever. While color from botanical and fruit sources are inherently more complex, vivid and bright shades are certainly possible. Compromise on shade and flavor is unnecessary.

One thing is clear though — consumers prefer naturally colored products in vibrant shades. Entice the conscious consumers of today with nature's brilliance in your application-specific color solutions from Sensient.

CLASS ACTIVITY

Summarize key points on the whiteboard/flip chart, highlighting differences between natural and synthetic colorants, and any interesting observations from the experiments. And Assign a task where students must choose a food product from their home, research its colorants, and write a short report on its benefits and potential risks.

SUMMARY

- The use of pesticides, irradiation, food additives, and even trace substances in the production and preservation of food, has generated a high concern among consumers, due to the possible short, medium, and long-term health effects.
- Organoleptic characteristics largely determine the acceptance, selection, and subsequent consumption of foods.
- Color can be considered one of the most impressive and charming attributes of foods, and although natural food products have their own color, the different processes they undergo and factors, such as the presence or absence of oxygen, metals, light, pH, and water activity, can produce undesirable modifications.
- It should be emphasized that, in general, the stability can be understood as one of the most important factors for the use of natural food colorants, since unpleasant organoleptic characteristics appear when unstable food colors are applied, leading to food rejection by consumers.
- Antimicrobial compounds can be naturally present in foods or added to delay and/or prevent the proliferation and growth of natural microorganisms responsible for the spoilage of foods (bacteria, yeasts, and molds) as well as to prevent/control contamination by pathogenic microorganisms, thus ensuring food safety and quality.
- Molecular oxygen oxidation is the primary degradation process in foods; it promotes a loss of nutritional value and produces an undesirable taste and smell.
- Color additives are used extensively in several food products, as to enhance their appearance, and their nutritional properties.
- Naturally occurring food colorants can be allocated in different sub-categories, namely anthocyanins, carotenoids, beet colorants, and phenolic compounds. In addition, annatto, carminic acid, and some curcuminoids have been studied, particularly curcumin.

REVIEW QUESTIONS

1. What are the food additives and their demands?
2. What are the food colorants? Explain in detail.
3. Discuss about the food preservatives.
4. Explain the toxicological aspects and regulatory framework of food colorants.
5. What are the analytical methodologies for the determination of food colorants?

REFERENCES

1. Ab Rashid, S., Tong, W. Y., Leong, C. R., Abdul Ghazali, N. M., Taher, M. A., Ahmad, N., et al. (2021). Anthocyanin microcapsule from *Clitoria ternatea*: Potential

bio-preservative and blue colorant for baked food products. *Arabian Journal for Science and Engineering, 46*(1), 65–72. https://doi.org/10.1007/s13369-020-04716-y.

2. Abd-Elhakim, Y. M., Hashem, M. M., El-Metwally, A. E., Anwar, A., Abo-EL-Sooud, K., Moustafa, G. G., et al. (2018). Comparative haemato-immunotoxic impacts of long-term exposure to tartrazine and chlorophyll in rats. *International Immunopharmacology, 63*, 145–154. https://doi.org/10.1016/j.intimp.2018.08.002.
3. Aberamound, A. (2011). A review article on edible pigments properties and sources as natural biocolorants in foodstuff and food industry. *World Journal of Dairy and Food Sciences, 6*, 71–78.
4. Abo-EL-Sooud, K., Hashem, M. M., Badr, Y. A., Eleiwa, M. M. E., Gab-Allaha, A. Q., Abd-Elhakim, Y. M., et al. (2018). Assessment of hepato-renal damage and genotoxicity induced by long-term exposure to five permitted food additives in rats. *Environmental Science and Pollution Research, 25*(26), 26341–26350. https://doi.org/10.1007/s11356-018-2665-z.
5. Aguayo, E., Requejo-Jackman, C., Stanley, R., & Woolf, A. (2010). Effects of calcium ascorbate treatments and storage atmosphere on antioxidant activity and quality of fresh-cut apple slices. *Postharvest Biology and Technology, 57*(1), 52–60. https://doi.org/10.1016/j.postharvbio.2010.03.001.
6. Alandes, L., Hernando, I., Quiles, A., Pérez-Munuera, I., & Lluch, M. A. (2006). Cell wall stability of fresh-cut Fuji apples treated with calcium lactate. *Journal of Food Science, 71*(9), S615–S620. https://doi.org/10.1111/j.1750-3841.2006.00180.x.
7. Albuquerque, B. R., Oliveira, M. B. P. P., Barros, L., & Ferreira, I. C. F. R. (2021). Could fruits be a reliable source of food colorants? Pros and cons of these natural additives. *Critical Reviews in Food Science and Nutrition, 61*(5), 805–835. https://doi.org/10.1080/10408398.2020.1746904.
8. Al-Habib, A., Al-Saleh, E., Safer, A.-M., & Afzal, M. (2010). Bactericidal effect of grape seed extract on methicillin-resistant *Staphylococcus aureus* (MRSA). *Journal of Toxicological Sciences, 35*(3), 357–364. https://doi.org/10.2131/jts.35.357.
9. Amchova, P., Kotolova, H., & Ruda-Kucerova, J. (2015). Health safety issues of synthetic food colorants. *Regulatory Toxicology and Pharmacology, 73*(3), 914–922. https://doi.org/10.1016/j.yrtph.2015.09.026.
10. Aparicio, I. M. (2017). *Aditivos alimentarios* (1st ed.). Dextra Editorial.
11. Arioglu-Tuncil, S., Voelker, A. L., Taylor, L. S., & Mauer, L. J. (2020). Amorphization of thiamine chloride hydrochloride: Effects of physical state and polymer type on the chemical stability of thiamine in solid dispersions. *International Journal of Molecular Sciences, 21*(16), 5935. https://doi.org/10.3390/ijms21165935.
12. Arqués, J. L., Rodríguez, E., Nuñez, M., & Medina, M. (2011). Combined effect of reuterin and lactic acid bacteria bacteriocins on the inactivation of food-borne pathogens in milk. *Food Control, 22*(3–4), 457–461. https://doi.org/10.1016/j.foodcont.2010.09.027.
13. Arvanitoyannis, I. S., & Van Houwelingen-Koukaliaroglou, M. (2005). Functional foods: A survey of health claims, pros and cons, and current legislation. *Critical Reviews in Food Science and Nutrition, 45*(5), 385–404. https://doi.org/10.1080/10408390590967667.

14. Atta, E. M., Mohamed, N. H., & Abdelgawad, A. A. M. (2017). Antioxidants: An overview on the natural and synthetic types. *European Chemical Bulletin, 6*(8), 365. https://doi.org/10.17628/ecb.2017.6.365-375.

15. Ayala, A., Muñoz, M. F., & Argüelles, S. (2014). Lipid peroxidation: Production, metabolism, and signaling mechanisms of malondialdehyde and 4-hydroxy-2-nonenal. *Oxidative Medicine and Cellular Longevity, 2014*, 360438. https://doi.org/10.1155/2014/360438.

16. Babu, D., Kushwaha, K., Sehgal, S., & Juneja, V. K. (2017). Antimicrobials of plant origin. In V. Juneja, H. Dwivedi, & J. Sofos (Eds.), *Microbial Control and Food Preservation* (pp. 85–100). Springer. https://doi.org/10.1007/978-1-4939-7556-3_5.

17. Baines, D., & Seal, R. (2012). *Natural Food Additives, Ingredients and Flavorings.* Woodhead Publishing, Cambridge, U.K.

18. Bakowska-Barczak, A. (2005). Acylated anthocyanins as stable, natural food colorants – A review. *Polish Journal of Food and Nutrition Sciences, 14*(55), 107–116.

19. Banipal, T. S., Kaur, H., Kaur, A., & Banipal, P. K. (2016). Effect of tartarate and citrate-based food additives on the micellar properties of sodium dodecylsulfate for prospective use as food emulsifier. *Food Chemistry, 190*, 599–606. https://doi.org/10.1016/j.foodchem.2015.05.130.

20. Baptista, R. C., Horita, C. N., & Sant'Ana, A. S. (2020). Natural products with preservative properties for enhancing the microbiological safety and extending the shelf-life of seafood: A review. *Food Research International, 127*, 108762. https://doi.org/10.1016/j.foodres.2019.108762.

21. Barbosa-Pereira, L., Cruz, J. M., Sendón, R., Rodríguez Bernaldo de Quirós, A., Ares, A., Castro-López, M., et al. (2013). Development of antioxidant active films containing tocopherols to extend the shelf life of fish. *Food Control, 31*(1), 236–243. https://doi.org/10.1016/j.foodcont.2012.09.036.

22. Belitz, H.-D., Groch, W., & Schieberle, P. (2009). *Food Chemistry* (4th ed.). Springer.

23. Bensid, A., El Abed, N., Houicher, A., Regenstein, J. M., & Ozogul, F. (2020). Antioxidant and antimicrobial preservatives: Properties, mechanism of action and applications in food–A review. *Critical Reviews in Food Science and Nutrition, 1.* https://doi.org/10.1080/10408398.2020.1862046.

24. Berardo, A., de Maere, H., Stavropoulou, D. A., Rysman, T., Leroy, F., & de Smet, S. (2016). Effect of sodium ascorbate and sodium nitrite on protein and lipid oxidation in dry fermented sausages. *Meat Science, 121*, 359–364. https://doi.org/10.1016/j.meatsci.2016.07.003.

25. Berdahl, D. R., Nahas, R. I., & Barren, J. P. (2010). Synthetic and natural antioxidant additives in food stabilization: Current applications and future research. In *Oxidation in Foods and Beverages and Antioxidant Applications: Understanding Mechanisms of Oxidation and Antioxidant Activity* (pp. 272–320). https://doi.org/10.1533/9780857090447.2.272.

26. Bernardi, R. C., Gomes, D. E. B., Gobato, R., Taft, C. A., Ota, A. T., & Pascutti, P. G. (2009). Molecular dynamics study of biomembrane/local anesthetics interactions. *Molecular Physics, 107*(14), 1437–1443. https://doi.org/10.1080/00268970902926238.

27. Bhatt, D., Vyas, K., Singh, S., John, P. J., & Soni, I. (2018). Tartrazine induced neurobiochemical alterations in rat brain sub-regions. *Food and Chemical Toxicology, 113*, 322–327. https://doi.org/10.1016/j.fct.2018.02.011.

28. Binstok, G., Campos, C., Varela, O., & Gerschenson, L. N. (1998). Sorbate–nitrite reactions in meat products. *Food Research International, 31*(8), 581–585. https://doi.org/10.1016/S0963-9969(99)00031-9.

29. Birti⊠, S., Dussort, P., Pierre, F. X., Bily, A. C., & Roller, M. (2015). Carnosic acid. *Phytochemistry, 115*(1), 9–19. https://doi.org/10.1016/j.phytochem.2014.12.026.

30. Błaszczyk, A., Augustyniak, A., & Skolimowski, J. (2013). Ethoxyquin: An antioxidant used in animal feed. *International Journal of Food Science, 2013*, 585931. https://doi.org/10.1155/2013/585931.

31. Borges, M. E., Tejera, R. L., Díaz, L., Esparza, P., & Ibáñez, E. (2012). Natural dyes extraction from cochineal (*Dactylopius coccus*): New extraction methods. *Food Chemistry, 132*(4), 1855–1860. https://doi.org/10.1016/j.foodchem.2011.12.018.

32. Boualem, K., Subirade, M., Desjardins, Y., & Saucier, L. (2013). Development of an encapsulation system for the protection and controlled release of antimicrobial nisin at meat cooking temperature. *Journal of Food Research, 2*, 36–45. https://doi.org/10.5539/jfr.v2n3p36.

33. Bourrier, T. (2006). Intolerance and allergy to colorants and additives. *Revue Française d'Allergologie et d'Immunologie Clinique, 46*(2), 68–79. https://doi.org/10.1016/j.allerg.2005.12.002.

34. Boziaris, I. S., Proestos, C., Kapsokefalou, M., & Komaitis, M. (2011). Antimicrobial effect of *Filipendula ulmaria* plant extract against selected foodborne pathogenic and spoilage bacteria in laboratory media, fish flesh and fish roe product. *Food Technology and Biotechnology, 49*(2), 263–270.

35. Branen, A. L., Davidson, P. M., Salminen, S., & Thorngate, J. (2001). *Food Additives* (2nd ed.). Taylor & Francis.

36. Burrows, A. J. D. (2009). Palette of our palates: A brief history of food coloring and its regulation. *Comprehensive Reviews in Food Science and Food Safety, 9*, 394–408.

37. Caflisch, A., & Paci, E. (2008). Molecular dynamics simulations to study protein folding and unfolding. In *Protein Science Encyclopedia: Online* (pp. 1143–1169). https://doi.org/10.1002/9783527610754.sf11.

38. Cammack, R., Joannou, C. L., Cui, X. Y., Torres Martinez, C., Maraj, S. R., & Hughes, M. N. (1999). Nitrite and nitrosyl compounds in food preservation. *Biochimica et Biophysica Acta (BBA)–Bioenergetics, 1411*(2–3), 475–488. https://doi.org/10.1016/S0005-2728(99)00033-X.

39. Capuano, E., Oliviero, T., & van Boekel, M. A. J. S. (2018). Modeling food matrix effects on chemical reactivity: Challenges and perspectives. *Critical Reviews in Food Science and Nutrition, 58*(16), 2814–2828. https://doi.org/10.1080/10408398.2017.1342595.

40. Carocho, M., & Ferreira, I. C. (2013). A review on antioxidants, prooxidants and related controversy: Natural and synthetic compounds, screening and analysis methodologies and future perspectives. *Food and Chemical Toxicology, 51*, 15–25.

41. Carocho, M., Barreiro, M. F., Morales, P., & Ferreira, I. C. F. R. (2014). Adding molecules to food, pros and cons: A review on synthetic and natural food additives. *Comprehensive Reviews in Food Science and Food Safety, 13*(4), 377–399. https://doi.org/10.1111/1541-4337.12065.

42. Carocho, M., Morales, P., & Ferreira, I. C. F. R. (2015). Natural food additives: Quo Vadis? *Trends in Food Science & Technology, 45*(2), 284–295. https://doi.org/10.1016/j.tifs.2015.06.007.

43. Carocho, M., Morales, P., & Ferreira, I. C. F. R. (2018). Antioxidants: Reviewing the chemistry, food applications, legislation and role as preservatives. *Trends in Food Science & Technology, 71*, 107–120. https://doi.org/10.1016/j.tifs.2017.11.008.

44. Chammem, N., Saoudi, S., Sifaoui, I., Sifi, S., de Person, M., Abderraba, M., et al. (2015). Improvement of vegetable oils quality in frying conditions by adding rosemary extract. *Industrial Crops and Products, 74*, 592–599. https://doi.org/10.1016/j.indcrop.2015.05.054.

45. Chaudhari, A. K., Singh, V. K., Das, S., Deepika, Prasad, J., Dwivedy, A. K., et al. (2020). Improvement of in vitro and in situ antifungal, AFB1 inhibitory and antioxidant activity of *Origanum majorana* L. essential oil through nanoemulsion and recommending as novel food preservative. *Food and Chemical Toxicology, 143*, 111536. https://doi.org/10.1016/j.fct.2020.111536.

46. Cheng, C., Arritt, F., & Stevenson, C. (2015). Controlling *Listeria monocytogenes* in cold smoked salmon with the antimicrobial peptide salmine. *Journal of Food Science, 80*(6), M1314–M1318. https://doi.org/10.1111/1750-3841.12886.

47. Codex Alimentarius. (2018). Retrieved from: http://www.codexalimentarius.org/standards/gsfa/ (accessed on 3 July 2024).

48. Crist, C. A., Williams, J. B., Schilling, M. W., Hood, A. F., Smith, B. S., & Campano, S. G. (2014). Impact of sodium lactate and vinegar derivatives on the quality of fresh Italian pork sausage links. *Meat Science, 96*(4), 1509–1516. https://doi.org/10.1016/j.meatsci.2013.11.016.

49. Cruzen, S. M., Kim, Y. H. B., Lonergan, S. M., Grubbs, J. K., Fritchen, A. N., & Huff-Lonergan, E. (2015). Effect of early postmortem enhancement of calcium lactate/phosphate on quality attributes of beef round muscles under different packaging systems. *Meat Science, 101*, 63–72. https://doi.org/10.1016/j.meatsci.2014.11.004.

50. da Cruz Cabral, L., Fernández Pinto, V., & Patriarca, A. (2013). Application of plant derived compounds to control fungal spoilage and mycotoxin production in foods. *International Journal of Food Microbiology, 166*(1), 1–14. https://doi.org/10.1016/j.ijfoodmicro.2013.05.026.

51. Dalié, D. K. D., Deschamps, A. M., & Richard-Forget, F. (2010). Lactic acid bacteria–Potential for control of mold growth and mycotoxins: A review. *Food Control, 21*(4), 370–380. https://doi.org/10.1016/j.foodcont.2009.07.011.

52. Damodaran, S., & Parkin, K. L. (2017). *Fennema's Food Chemistry* (5th ed.). CRC Press.

53. Davidson, P. M., & Zivanovic, S. (2003). The use of natural antimicrobials. In M. S. Borad & G. W. Gould (Eds.), *Food Preservation Techniques* (pp. 5–30). https://doi.org/10.1533/9781855737143.1.5.
54. Davidson, P. M., Taylor, T. M., & Schmidt, S. E. (2013). Chemical Preservatives and Natural Antimicrobial Compounds. In M. P. Doyle & R. L. Buchanan (Eds.), *Food Microbiology: Fundamentals and Frontiers* (4th ed., pp. 765–802). American Society for Microbiology.
55. de Mejia, E. G., Zhang, Q., Penta, K., Eroglu, A., & Lila, M. A. (2020). The Colors of Health: Chemistry, Bioactivity, and Market Demand for Colorful Foods and Natural Food Sources of Colorants. *Annual Review of Food Science and Technology, 11*, 145–182. https://doi.org/10.1146/annurev-food-032519-051729.
56. De'Nobili, M. D., Soria, M., Martinefski, M. R., Tripodi, V. P., Fissore, E. N., & Rojas, A. M. (2016). Stability of L- (+)-ascorbic acid in alginate edible films loaded with citric acid for antioxidant food preservation. *Journal of Food Engineering, 175*, 1–7. https://doi.org/10.1016/j.jfoodeng.2015.11.015.
57. Del Campo, J. A., Garcia-Gonzalez, M., & Guerrero, M. G. (2007). Outdoor cultivation of microalgae for carotenoid production: Current state and perspectives. *Applied Microbiology and Biotechnology, 74*, 1163–1174. https://doi.org/10.1007/s00253-007-0844-9.
58. Delgado-Vargas, F. (2002). *Natural Colorants for Food and Nutraceutical Uses*. CRC Press. https://doi.org/10.1201/9781420031713.
59. Devlieghere, F., & Debevere, J. (2000). Influence of dissolved carbon dioxide on the growth of spoilage bacteria. *LWT–Food Science and Technology, 33*(8), 531–537. https://doi.org/10.1006/fstl.2000.0705.
60. Dias, M. G., Camoes, M. F. G., & Oliveira, L. (2009). Carotenoids in traditional Portuguese fruits and vegetables. *Food Chemistry, 113*, 808–815.
61. Dias, M. I., Ferreira, I. C. F. R., & Barreiro, M. F. (2015). Microencapsulation of bioactives for food applications. *Food & Function, 6*(4), 1035–1052. https://doi.org/10.1039/C4FO01175A.
62. Dufossé, L., Fouillaud, M., Caro, Y., Mapari, S. A. S., & Sutthiwong, N. (2014). Filamentous fungi are large-scale producers of pigments and colorants for the food industry. *Current Opinion in Biotechnology, 26*, 56–61. https://doi.org/10.1016/j.copbio.2013.09.007.
63. EFSA Panel on Food Additives and Nutrient Sources added to Food (ANS). (2015). Scientific opinion on the re-evaluation of ascorbic acid (E300), sodium ascorbate (E301) and calcium ascorbate (E302) as food additives. *EFSA Journal, 13*(5), 4087. https://doi.org/10.2903/j.efsa.2015.4087.
64. EFSA Panel on Food Additives and Nutrient Sources added to Food (ANS). (2015). Scientific opinion on the re-evaluation of ascorbyl palmitate (E304(i)) and ascorbyl stearate (E304(ii)) as food additives. *EFSA Journal, 13*, 4289–4346. https://doi.org/10.2903/j.efsa.2015.4289.
65. EFSA Panel on Food Additives and Nutrient Sources added to Food (ANS). (2011). Scientific opinion on the re-evaluation of butylated hydroxyanisole–BHA (E320) as a food additive. *EFSA Journal, 9*, 2392–2441. https://doi.org/10.2903/j.efsa.2011.2392.

66. EFSA Panel on Food Additives and Nutrient Sources added to Food (ANS). (2012). Scientific opinion on the re-evaluation of mixed carotenes (E160a(i)) and beta-carotene (E160a(ii)) as a food additive. *EFSA Journal, 10*(3), 2593. https://doi.org/10.2903/j.efsa.2012.2593.

67. Eriksen, N. T. (2008). Production of phycocyanin–A pigment with applications in biology, biotechnology, foods and medicine. *Applied Microbiology and Biotechnology, 80*(1), 1–14. https://doi.org/10.1007/s00253-008-1542-y.

68. Esatbeyoglu, T., Wagner, A. E., Schini-Kerth, V. B., & Rimbach, G. (2015). Betanin—A food colorant with biological activity. *Molecular Nutrition & Food Research, 59*(1), 36–47. https://doi.org/10.1002/mnfr.201400484.

69. European Union. (2011). Commission Regulation No. 1129/2011 amending Annex II to Regulation (EC) No 1333/2008 of the European parliament and of the council by establishing a union list of food additives.

70. Fennema, O. R. (1987). Food additives—An unending controversy. *American Journal of Clinical Nutrition, 46*, 201–203. https://doi.org/10.1093/ajcn/46.1.201.

71. Ferreira, I. C. F. R., Morales, P., & Barros, L. (Eds.). (2017). *Wild Plants, Mushrooms and Nuts: Functional Food Properties and Applications* (1st ed.). Wiley.

72. Finamore, A., Palmery, M., Bensehaila, S., & Peluso, I. (2017). Antioxidant, immunomodulating, and microbial-modulating activities of the sustainable and ecofriendly *Spirulina*. *Oxidative Medicine and Cellular Longevity, 2017*, 3247528. https://doi.org/10.1155/2017/3247528.

73. Fundo, J. F., Quintas, M. A. C., & Silva, C. L. M. (2015). Molecular dynamics and structure in physical properties and stability of food systems. *Food Engineering Reviews, 7*(4), 384–392. https://doi.org/10.1007/s12393-015-9109-z.

74. Gallen, C., & Pla, J. (2013). Allergie et intolérance aux additifs alimentaires. *Revue Française d'Allergologie, 53*(S1), 9–18. https://doi.org/10.1016/S1877-0320(13)70044-7.

75. Galvez, A., Grande Burgos, M. J., Lucas Lopez, R., & Perez Pulido, R. (2014). Natural antimicrobials for food bio preservation. In *Food Bio Preservation* (Springer Briefs in Food, Health, and Nutrition, pp. 2–14). Springer.

76. Gao, M., Ling, B., Yuan, Q., Xiong, Q., & Yang, L. (2014). A robust collaborative filtering approach based on user relationships for recommendation systems. *Mathematical Problems in Engineering, 2014*, 162521. https://doi.org/10.1155/2014/162521.

77. Gengatharan, A., Dykes, G. A., & Choo, W. S. (2015). Betalains: Natural plant pigments with potential application in functional foods. *LWT–Food Science and Technology, 64*(2), 645–649. https://doi.org/10.1016/j.lwt.2015.06.052.

78. Ghanbari, M., Jami, M., Domig, K. J., & Kneifel, W. (2013). Seafood bio-preservation by lactic acid bacteria–A review. *LWT–Food Science and Technology, 54*(2), 315–324. https://doi.org/10.1016/j.lwt.2013.05.039.

79. Gimeno, O., Astiasarán, I., & Bello, J. (2001). Calcium ascorbate as a potential partial substitute for NaCl in dry fermented sausages: Effect on color, texture and hygienic quality at different concentrations. *Meat Science, 57*(1), 23–29. https://doi.org/10.1016/S0309-1740(00)00070-X.

80. Gokoglu, N. (2019). Novel natural food preservatives and applications in seafood preservation: A review. *Journal of the Science of Food and Agriculture, 99*(5), 2068–2077. https://doi.org/10.1002/jsfa.9416.

81. Gøtterup, J., Olsen, K., Knöchel, S., Tjener, K., Stahnke, L. H., & Møller, J. K. S. (2007). Relationship between nitrate/nitrite reductase activities in meat associated staphylococci and nitrosyl myoglobin formation in a cured meat model system. *International Journal of Food Microbiology, 120*(3), 303–310. https://doi.org/10.1016/j.ijfoodmicro.2007.08.034.

82. Gulcin, ⊠. (2020). Antioxidants and antioxidant methods: An updated overview. *Archives of Toxicology, 94*(3), 651–715. https://doi.org/10.1007/s00204-020-02689-3.

83. Guo, Q., Gao, S., Sun, Y., Gao, Y., Wang, X., & Zhang, Z. (2016). Antioxidant efficacy of rosemary ethanol extract in palm oil during frying and accelerated storage. *Industrial Crops and Products, 94*, 82–88. https://doi.org/10.1016/j.indcrop.2016.08.032.

84. Gyawali, R., & Ibrahim, S. A. (2014). Natural products as antimicrobial agents. *Food Control, 46*, 412–429. https://doi.org/10.1016/j.foodcont.2014.05.047.

85. Hassoun, A., & Emir Çoban, Ö. (2017). Essential oils for antimicrobial and antioxidant applications in fish and other seafood products. *Trends in Food Science & Technology, 68*, 26–36. https://doi.org/10.1016/j.tifs.2017.07.016.

86. Hayek, S. A., Gyawali, R., & Ibrahim, S. A. (2013). Antimicrobial natural products. In Méndez-Vilas, A. (Ed.), *Microbial Pathogens and Strategies for Combating Them: Science, Technology and Education* (pp. 910–921). Formatex Research Center.

87. Hintz, T., Matthews, K. K., & Di, R. (2015). The use of plant antimicrobial compounds for food preservation. *BioMed Research International, 2015*, 246264. https://doi.org/10.1155/2015/246264.

88. Honikel, K. O. (2008). The use and control of nitrate and nitrite for the processing of meat products. *Meat Science, 78*(1–2), 68–76. https://doi.org/10.1016/j.meatsci.2007.05.030.

89. Hui, Y. H. (2006). *Handbook of Food Science, Technology, and Engineering* (Vol. 1, 1st ed.). CRC Press, Taylor & Francis.

90. Hussain, N., Irshad, F., Jabeen, Z., Shamsi, I. H., Li, Z., & Jiang, L. (2013). Biosynthesis, structural, and functional attributes of tocopherol in plants; past, present, and future perspectives. *Journal of Agricultural and Food Chemistry, 61*, 6137–6149. https://doi.org/10.1021/jf4010302.

91. Iammarino, M., di Taranto, A., & Cristino, M. (2013). Endogenous levels of nitrites and nitrates in wide consumption foodstuffs: Results of five years of official controls and monitoring. *Food Chemistry, 140*(4), 763–771. https://doi.org/10.1016/j.foodchem.2012.10.094.

92. Iammarino, M., di Taranto, A., & Muscarella, M. (2012). Investigation on the presence of sulphites in fresh meat preparations: Estimation of an allowable maximum limit. *Meat Science, 90*(2), 304–308. https://doi.org/10.1016/j.meatsci.2011.07.015

93. Ilhak, O. I., & Guran, H. S. (2014). Combined antimicrobial effect of thymol and sodium lactate against *Listeria monocytogenes* and *Salmonella typhimurium* in fish patty. *Journal of Food Safety, 34*(3), 211–217. https://doi.org/10.1111/jfs.12115.

94. Inanli, A. G., Tümerkan, E. T. A., Abed, N. el, Regenstein, J. M., & Özogul, F. (2020). The impact of chitosan on seafood quality and human health: A review. *Trends in Food Science & Technology, 97*, 404–416. https://doi.org/10.1016/j.tifs.2020.01.029.

95. Jane Vickers, P., Braybrook, J., Lawrence, P., & Gray, K. (2007). Detecting tartrate additives in foods: Evaluating the use of capillary electrophoresis. *Journal of Food Composition and Analysis, 20*(3–4), 252–256. https://doi.org/10.1016/j.jfca.2006.05.002

96. Jasour, M. S., Ehsani, A., Mehryar, L., & Naghibi, S. S. (2015). Chitosan coating incorporated with the lactoperoxidase system: An active edible coating for fish preservation. *Journal of the Science of Food and Agriculture, 95*(6), 1373–1378. https://doi.org/10.1002/jsfa.6838.

97. Jones, J. A., Vernacchio, V. R., Collins, S. M., Shirke, A. N., Xiu, Y., Englaender, J. A., Cress, B. F., McCutcheon, C. C., Linhardt, R. J., Gross, R. A., & Koffas, M. A. G. (2017). Complete biosynthesis of anthocyanins using *E. coli* polycultures. *mBio, 8*(3), e00621–17. https://doi.org/10.1128/mBio.00621-17.

98. Jongberg, S., Tørngren, M. A., Gunvig, A., Skibsted, L. H., & Lund, M. N. (2013). Effect of green tea or rosemary extract on protein oxidation in Bologna type sausages prepared from oxidatively stressed pork. *Meat Science, 93*(3), 538–546. https://doi.org/10.1016/j.meatsci.2012.11.005.

99. Joo, N. E., Ritchie, K., Kamarajan, P., Miao, D., & Kapila, Y. (2012). Nisin, an apoptogenic bacteriocin and food preservative, attenuates HNSCC tumorigenesis via CHAC1. *Cancer Medicine, 1*, 295–305. https://doi.org/10.1002/cam4.35.

100. Junqueira-Goncalves, M., Yanez, L., Morales, C., Navarro, M., Contreras, R. A., & Zuniga, G. (2015). Isolation and characterization of phenolic compounds and anthocyanins from Murta (*Ugni molinae Turcz.*) fruits. Assessment of antioxidant and antibacterial activity. *Molecules, 20*(4), 5698–5713. https://doi.org/10.3390/molecules20045698.

101. Kaewprachu, P., Osako, K., Benjakul, S., & Rawdkuen, S. (2015). Quality attributes of minced pork wrapped with catechin–lysozyme incorporated gelatin film. *Food Packaging and Shelf Life, 3*, 88–96. https://doi.org/10.1016/j.fpsl.2014.11.002.

102. Kallinteri, L. D., Kostoula, O. K., & Savvaidis, I. N. (2013). Efficacy of nisin and/or natamycin to improve the shelf-life of Galotyri cheese. *Food Microbiology, 36*, 176–181. https://doi.org/10.1016/j.fm.2013.05.006.

103. Kalpana, V. N., & Rajeswari, V. D. (2019). Preservatives in beverages: Perception and needs. In *Preservatives and Preservation Approaches in Beverages* (pp. 1–30). The Science of Beverages, 15. https://doi.org/10.1016/B978-0-12-816685-7.00001-X.

104. Kamal, A. A., & Fawzia, S. E.-S. (2018). Toxicological and safety assessment of tartrazine as a synthetic food additive on health biomarkers: A review. *African Journal of Biotechnology, 17*(6), 139–149. https://doi.org/10.5897/AJB2017.16300.

105. Karplus, M., & McCammon, J. A. (2002). Molecular dynamics simulations of biomolecules. *Nature Structural Biology, 9*(9), 646–652. https://doi.org/10.1038/nsb0902-646.

106. Karpuzoglu, E., Holladay, S. D., & Gogal, R. M. (2013). Parabens: Potential impact of low-affinity estrogen receptor binding chemicals on human health. *Journal of*

Toxicology and Environmental Health, Part B, 16(5), 321–335. https://doi.org/10.1080/10937404.2013.809252.

107. Khan, M. I. (2016a). Plant betalains: Safety, antioxidant activity, clinical efficacy, and bioavailability. *Comprehensive Reviews in Food Science and Food Safety, 15*(2), 316–330. https://doi.org/10.1111/1541-4337.12185.
108. Khan, M. I. (2016b). Stabilization of betalains: A review. *Food Chemistry, 197*, 1280–1285. https://doi.org/10.1016/j.foodchem.2015.11.043.
109. Khan, M. I., Anjum, F. M., Sohaib, M., & Sameen, A. (2013). Tackling metabolic syndrome by functional foods. *Reviews in Endocrine and Metabolic Disorders, 14*(3), 287–297. https://doi.org/10.1007/s11154-013-9270-8.
110. Khanna, S., Dash, P. R., & Darbre, P. D. (2014). Exposure to parabens at the concentration of maximal proliferative response increases migratory and invasive activity of human breast cancer cells in vitro. *Journal of Applied Toxicology, 34*(9), 1051–1059. https://doi.org/10.1002/jat.3003.
111. Kiokias, S., & Gordon, M. H. (2003). Antioxidant properties of annatto carotenoids. *Food Chemistry, 83*(4), 523–529. https://doi.org/10.1016/S0308-8146(03)00148-1.
112. Klaenhammer, T. R. (1993). Genetics of bacteriocins produced by lactic acid bacteria. *FEMS Microbiology Reviews, 12*(1–3), 39–85. https://doi.org/10.1111/j.1574-6976.1993.tb00012.x.
113. Krishna, S., Sreedhar, I., & Patel, C. M. (2021). Molecular dynamics simulation of polyamide-based materials–A review. *Computational Materials Science, 200*, 110853. https://doi.org/10.1016/j.commatsci.2021.110853.
114. Kumar, N., & Pruthi, V. (2014). Potential applications of ferulic acid from natural sources. *Biotechnology Reports, 4*(1), 86–93. https://doi.org/10.1016/j.btre.2014.09.002.
115. Lambert, J. D., & Elias, R. J. (2010). The antioxidant and pro-oxidant activities of green tea polyphenols: A role in cancer prevention. *Archives of Biochemistry and Biophysics, 501*, 65–72. https://doi.org/10.1016/j.abb.2010.06.013.
116. Laokuldilok, N., Thakeow, P., Kopermsub, P., & Utama-Ang, N. (2016). Optimization of microencapsulation of turmeric extract for masking flavor. *Food Chemistry, 194*, 695–704. https://doi.org/10.1016/j.foodchem.2015.07.150.
117. Le Meste, M., Champion, D., Roudaut, G., Blond, G., & Simatos, D. (2002). Glass transition and food technology: A critical appraisal. *Journal of Food Science, 67*(7), 2444–2458. https://doi.org/10.1111/j.1365-2621.2002.tb08758.x.
118. Lee, D. (2010). *Nature⍰s Palette: The Science of Plant Color* (Reprint). University of Chicago Press.
119. Lennerz, B. S., Vafai, S. B., Delaney, N. F., Clish, C. B., Deik, A. A., Pierce, K. A., et al. (2015). Effects of sodium benzoate, a widely used food preservative, on glucose homeostasis and metabolic profiles in humans. *Molecular Genetics and Metabolism, 114*(1), 73–79. https://doi.org/10.1016/j.ymgme.2014.11.010.
120. Leo, L., Loong, C., Ho, X. L., Raman, M. F. B., Suan, M. Y. T., & Loke, W. M. (2018). Occurrence of azo food dyes and their effects on cellular inflammatory responses. *Nutrition, 46*, 36–40. https://doi.org/10.1016/j.nut.2017.08.010.

121. Li, D., Wang, P., Luo, Y., Zhao, M., & Chen, F. (2017). Health benefits of anthocyanins and molecular mechanisms: Update from recent decade. *Critical Reviews in Food Science and Nutrition, 57*(8), 1729–1741. https://doi.org/10.1080/10408398.2015.1030064.

122. Lin, S. J., & Pascall, M. A. (2014). Incorporation of vitamin E into chitosan and its effect on the film forming solution (viscosity and drying rate) and the solubility and thermal properties of the dried film. *Food Hydrocolloids, 35*, 78–84. https://doi.org/10.1016/j.foodhyd.2013.04.015.

123. Lin, S.-M., Lin, B.-H., Hsieh, W.-M., Ko, H.-J., Liu, C.-D., Chen, L.-G., & Chiou, R. Y.-Y. (2010). Structural identification and bioactivities of red-violet pigments present in *Basella alba* fruits. *Journal of Agricultural and Food Chemistry, 58*(19), 10364–10372. https://doi.org/10.1021/jf1017719.

124. Liu, Y., Strümpfer, J., Freddolino, P. L., Gruebele, M., & Schulten, K. (2012). Structural characterization of λ-repressor folding from all-atom molecular dynamics simulations. *Journal of Physical Chemistry Letters, 3*(9), 1117–1123. https://doi.org/10.1021/jz300017c.

125. Liu, Z.-Q. (2006). Kinetic study on the prooxidative effect of vitamin C on the autoxidation of glycerol trioleate in micelles. *Journal of Physical Organic Chemistry, 19*(2), 136–142. https://doi.org/10.1002/poc.1011.

126. MacDougall, D. B. (2002). *Color in Food*. Woodhead Publishing Limited.

127. Malheiros, P. d. S., Sant'Anna, V., Barbosa, M. d. S., Brandelli, A., & Franco, B. D. G. d. M. (2012). Effect of liposome-encapsulated nisin and bacteriocin-like substance P34 on *Listeria monocytogenes* growth in *Minas frescal* cheese. *International Journal of Food Microbiology, 156*, 272–277. https://doi.org/10.1016/j.ijfoodmicro.2012.04.004.

128. Malhotra, B., Keshwani, A., & Kharkwal, H. (2015). Antimicrobial food packaging: Potential and pitfalls. *Frontiers in Microbiology, 6*, 611. https://doi.org/10.3389/fmicb.2015.00611.

129. Mamur, S., Yüzbaşıoğlu, D., Ünal, F., & Aksoy, H. (2012). Genotoxicity of food preservative sodium sorbate in human lymphocytes in vitro. *Cytotechnology, 64*(5), 553–562. https://doi.org/10.1007/s10616-012-9434-5.

130. Mamur, S., Yüzbaşiolu, D., Ünal, F., & Yilmaz, S. (2010). Does potassium sorbate induce genotoxic or mutagenic effects in lymphocytes? *Toxicology in Vitro, 24*(3), 790–794. https://doi.org/10.1016/j.tiv.2009.12.021.

131. Marcos, B., Sárraga, C., Castellari, M., Kappen, F., Schennink, G., & Arnau, J. (2014). Development of biodegradable films with antioxidant properties based on polyesters containing α-tocopherol and olive leaf extract for food packaging applications. *Food Packaging and Shelf Life, 1*(2), 140–150. https://doi.org/10.1016/j.fpsl.2014.04.002.

132. Martinez-Delgado, A. A., Khandual, S., & Villanueva–Rodriguez, S. J. (2017). Chemical stability of astaxanthin integrated into a food matrix: Effects of food processing and methods for preservation. *Food Chemistry, 225*, 23–30. https://doi.org/10.1016/j.foodchem.2016.11.092.

133. Martins, F. C. O. L., Sentanin, M. A., & De Souza, D. (2019). Analytical methods in food additives determination: Compounds with functional applications. *Food Chemistry, 272*, 732–750. https://doi.org/10.1016/j.foodchem.2018.08.060.

134. Martins, N., & Ferreira, I. C. F. R. (2017). Mountain food products: A broad spectrum of market potential to be exploited. *Trends in Food Science & Technology, 67*, 12–18. https://doi.org/10.1016/j.tifs.2017.06.013.

135. Martins, N., Roriz, C. L., Morales, P., Barros, L., & Ferreira, I. C. F. R. (2017). Coloring attributes of betalains: A key emphasis on stability and future applications. *Food & Function, 8*, 1357–1372. https://doi.org/10.1039/C7FO00144D.

136. Martins, N., Roriz, C. L., Morales, P., Barros, L., & Ferreira, I. C. F. R. (2016). Food colorants: Challenges, opportunities and current desires of agro-industries to ensure consumer expectations and regulatory practices. *Trends in Food Science & Technology, 52*, 1–15. https://doi.org/10.1016/j.tifs.2016.03.009.

137. Moral Turiel, M. C. (2017). *Estudio de los colorantes alimentarios para su aplicación en las bellas artes* (Thesis). Universidad Complutense de Madrid, Madrid, Spain.

138. Mpountoukas, P., Vantarakis, A., Sivridis, E., & Lialiaris, T. (2008). Cytogenetic study in cultured human lymphocytes treated with three commonly used preservatives. *Food and Chemical Toxicology, 46*(7), 2390–2393. https://doi.org/10.1016/j.fct.2008.03.021.

139. Msagati, T. A. M. (2012). *The Chemistry of Food Additives and Preservatives*. Wiley-Blackwell.

140. Mumford, P. W., Kephart, W. C., Romero, M. A., Haun, C. T., Mobley, C. B., Osburn, S. C., Healy, J. C., Moore, A. N., Pascoe, D. D., Ruffin, W. C., Beck, D. T., Martin, J. S., Roberts, M. D., & Young, K. C. (2018). Effect of 1-week betalain-rich beetroot concentrate supplementation on cycling performance and select physiological parameters. *European Journal of Applied Physiology, 118*(11), 2465–2476. https://doi.org/10.1007/s00421-018-3973-1.

141. Narsing Rao, M. P., Xiao, M., & Li, W. J. (2017). Fungal and bacterial pigments: Secondary metabolites with wide applications. *Frontiers in Microbiology, 8*, 1113. https://doi.org/10.3389/fmicb.2017.01113.

142. Naveena, B. M., Vaithiyanathan, S., Muthukumar, M., Sen, A. R., Kumar, Y. P., Kiran, M., et al. (2013). Relationship between the solubility, dosage and antioxidant capacity of carnosic acid in raw and cooked ground buffalo meat patties and chicken patties. *Meat Science, 95*(2), 195–202. https://doi.org/10.1016/j.meatsci.2013.04.043.

143. Newsome, A. G., Culver, C. A., & van Breemen, R. B. (2014). Nature's palette: The search for natural blue colorants. *Journal of Agricultural and Food Chemistry, 62*(28), 6498–6511. https://doi.org/10.1021/jf501419q.

144. Oms-Oliu, G., Rojas-Grau, M. A., Gonzalez, L. A., Varela, P., Soliva-Fortuny, R., Hernando, M. I. H., Munuera, I. P., Fiszman, S., Martin-Belloso, O., et al. (2010). Recent approaches using chemical treatments to preserve quality of fresh-cut fruit: A review. *Postharvest Biology and Technology, 57*(3), 139–148. https://doi.org/10.1016/j.postharvbio.2010.04.001.

145. Østerlie, M., & Lerfall, J. (2005). Lycopene from tomato products added to minced meat: Effect on storage quality and color. *Food Research International, 38*, 925–929. https://doi.org/10.1016/j.foodres.2004.12.003.

146. Ou, S., & Kwok, K.-C. (2004). Ferulic acid: Pharmaceutical functions, preparation and applications in foods. *Journal of the Science of Food and Agriculture, 84*(11), 1261–1269. https://doi.org/10.1002/jsfa.1873.

147. Oyinloye, B. E., Adenowo, A. F., & Kappo, A. P. (2015). Reactive oxygen species, apoptosis, antimicrobial peptides and human inflammatory diseases. *Pharmaceuticals, 8*(2), 151–175. https://doi.org/10.3390/ph8020151.

148. Ozogul, Y., Yuvka, İ., Ucar, Y., Durmus, M., Kösker, A. R., Öz, M., et al. (2017). Evaluation of effects of nanoemulsion based on herb essential oils (rosemary, laurel, thyme and sage) on sensory, chemical and microbiological quality of rainbow trout (*Oncorhynchus mykiss*) fillets during ice storage. *LWT–Food Science and Technology, 75*, 677–684. https://doi.org/10.1016/j.lwt.2016.10.009.

149. Pandey, V. D., Pandey, A., & Sharma, V. (2013). Biotechnological applications of cyanobacterial phycobiliproteins. *International Journal of Current Microbiology and Applied Sciences, 2*(9), 89–97.

150. Paramita, V. D., & Kasapis, S. (2019). Molecular dynamics of the diffusion of natural bioactive compounds from high-solid biopolymer matrices for the design of functional foods. *Food Hydrocolloids, 88*, 301–319. https://doi.org/10.1016/j.foodhyd.2018.09.007.

151. Paul Ross, R., Morgan, S., & Hill, C. (2002). Preservation and fermentation: Past, present and future. *International Journal of Food Microbiology, 79*(1–2), 3–16. https://doi.org/10.1016/S0168-1605(02)00174-5.

152. Pisoschi, A. M., Pop, A., Georgescu, C., Turcuş, V., Olah, N. K., & Mathe, E. (2018). An overview of natural antimicrobials role in food. *European Journal of Medicinal Chemistry, 143*, 922–935. https://doi.org/10.1016/j.ejmech.2017.11.095.

153. Polturak, G., Grossman, N., Vela-Corcia, D., Dong, Y., Nudel, A., Pliner, M., Levy, M., Rogachev, I., & Aharoni, A. (2017). Engineered gray mold resistance, antioxidant capacity, and pigmentation in betalain-producing crops and ornamentals. *Proceedings of the National Academy of Sciences of the United States of America, 114*(34), 9062–9067. https://doi.org/10.1073/pnas.1707176114.

154. Prabhakara Rao, P. G., Jyothirmayi, T., Balaswamy, K., Satyanarayana, A., & Rao, D. G. (2005). Effect of processing conditions on the stability of annatto (*Bixa orellana* L.) dye incorporated into some foods. *LWT–Food Science and Technology, 38*(7), 779–784. https://doi.org/10.1016/j.lwt.2004.08.015.

155. Quinto, E. J., Caro, I., Villalobos-Delgado, L. H., Mateo, J., De-Mateo-Silleras, B., & Redondo-Del-Río, M. P. (2019). Food safety through natural antimicrobials. *Antibiotics, 8*, 208. https://doi.org/10.3390/antibiotics8040208.

156. Rai, M., Pandit, R., Gaikwad, S., & Kövics, G. (2016). Antimicrobial peptides as natural bio-preservative to enhance the shelf-life of food. *Journal of Food Science and Technology, 53*, 3381–3394. https://doi.org/10.1007/s13197-016-2318-5.

157. Ray, R. C., & Joshi, V. K. (2015). Fermented foods: Past, present and future. In D. Montet & R. C. Ray (Eds.), *Microorganisms and Fermentation of Traditional Food* (pp. 1–36). CRC Press.

158. Ray, S., Raychaudhuri, U., & Chakraborty, R. (2016). An overview of encapsulation of active compounds used in food products by drying technology. *Food Bioscience, 13*, 76–83. https://doi.org/10.1016/j.fbio.2015.12.009.

159. Rodriguez-Amaya, D. B. (2016). Natural food pigments and colorants. *Current Opinion in Food Science, 7*, 20–26. https://doi.org/10.1016/j.cofs.2015.08.004.

160. Sajid, S., & Akbar, N. (2018). Applications of fungal pigments in biotechnology. *Pure and Applied Biology, 7*(3), 922–930. https://doi.org/10.19045/bspab.2018.700111.

161. Saltmarsh, M., & Insall, L. (2013). *Essential Guide to Food Additives* (4th ed.). RSC Publishing.

162. Samaranayaka, A. G. P., & Li-Chan, E. C. Y. (2011). Food-derived peptide antioxidants: A review of their production, assessment, and potential applications. *Journal of Functional Foods, 3*(4), 229–254. https://doi.org/10.1016/j.jff.2011.05.006.

163. Sarıkaya, R., Selvi, M., & Erkoç, F. (2012). Evaluation of potential genotoxicity of five food dyes using the somatic mutation and recombination test. *Chemosphere, 88*(8), 974–979. https://doi.org/10.1016/j.chemosphere.2012.03.032.

164. Schelegueda, L., Gliemmo, M., & Campos, C. (2012). Antimicrobial synergic effect of chitosan with sodium lactate, nisin or potassium sorbate against the bacterial flora of fish. *Journal of Food Research, 1*(3), 272. https://doi.org/10.5539/jfr.v1n3p272.

165. Scotter, M. (2009). The chemistry and analysis of annatto food coloring: A review. *Food Additives & Contaminants: Part A, 26*(8), 1123–1145. https://doi.org/10.1080/02652030902942873.

166. Sebranek, J. G., & Bacus, J. N. (2007). Cured meat products without direct addition of nitrate or nitrite: What are the issues? *Meat Science, 77*(1), 136–147. https://doi.org/10.1016/j.meatsci.2007.03.025.

167. Serra, A. T., Poejo, J., Matias, A. A., Bronze, M. R., & Duarte, C. M. M. (2013). Evaluation of *Opuntia spp.* derived products as antiproliferative agents in human colon cancer cell line (HT29). *Food Research International, 54*(1), 892–901. https://doi.org/10.1016/j.foodres.2013.08.043.

168. Shahid, M., Shahid-ul-Islam, & Mohammad, F. (2013). Recent advancements in natural dye applications: A review. *Journal of Cleaner Production, 53*, 310–331. https://doi.org/10.1016/j.jclepro.2013.03.031.

169. Shahidi, F. (2015). *Handbook of Antioxidants and Food Preservation*. Woodhead Publishing Limited.

170. Sharifi, F., Khanzadi, S., Hashemi, M., & Azizzadeh, M. (2017). Control of *Listeria monocytogenes* and *Escherichia coli* O157 inoculated on fish fillets using alginate coating containing lactoperoxidase system and *Zataria multiflora* Boiss essential oil. *Journal of Aquatic Food Product Technology, 26*(9), 1014–1021. https://doi.org/10.1080/10498850.2017.1375057.

171. Shea, J. E., & Brooks, C. L. III. (2001). From folding theories to folding proteins: A review and assessment of simulation studies of protein folding and unfolding. *Annual Review of Physical Chemistry, 52*, 499–535. https://doi.org/10.1146/annurev.physchem.52.1.499.

172. Sigurdson, G. T., Tang, P., & Giusti, M. M. (2017). Natural colorants: Food colorants from natural sources. *Annual Review of Food Science and Technology, 8*, 261–280.

173. Silva, C. C. G., Silva, S. P. M., & Ribeiro, S. C. (2018). Application of bacteriocins

and protective cultures in dairy food preservation. *Frontiers in Microbiology, 9*, 594. https://doi.org/10.3389/fmicb.2018.00594.

174. Sindelar, J. J., & Milkowski, A. L. (2012). Human safety controversies surrounding nitrate and nitrite in the diet. *Nitric Oxide, 26*(4), 259–266. https://doi.org/10.1016/j.niox.2012.03.011.
175. Siva, R., Palackan, M. G., Maimoon, L., Geetha, T., Bhakta, D., Balamurugan, P., & Rajanarayanan, S. (2011). Evaluation of antibacterial, antifungal, and antioxidant properties of some food dyes. *Food Science and Biotechnology, 20*(1), 7–13. https://doi.org/10.1007/s10068-011-0002-0.
176. Sivakanthan, S., Rajendran, S., Gamage, A., Madhujith, T., & Mani, S. (2020). Antioxidant and antimicrobial applications of biopolymers: A review. *Food Research International, 136*, 109327. https://doi.org/10.1016/j.foodres.2020.109327.
177. Smeriglio, A., Barreca, D., Bellocco, E., & Trombetta, D. (2016). Chemistry, pharmacology, and health benefits of anthocyanins. *Phytotherapy Research, 30*(8), 1265–1286. https://doi.org/10.1002/ptr.5642.
178. Smith, J., & Hong-Shum, L. (2011). *Food Additives Data Book*. Blackwell Publishing.
179. Sothiselvam, S., Liu, B., Han, W., Ramu, H., Klepacki, D., Atkinson, G. C., Brauer, A., Remm, M., Tenson, T., Schulten, K., Vazquez-Laslop, N., & Mankin, A. S. (2014). Macrolide antibiotics allosterically predispose the ribosome for translation arrest. *Proceedings of the National Academy of Sciences of the United States of America, 111*(27), 9804–9809. https://doi.org/10.1073/pnas.1403586111.
180. Sreekanth, D., Arunasree, M. K., Roy, K. R., Chandramohan Reddy, T., Reddy, G. V., & Reddanna, P. (2007). Betanin, a betacyanin pigment purified from fruits of *Opuntia ficus-indica* induces apoptosis in human chronic myeloid leukemia cell line-K562. *Phytomedicine, 14*(11), 739–746. https://doi.org/10.1016/j.phymed.2007.03.017.
181. Suh, H. J., Cho, Y. H., Chung, M. S., & Kim, B. H. (2007). Preliminary data on sulphite intake from the Korean diet. *Journal of Food Composition and Analysis, 20*(3–4), 212–219. https://doi.org/10.1016/j.jfca.2006.04.012.
182. Tamme, T., Reinik, M., & Roasto, M. (2010). Nitrates and nitrites in vegetables: Occurrence and health risks. In R. R. Watson & V. R. Preedy (Eds.), *Bioactive Foods in Promoting Health: Fruits and Vegetables* (pp. 307–321). Academic Press. https://doi.org/10.1016/B978-0-12-374628-3.00021-9.
183. The European Parliament and the Council of the European Union. (2008). Regulation (EC) No. 1333/2008 of the European Parliament and of the Council of 16 December 2008 on food additives. *Official Journal of the European Union, L354/16*. https://eur-lex.europa.eu/legal-content/EN/TXT/?uri=celex%3A32008R1333 (accessed on 3 July 2024).
184. Tian, P. (2008). Molecular dynamics simulations of nanoparticles. *Annual Reports on the Progress of Chemistry–Section C: Physical Chemistry, 104*, 142–164. https://doi.org/10.1039/b703897f.
185. Tiwari, B. K., Valdramidis, V. P., O'Donnell, C. P., Muthukumarappan, K., Bourke, P., & Cullen, P. J. (2009). Application of natural antimicrobials for food preservation. *Journal of Agricultural and Food Chemistry, 57*, 5987–6000. https://doi.org/10.1021/jf900668n.

186. Trouillas, P., Sancho-Garcia, J. C., De Freitas, V., Gierschner, J., Otyepka, M., & Dangles, O. (2016). Stabilizing and modulating color by copigmentation: Insights from theory and experiment. *Chemical Reviews, 116*(9), 4937–4982. https://doi.org/10.1021/acs.chemrev.5b00507.

187. Turgis, M., Vu, K. D., Dupont, C., & Lacroix, M. (2012). Combined antimicrobial effect of essential oils and bacteriocins against foodborne pathogens and food spoilage bacteria. *Food Research International, 48*(2), 696–702. https://doi.org/10.1016/j.foodres.2012.06.016.

188. Ultee, A., & Smid, E. J. (2001). Influence of carvacrol on growth and toxin production by *Bacillus cereus*. *International Journal of Food Microbiology, 64*, 373–378. https://doi.org/10.1016/S0168-1605(00)00480-3.

189. Velmurugan, P., Lee, Y. H., Nanthakumar, K., Kamala-Kannan, S., Dufosse, L., Mapari, S. A., & Oh, B.-T. (2010). Water-soluble red pigments from *Isaria farinosa* and structural characterization of the main-colored component. *Journal of Basic Microbiology, 50*(6), 581–590. https://doi.org/10.1002/jobm.201000097.

190. Vieira Teixeira da Silva, D., dos Santos Baiao, D., de Oliveira Silva, F., Alves, G., Perrone, D., Mere Del Aguila, E., & Flosi Paschoalin, V. M. (2019). Betanin, a natural food additive: Stability, bioavailability, antioxidant and preservative ability assessments. *Molecules, 24*(3), 458. https://doi.org/10.3390/molecules24030458.

191. Vulic, J. J., Cebovic, T. N., Canadanovic, V. M., Cetkovic, G. S., Djilas, S. M., Canadanovic-Brunet, J. M., Velicanski, A. S., Cvetkovic, D. D., & Tumbas, V. T. (2013). Antiradical, antimicrobial and cytotoxic activities of commercial beetroot pomace. *Food Function, 4*(5), 713–721. https://doi.org/10.1039/c3fo30315b.

192. Wang, C. Q., & Yang, G. Q. (2010). Betacyanins from *Portulaca oleracea* L. ameliorate cognition deficits and attenuate oxidative damage induced by D-galactose in the brains of senescent mice. *Phytomedicine, 17*(7), 527–532. https://doi.org/10.1016/j.phymed.2009.09.006.

193. Wang, J. C., & Liapis, A. I. (2012). Water–water and water–macromolecule interactions in food dehydration and the effects of the pore structures of food on the energetics of the interactions. *Journal of Food Engineering, 110*(4), 514–524. https://doi.org/10.1016/j.jfoodeng.2012.01.008.

194. Wang, Y., Li, F., Zhuang, H., Chen, X., Li, L., Qiao, W., et al. (2015). Effects of plant polyphenols and ⊠-tocopherol on lipid oxidation, residual nitrites, biogenic amines, and N-nitrosamines formation during ripening and storage of dry-cured bacon. *LWT–Food Science and Technology, 60*(1), 199–206. https://doi.org/10.1016/j.lwt.2014.09.022.

195. Wang, Y., Lu, Z., Wu, H., & Lv, F. (2009). Study on the antibiotic activity of microcapsule curcumin against foodborne pathogens. *International Journal of Food Microbiology, 136*(1), 71–74. https://doi.org/10.1016/j.ijfoodmicro.2009.09.001.

196. WHO/FAO. (2018). *Codex Alimentarius: General Standard for Food Additives* (2018th ed.). WHO/FAO, Rome.

197. Xue, J., Davidson, P. M., & Zhong, Q. (2013). Thymol nanoemulsified by whey protein-maltodextrin conjugates: The enhanced emulsifying capacity and antilisterial

properties in milk by propylene glycol. *Journal of Agricultural and Food Chemistry, 61*(51), 12720–12726. https://doi.org/10.1021/jf4043437.

198. Yang, Y., Song, X., Sui, X., Qi, B., Wang, Z., Li, Y., et al. (2016). Rosemary extract can be used as a synthetic antioxidant to improve vegetable oil oxidative stability. *Industrial Crops and Products, 80*, 141–147. https://doi.org/10.1016/j.indcrop.2015.11.044.

199. Zhang, H., Zhan, J., Su, K., & Zhang, Y. (2006). A kind of potential food additive produced by *Streptomyces coelicolor*: Characteristics of blue pigment and identification of a novel compound, ⊠-actinorhodin. *Food Chemistry, 95*(2), 186–192. https://doi.org/10.1016/j.foodchem.2004.12.028.

200. Zhang, J., Celli, G. B., & Brooks, M. S. (2019). Chapter 1: Natural sources of anthocyanins. In *Anthocyanins from Natural Sources: Exploiting Targeted Delivery for Improved Health* (pp. 1–33). Royal Society of Chemistry. https://doi.org/10.1039/9781788012614.

201. Zhang, J., Hou, X., Ahmad, H., Zhang, H., Zhang, L., & Wang, T. (2014). Assessment of free radicals scavenging activity of seven natural pigments and protective effects in AAPH-challenged chicken erythrocytes. *Food Chemistry, 145*, 57–65. https://doi.org/10.1016/j.foodchem.2013.08.025.

202. Zhao, G., Perilla, J. R., Yufenyuy, E. L., Meng, X., Chen, B., Ning, J., Ahn, J., Gronenborn, A. M., Schulten, K., Aiken, C., & Zhang, P. (2013). Mature HIV-1 capsid structure by cryo-electron microscopy and all-atom molecular dynamics. *Nature, 497*(7451), 643–646. https://doi.org/10.1038/nature12162.

CHAPTER

9 Flavors

LEARNING OBJECTIVES

After studying this chapter, you will be able to:

- Understand the flavor chemistry
- Learn about taste and other saporous substances
- Discuss the vegetable, fruit, and spice flavors

KEY TERMS FROM THIS CHAPTER

Smell	Sugar	Methyl salicylate
Flavor chemistry	Inorganic compounds	Aromatic organic chemicals
Taste	Sweet	Sour
Bitter	Salty	Umami
Natural flavors	Food flavors	Sensory assessments
Odor units	Sweet substances	Chloroform
Saccharin	Stereochemistry	Hydrophobic group
Quinine		

INTRODUCTION

Food flavor has a significant impact on how food tastes and smells to the consumer, particularly when it comes to sensory analysis. While some of these products, like sugar and salt, are naturally occurring, many of these flavors are created for food products by flavor chemists, also known as flavorists. These artificial flavors include lactic acid, which gives milk its tart flavor, and methyl salicylate, which gives wintergreen its scent.

9.1. FLAVOR CHEMISTRY

The integrated reaction to a complex mixture of stimuli primarily on the senses of smell and taste, but also on those associated with sight (color and appearance), tactile sensations (texture and mouth-feel), and pain (pungency), is the universal, everyday experience of appreciating flavor. Thus, in its broadest sense, flavor chemistry is a specialized area of organic chemistry that influences numerous other fields, such as zoology, botany, anatomy, physiology, histology, cytology, etc., for its thorough comprehension.

Here, "flavor chemistry" refers to the objective study of the biogenesis, biosynthesis, and deterioration of the constituents that give natural products of plant and animal origin their distinct odor and flavor, as well as their chemical makeup. Discussing the entire spectrum of organic chemical reactions involved in the synthesis of odorants and flavorants in a laboratory or on a commercial scale is obviously impractical in this context. Hundreds of aromatic chemicals are synthesized annually, either as intermediates or as final products (nearly 6,000 compounds have been identified to date).

Some of these have no such counterparts, while the majority are exactly like those found in nature. While most of these synthetics never become commercially significant, a small number do reach full production scale and are used in flavorings and fragrances once their safety in use is proven. The literature provides a thorough description of the chemistry of synthetic aromatic chemicals (Bedoukian, 1991).

It is highly desirable for flavorists and anyone involved in flavor applications to understand the sensory mechanisms by which these substances are perceived and identified. This is also a highly specialized area of structural chemistry, psychology, and physiology. Issues pertaining to sensory evaluation, reaching consensus on examination techniques, and conveying sensory reactions.

Food and related consumables have flavor because of chemicals that include a wide range of aliphatic and aromatic organic compounds as well as some inorganic ones. Flavor can be added purposefully or arise naturally. Flavor (e.g., sweetness, bitterness, etc.) and a few additional qualities (e.g., astringency, pungency, etc.) are caused by nonvolatile elements. Foods' unique and frequently distinguishable flavor profile is primarily influenced by their odor. Therefore, while taste is important, studying the volatile components is more important to flavor chemists and flavorists who want to match or reconstruct a specific natural profile. With fruit flavors, however, the perception of the aromatic impact can be greatly influenced by the acid/sugar ratio.

The amount of articles and research papers on the biochemistry of natural products in technical literature is constantly growing. Long lists of components that have been isolated and identified can be found in the majority of these qualitative studies (Maarse & Visscher, 1989). Determining the sensory contribution of the identified flavor compounds is becoming increasingly important, after years of ignoring their sensory properties (Acree, Bernard, & Cunningham, 1984, Ullrich & Grosch, 1987; Ney, 1990).

9.1.1. Flavor Classification

There are many ways to categorize flavors, but from the perspective of their chemistry, the most practical way is to group them according to how they are formed—either spontaneously through biosynthetic pathways from recognized precursors, or through processing that modifies the biological, chemical, or physical properties of natural or synthetic starting materials. Since natural flavors are primarily metabolites, how they form in living tissues is complicated, genetically determined, and affected by growth conditions that occur naturally. Processing can yield flavors that are either complex interaction products or direct breakdown products, depending on the raw materials used and the processing conditions applied.

In this section, the chemistry of odor and taste-stimulating components will be discussed in the following groups:

a. The biochemistry of volatile and nonvolatile constituents that are formed during normal plant metabolism and remain in situ when the plant is harvested (e.g., essential oils, fruit and vegetable flavors).
b. Flavoring components resulting from enzyme-catalyzed reactions (e.g., enzyme-modified dairy products).
c. Flavoring constituents produced as a result of microbiological action and fermentation (e.g., wines, dairy products, tea).
d. Flavors resulting from heat processing or cooking (e.g., cooked meat flavors, roasted coffee).

9.1.2. General Philosophy

With the introduction of gas chromatography and fast-scan mass spectrometry in the late 1950s, the field of food chemistry has progressed, and understanding the chemistry of flavors is generally regarded as a relatively recent development. Even though these analytical instruments have made it possible to conduct a thorough investigation of the whole spectrum of flavor compounds, much earlier research skillfully employed classical chemical techniques, particularly with regard to essential oils and spice extractives. A discipline-oriented identity for the field of flavors has slowly evolved as a result of the broad and somewhat isolated attention that perfumery has received, as well as the quick and seemingly chaotic development of knowledge about the chemistry of food flavors.

The only necessary condition for a molecule to be included in the flavor category of food chemistry is that it must have the ability to "stimulate taste or aroma receptors or specialized nerves to produce an integrated psychological response known as flavor." This is true even though flavor substances represent an incredibly broad range of chemical structures derived from almost every notable constituent of foods. From a wider viewpoint, however, the term "flavor" now refers to an integrated overall perception of all contributing senses (taste, smell, sight, feeling, and sound) at the moment of food consumption. Consequently, even though the nonchemical or indirect senses—sound, touch, and sight—often have a significant impact on how flavors are perceived.

Chemicals that elicit specific odor- and/or taste-related responses are the focus of attention, though it is not always possible to distinguish between the meaning of these terms and that of flavor. The chemistry of significant flavor systems and character-impact compounds—which have been chosen to highlight the chemistry of pertinent food systems and the molecular underpinnings of flavor compounds in foods—will be covered. Relationships between flavor compounds' structure and activity are noted when applicable and information is available.

Listings of flavor compound profiles found in various foods are given limited notice here. You can find extensive lists of food-specific flavor compounds and tabulations of threshold concentrations for individual compounds elsewhere. Lastly, it is up to the reader to decide whether or not to include information about the flavor chemistry of important food ingredients.

9.1.3. Methods for Flavor Analysis

The analysis of volatile compounds using gas chromatography in conjunction with fast-scan mass spectrometry has traditionally been associated with flavor chemistry; however, this perspective is overly limited given the wide range of techniques that have been developed for the analysis of flavor compounds. However, since in-depth discussions about flavor analysis can be found elsewhere, only a limited amount of attention will be focused on it here.

A number of elements contribute to the difficulty of flavor analysis: flavors exist in low concentrations (ppm, 1×10^6; ppb, 1×10^9; ppt, 1×10^{12}), mixtures are complex (e.g., coffee contains more than 450 different volatile compounds), some flavor compounds are unstable, and their volatility can vary greatly (high vapor pressures) or drastically (low vapor pressures). It is usually necessary to first isolate flavor compounds from the bulky food ingredients and then combine that isolation with a significant concentration (e.g., distillation), but particularly in the case of studying flavor quality, this ought to happen with the least amount of distortion of the original composition. Minimizing the destruction of sensitive compounds during isolation has been made possible by the adsorption of equilibrium gas-phase flavor compounds over foods onto porous polymers, followed by either solvent elution or thermal desorption. To ensure sufficient recovery for analysis, distillation techniques may still be necessary for compounds with higher boiling points and for certain compounds present in extremely low concentrations.

A primary objective of flavor analysis is the identification of precursor substances and flavor-active compounds; however, an equally important goal is the precise measurement of these compounds' concentrations in food. When attempting to establish correlations between the occurrences of flavor compounds and the ensuing sensory perceptions, quantitative information is particularly useful. Even though a lot of information has been gathered about the quantitative presence of flavor compounds in headspace gases over a variety of foods and differently obtained isolates from foods, it is frequently challenging to reconstruct high-quality facsimile flavors from these data due to missing compounds or distorted quantitative values.

Furthermore, efforts to replicate flavors in foods with altered nutritional values (e.g., low-fat formulations) by modifying the taste profiles of foods that haven't been altered (e.g., full-fat formulations) have also typically had only patchy success. In reaction to concerns that the modified food matrix's individual flavor compounds would release at varying relative rates (e.g., low-fat formulation) in the mouth explained these challenges, a great deal of work has gone into using atmospheric pressure ionization mass spectrometry techniques to measure the release of flavor compounds in the oral cavity in real-time. The concept of time-intensity release rate for individual compounds from foods in the mouth has been validated by elegant experiments; however, the observed differences primarily pertain to flavor intensity effects and do not appear to be correlated with variations in perceived flavor qualities between unmodified and modified food matrix compositions. These results have caused the focus to shift toward the identification of flavor-modifying compounds found in unmodified foods (e.g., full-fat recipes), but which are left out or omitted from altered foods (e.g., reduced-fat recipes).

Overall, routine flavor assessment by purely analytical means is still limited, despite advancements in the development and application of methods that correlate objective flavor chemical data with subjective sensory information. The long-standing need for quick ways to measure chemical parameters that offer accurate information about food quality and flavor intensity led to the development and commercialization of electronic nose devices. While there have been some documented applications successes for these devices, particularly in the screening of food for deteriorating flavors (e.g., oxidative rancidity), they are still typically considered during the initial phases of development.

9.1.4. Sensory Assessment of Flavors

Regardless of the ultimate goals, sensory evaluations of food ingredients and flavor compounds are critical to meeting the objectives of flavor investigations. In certain cases, samples must be sensory characterized by qualified personnel (such as trained flavorists or researchers). Nonetheless, formal sensory panels and statistical analysis are often required to identify flavor differences, gather descriptive flavor data, or ascertain consumer flavor preferences.

Both qualitative and quantitative flavor or aroma sensations for individual or combinations of chemicals vary with concentration, and sensory assessments are used to record the qualitative flavor characteristics of flavor and aroma chemicals. A single representative of the general population is typically used to determine detection threshold values for flavor chemicals. This process gives a measure of the potency of flavor that each compound provides. Sensory panelists show a range of concentrations of a chosen flavor compound in a specified medium (water, milk, air, etc.), and each panelist indicates whether or not the compound can be detected. The term "taste threshold" refers to the concentration range at which the compound is detected by at least half (and sometimes more) of the panelists. The potency of a compound's flavor or odor varies widely, so a small quantity of a compound with a very low threshold is thought to have a significantly bigger impact on a food's flavor than a compound that is fairly abundant but has a high threshold.

Odor units (OUs) are a measure of the contribution made by each flavor compound. They are computed by dividing the concentration of the flavor compound by its flavor threshold (OU = concentration present/threshold concentration). The most potent odorants in food have recently been found using a technique called aroma extract dilution analysis, which entails sensory detection of individual compounds (flavor dilution factor) in gas chromatographic effluents produced by serial dilutions of aroma extracts from food. These techniques yield quantitative data regarding the relative potency or intensity of flavoring compounds found in various foods and beverages. Nevertheless, these assessments frequently overlook or significantly undervalue the qualitative flavor characteristics of substances. This is particularly true when specific compounds give a flavor it's distinctive, defining, or character-impacting quality. These most effective techniques for detecting aroma compounds are also criticized because they yield data on flavor potency that is calculated without taking into account the effects of food matrices or the interactive psychophysical effects of perceiving flavor compound mixtures. Consequently, there is very little chance of extrapolating such data to real food systems.

9.1.5. Molecular Mechanisms of Flavor Perception

Research in this area has only recently started to replace theories with accepted facts and principles, although comprehending the molecular basis of flavor perception has long been sought as a significant goal with numerous practical applications. Recent advances in our understanding of taste perception have largely come from the application of methods found in many modern molecular biology studies. A compelling motivation for furthering fundamental research on the mechanisms underlying flavor perception is the aim of making greater use of the structure–activity characteristics of molecules (e.g., a well-defined structural characteristic produces a consistent scent or flavor) to direct the creation of more beneficial and efficient flavor compounds (like strong sweeteners). In a similar vein, there is a great demand for compounds that neutralize or mask off taste characteristics of certain food ingredients (e.g., soy protein derivatives), and particularly the unwelcome bitterness that some medications and nutraceutical ingredients contain by nature. The olfactory epithelium in the nasal cavity contains specialized cells that are capable of detecting minute amounts of volatile odorants, which explains the almost infinite variations in flavor and odor intensity. Humans can taste umami, sweetness, sourness, saltiness, and bitterness thanks to taste buds on the tongue and back of the oral cavity. These sensations help create the flavor components of food. All things considered, the molecular process of odor and taste perception generally consists of three stages that build up to the taster's sensory experience. These are the three processes by which electrical impulse information is received, transduced, and processed by the brain. Initial perception event for odorants and some tastants (sweet, bitter, and umami) is the selective binding (possibly adhering to a structural lock-and-key conceptual model) of a flavor molecule to a particular receptor protein in the membrane of a suitable receptor cell.

One of several highly specialized biochemical reaction cascades is used to convert the chemical energy of the flavor molecule's binding to the receptor protein into electrical energy. When a flavor molecule and a receptor protein first bind, the G protein coupled receptor is stimulated, which in turn triggers enzymatic reactions that result in reaction cascade products (e.g., inositol triphosphate, or cyclic adenosine 5′-monophosphate [AMP]) that interact with and open the Na^+ or Ca^{2+} ion channels in the receptor cell membrane. The rapid passage of charged ions through the receptor cell membrane that follows depolarizes the cell and generates a distinct set of electrical charges known as action potentials or nerve impulses, which indicate the concentration of odorant stimulating the cell and reveal some details about the type of flavor molecule. At this point, the crucial mechanisms underlying the coding of electrical information are mostly theoretical, but genetic and physiological data lend support to the idea that the olfactory bulb and other brain structures produce spatial maps—that is, different neural firing rates and intensities—to accomplish this coding process. The molecular-level processes that underlie the perception of sour (H^+) and salt (Na^+) ionic tastants differ from those of odorants as well as sweet, bitter, and umami tastants. Nonetheless, ionic channels in taste receptor cell membranes are directly contacted by both salty and sour ionic tastants. H^+ ions directly bind to ion channels in sour tastants, closing them to Na^+ flow and triggering a neural impulse and membrane depolarization. On the other hand, because the ion channels are permeable to the salt cation (Na^+), saltiness perception is started by the direct entry of Na^+ ions from the external environment into tastant receptor cells. Therefore, in response to the presence of salt (NaCl) in the surrounding environment, receptor cells depolarize and produce a neural impulse when Na^+ ions enter and alter the electrical potential across the cell membrane. Certain flavor molecules have distinct sensory qualities that greatly influence the flavors of particular foods and beverages. These qualities include heat or pungency, cooling, and tingling sensations. Since these feelings are the result of influences on particular nerve fibers rather than the involvement of a particular receptor (i.e., taste, or smell) cells in their generation; historically, they were frequently referred to as non-specific saporous experiences. The cutaneous (skin) chemosensory systems detect similar sensations in the oral and nasal tissues, such as pain, irritation, heat, and cold. But in order to differentiate between these flavor-related experiences coming from the nasal and oral cavities' innervating systems (i.e., trigeminal, glossopharyngeal, and vagus), and more recently, the word "chemesthesis" was created to describe all of these saporous feelings combined. Other vague, artificially produced flavor-influencing feelings (complexity, fullness, etc.) are thought to be detected by the trigeminal neural system; however, neither the substances responsible for the effects nor the perception processes are fully understood.

9.2. TASTE AND OTHER SAPOROUS SUBSTANCES

When it comes to these aspects of taste perception, water soluble and relatively nonvolatile substances are usually, but not always, responsible. They have frequently been overlooked in flavor coverages, and they are typically found in foods in greater quantities than those that give them their scents. The chemistry of substances responsible for taste sensations and some of the less defined flavor sensations is examined here because of their extremely influential role in the acceptance of foods.

9.2.1. Sweet Taste Substances

Due to the interest in sugar substitutes and the need to find appropriate substitutes for specific low-calorie sweeteners, such as cyclamate and saccharin, sweet substances have received a lot of attention. Prior to the development of contemporary theories of sweetness, it was common knowledge that since hydroxyl (-OH) groups predominate in sugar molecules, sweetness was linked to them. This theory was quickly refuted, though, since polyhydroxy compounds come in a wide range of sweetness levels, in addition to the sweetness of many amino acids, certain metallic salts, and unrelated substances like saccharin and chloroform ($CHCl_3$). Nevertheless, it became clear that sweet substances shared certain traits, and over the next 75 years, a theory linking molecular structure to sweet taste developed that satisfactorily explains why some compounds are sweet.

The AH/B theory was initially put forth by Shallenberger and Acree (Shallenberger, & Acree, 1967) for the saporous (taste-eliciting) unit that is shared by all substances that elicit a sweet sensation (Figure 9.1). Initially, the saporous unit was thought to consist of an electronegative orbital situated approximately 3 Å away from the proton and a covalently bound H-bonding proton.

Therefore, a molecule's vicinal electronegative atoms are necessary for sweetness. Moreover, an H-bonding proton needs to be present in one of the atoms. In sweet molecules, oxygen, nitrogen, and chlorine atoms usually play these roles. Additionally, oxygen atoms in hydroxyl groups can play either the AH or B function. For the chemicals chloroform (I), saccharin (II), and glucose (III), basic AH/B relationships are displayed.

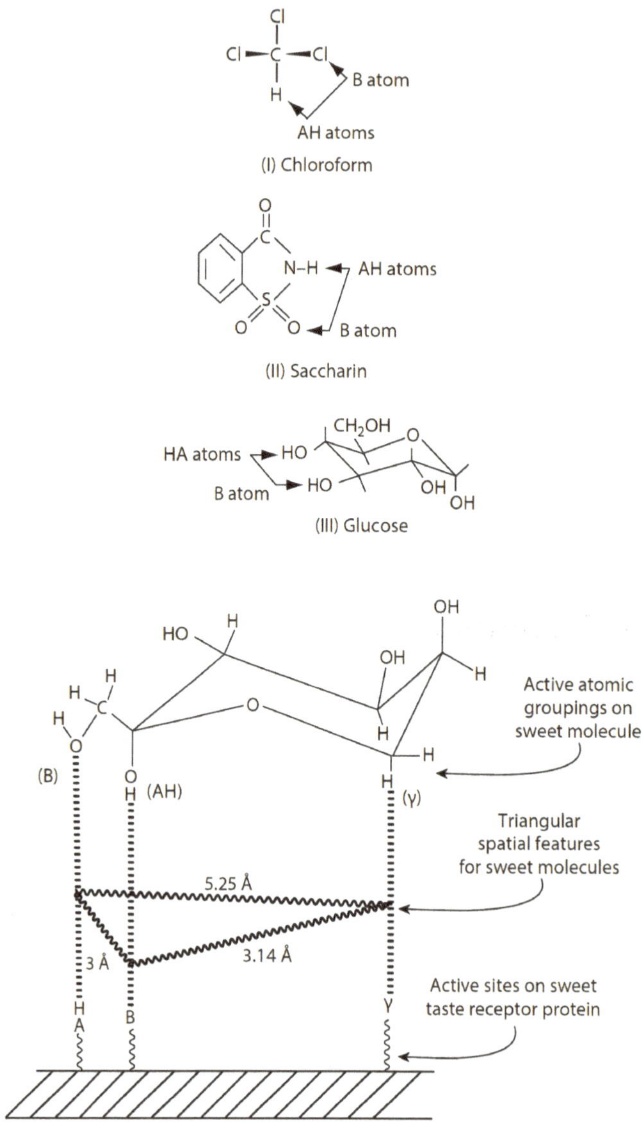

Figure 9.1. *Schematic showing the relationship between AH/B and γ-sites in the saporous sweet unit for β-d-fructopyranose.*

Source: Srinivasan Damodaran and Kirk L. Parkin. Fennema's Food Chemistry, Fifth Edition. CRC Press. ISBN: 9781482208122.

However, as shown in Figure 9.1, the AH/B parts of the saporous unit are also subject to stereochemical constraints in order to ensure proper alignment with the receptor site. It is currently thought that the H-bonding of the AH/B components to comparable structures in the taste receptor mediates the interaction between the active groups of the sweet molecule and the receptor.

In order to broaden the theory's applicability to extremely sweet substances, a third feature has also been added. This addition includes the proper stereochemical arrangement of sweet molecules' lipophilic regions, usually denoted by γ, which are drawn to the taste receptor's lipophilic regions in common. Sweet molecules often contain lipophilic regions, such as methylene ($-CH_2-$), methyl ($-CH_3$), or phenyl ($-C_6H_5$) groups. The rationale behind the tripartite structure theory of sweetness is that the entire sweet saporous structure is geometrically positioned so that, for highly sweet substances, triangular contact of all active units (AH, B, and γ) with the receptor molecule occurs.

Although it has less of an impact on sugar sweetness, the γ-site is a crucial component of highly sweet substances (Birch, 1981). It seems to work by making it easier for specific molecules to bind to the taste receptor site, which in turn modifies how sweet something tastes. Due to the general hydrophilic nature of sugars, this property is only partially relevant for certain sugars, like fructose.

A significant amount of the observed variation in sweetness quality between various sweet substances is likely explained by this component of the saporous sweet unit. It seems to be related to some of the interactions between sweet and bitter tastes that have been noted for specific compounds, in addition to being significant in the temporal or time-intensity aspects of sweetness perception.

A combined taste sensation is produced by the properties of sweet-bitter sugar structures that appear to enable them to interact with either or both types of receptors. Even when the concentration of the bitter sensation in a test solution is lower than the bitterness properties in the structure, the sweetness is still affected. Anomeric center configuration, ring oxygen, hexoses' primary alcohol group, and any substituents' nature all seem to have a role in imparting bitterness in sugars. A sweet molecule's sweetness can often be lost or suppressed, or bitterness can be introduced, as a result of modifications to its stereochemistry.

9.2.2. Bitter Taste Substances

Based on the relationship between molecular structure and receptors, sweetness and the bitterness sensation seem to be closely related. Due to its reliance on the stereochemistry of stimulus molecules, bitterness and sweetness are similar in that they are both elicited by similar features in molecules, some of which can produce both bitter and sweet sensations. Bitter molecules seem to require only one polar group and a hydrophobic group, whereas sweet molecules need two polar groups that can be supplemented with a nonpolar group.

Some, however, think that most bitter compounds have a hydrophobic group and an AH/B entity that are the same as those in sweet molecules. According to this theory, for molecules with the necessary molecular characteristics, the orientation of AH/B units within particular receptor sites—which are found on the flat bottom of receptor cavities—provides the ability to distinguish between sweetness and bitterness. Molecules aligned with the orientation of bitter compounds produce a bitter response, while those aligned with the orientation of sweetness produce a sweet response. A molecule would react in a bitter-sweet way if its geometry allowed it to orient in either direction.

For amino acids where the d-isomers are sweet and the l-isomers are bitter, such a model seems particularly appealing.

Due to its nondirectional lipophilicity, the hydrophobic or γ-site of the sweet receptor can contribute to either a sweet or bitter response. Stereochemical selectivity to the receptor sites inside each receptor cavity is provided by molecular bulkiness factors. We conclude that while the bitter taste modality has a very broad structural basis, current theories can account for the majority of empirical observations about molecular structure and bitterness.

It is widely acknowledged that quinine, an alkaloid, is the standard for the bitter taste sensation. For quinine hydrochloride (IV), the detection threshold is roughly 10 ppm. In general, bitter substances tend to be less soluble in water than other taste-active materials and have lower taste thresholds than other substances. Quinine is allowed as an additive in certain types of beverages, including soft drinks with tart-sweet characteristics. In these drinks, the bitterness offers a refreshing gustatory stimulation that melds well with the other flavors. The act of incorporating quinine into carbonated drinks appears to have originated from attempts to reduce or conceal quinine's acrid taste during its prescription for malaria.

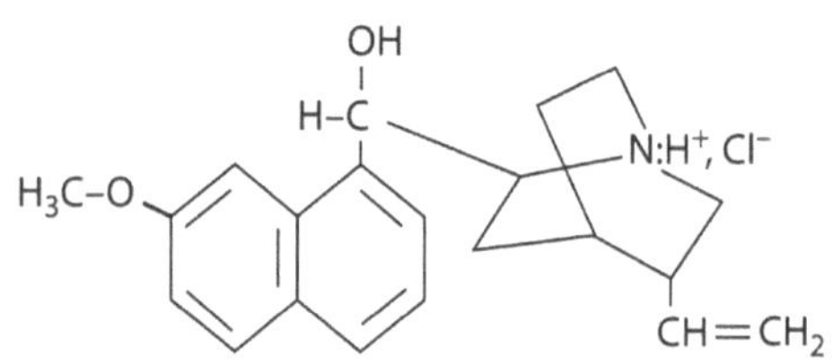

(IV) Quinine hydrochloride

Apart from carbonated drinks, bitterness is a significant flavor component of numerous other beverages that are ingested in large quantities, such as tea, coffee, and cocoa. Coffee, tea, and cola nuts contain caffeine (V), which has a moderately bitter taste at 150–200 parts per milliliter in water. Similar to caffeine, theobromine (VI) is primarily found in cocoa, where it adds to the bitter taste. Soft cola drinks can contain up to 200 parts per milligram of caffeine. The majority of the caffeine used for this purpose comes from green coffee bean extractions used to make decaffeinated coffee.

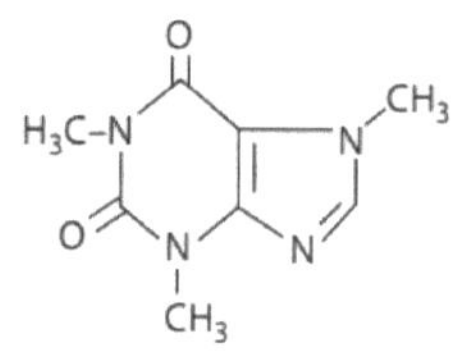

(V) Caffeine

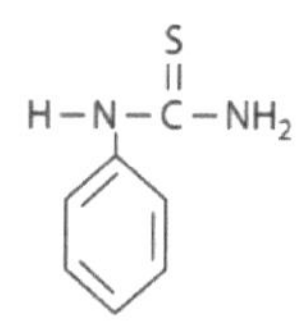

(VIII) Phenylthiocarbamide (PTC)

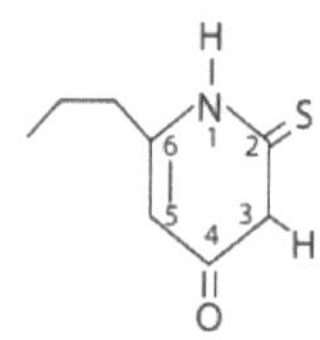

(IX) 6-*n*-Propyl-2-thiouracil (PROP or PTU)

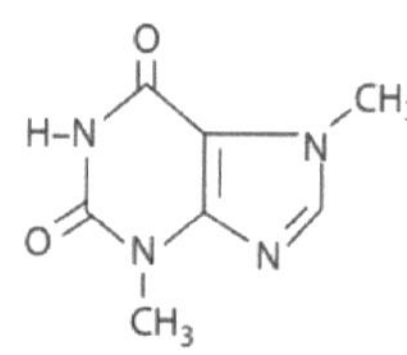

(VI) Theobromine

In order to give beer distinctive flavors, the brewing industry uses a lot of hops. Hop flavor is heavily influenced by bitterness, which is provided by certain peculiar compounds derived from isoprenoid compounds. These inert, bitter compounds are generally classified as humulone or lupulone derivatives, or α- or β-acids, as they are called in the brewing business. The most common material, humulone, undergoes an isomerization reaction during wort boiling to become isohumulone (Figure 9.2).

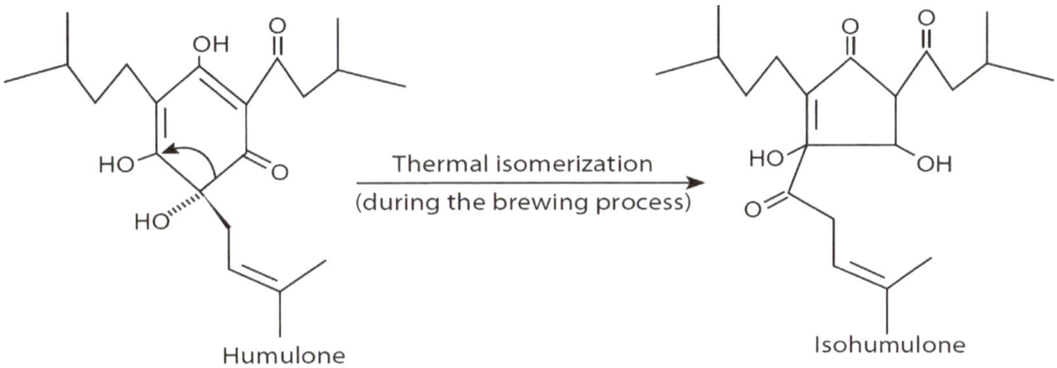

Figure 9.2. *Thermal isomerization of humulone to isohumulone occurring during wort boiling in the traditional brewing process.*

Source: Srinivasan Damodaran and Kirk L. Parkin. Fennema's Food Chemistry, Fifth Edition. CRC Press. ISBN: 9781482208122.

The precursor of the compound that gives light-exposed beer its "sunstruck" or "skunky" flavor is isohumulone. The carbon next to the keto group in the iso-hexenyl chain undergoes a photocatalyzed reaction when hydrogen sulfide from yeast fermentation is present. Prenylmercaptan, or 3-methyl-2-butene-l-thiol, is the result, and it smells like skunk.

This reaction can be avoided by selectively reducing the ketone in preisomerized hop extracts, allowing beer to be packaged in clear glass without developing a skunky or sunstruck flavor. For many years, it was debatable whether volatile hop aroma compounds made during the wort-boiling process survived.

Nevertheless, it is now well known that some important compounds do survive the wort-boiling process and that other compounds are formed from bitter hops; these compounds work together to give beer its distinctive kettle-hop aroma.

While some foods and beverages—such as those containing novel nutraceutical ingredients and pharmaceutical preparations—benefit from bitterness, unwanted bitterness is often an issue in other foods and beverages.

Many resources have been used in an attempt to find compounds that can hide bitter tastes, but the results have been mixed thus far. Although some gums and viscosity-enhancing polymers can reduce bitter tastes, using them usually only makes the issue slightly better.

Yet, it has recently been found that AMP (VII), which is linked to the metabolism of intermediary energy, has strong bitter-blocking abilities. This suggests that it may find use in the suppression of bitter flavors.

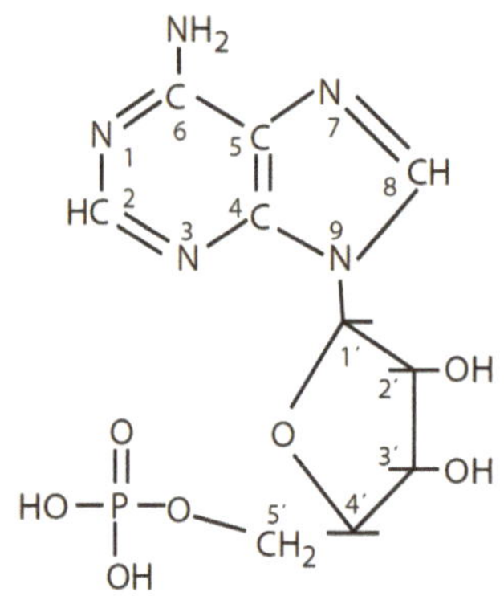

(VII) Adenosine-5′-monophosphate
(AMP, a potent bitter blocker)

One of the main issues facing the citrus industry is the emergence of excessive bitterness, particularly in processed products. While some bitterness is expected and desirable in grapefruit, many consumers find that both fresh and processed fruits have more bitterness than they would like. Numerous flavonone glycosides are present in citrus fruits, with naringin being the main flavonone present in grapefruit and bitter orange (Citrus aurantium). Higher naringin content juices are very bitter and have little commercial value unless they can be heavily diluted with lower naringin content juices. The structure of the molecule that results from the 1 to 2-linkage between rhamnose and glucose is linked to the bitterness of naringin.

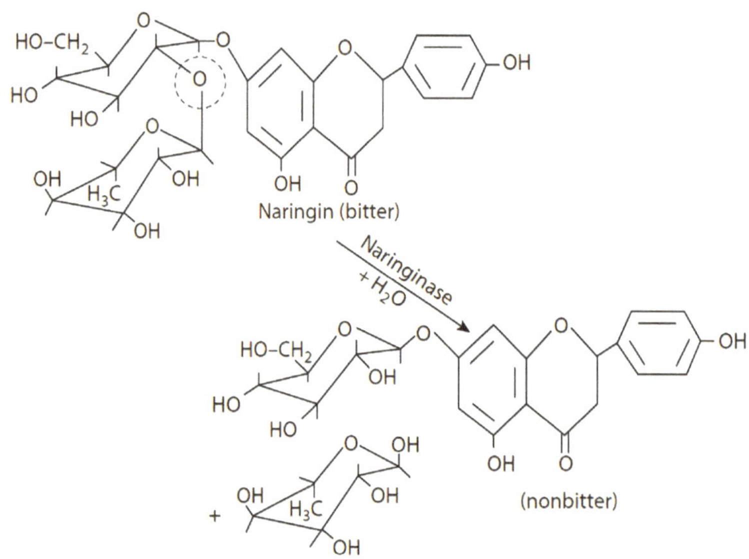

Figure 9.3. *Reaction showing the enzymatic hydrolysis of naringin by naringinase that is used in the debittering of citrus products.*

Source: Srinivasan Damodaran and Kirk L. Parkin. Fennema's Food Chemistry, Fifth Edition. CRC Press. ISBN: 9781482208122.

The enzyme naringinase, which has been isolated from Aspergillus sp., and commercial citrus pectin preparations, hydrolyzes the 1 to 2-linkage (Figure 9.3) to produce nonbitter products. Moreover, immobilized enzyme systems have been created to denature grapefruit juices with high naringin concentrations. In some food applications, naringin — which is extracted from grapefruit peels and used commercially — replaces caffeine for bitterness.

A triterpenoid dilactone (A- and D-rings) known as limonin is the main bittering ingredient in navel and Valencia oranges. It also functions as a bittering agent in grapefruit. The main form of limonin in intact fruits is a flavorless derivative created by the enzymatic hydrolysis of limonin's D-ring lactone; limonin is not present at all in intact fruits (Figure 9.4). Acidic conditions after juice extraction encourage the D-ring to close, forming limonin. This phenomenon of delayed bitterness has major economic ramifications.

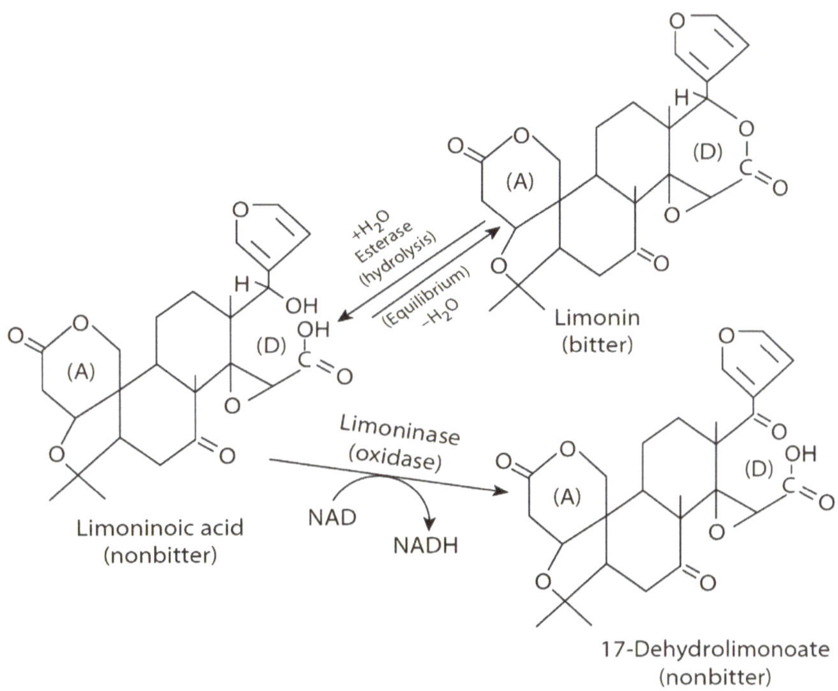

Figure 9.4. *Equilibrium reaction leading to the formation limonin and enzymatic debittering reactions that reverse bitterness development in citrus juices.*

Source: Srinivasan Damodaran and Kirk L. Parkin. Fennema's Food Chemistry, Fifth Edition. CRC Press. ISBN: 9781482208122.

Immobilized enzymes from *Arthrobacter* and *Acinetobacter* Sp have been used to develop debittering techniques for orange juice (Hasegawa, Patel, & Snyder, 1982). Because the D-ring lactone closes again in acidic environments, enzymes that merely open it can only address the issue momentarily. Nonetheless, an irreversible method of debittering orange juice is provided by the use of limonoate dehydrogenase to change the open D-ring compound into nonbitter 17-dehydrolimonoate A-ring lactone (Figure 9.4). Citrus juice debittering techniques also involve the use of polymeric adsorbents, which are presently the method of choice for commercial processors.

The overall hydrophobicity of amino acid side chains in peptides produces the unpleasant bitterness that is commonly detected in protein hydrolysates and aged cheeses. The ability of individual peptides to interact with the essential hydrophobic sites of bitterness receptors is largely dependent on the size and nature of their hydrophobic groupings. Despite this, all peptides contain an adequate number of AH-type polar groups that can fit the polar receptor site. Ney has demonstrated that the mean hydrophobicity value, or Q, of peptides can be calculated to predict their bitter flavor. The total hydrophobic contributions from the nonpolar amino acid side chains determine a protein's capacity for hydrophobic associations, which primarily contribute to the free energy (ΔG) involved in protein unfolding. Equation 9.1 can therefore be used to determine the mean hydrophobicity Q by adding the ΔG values for each side chain of an amino acid in a peptide.

$$Q = \frac{\Sigma \Delta G}{n}, \quad (9.1)$$

where n is the number of amino acid residues. Based on solubility data, individual ΔG values for amino acids have been calculated and are listed in Table 9.1. The peptide will be bitter if the Q-value is greater than 5,855 based on joules/mol (1,400 based on calories/mol); it will not be bitter if the Q-value is less than 5,436 based on joules/mol (1,300 based on calories/mol). A peptide's capacity to produce bitterness is also influenced by its molecular weight; only peptides with molecular weights less than 6,000 have this potential. It appears that peptides larger than this are blocked from binding to receptor sites due to their bulk.

Table 9.1. Calculated ΔG Values for Individual Amino Acids

Amino Acids	ΔG Value (kJ/mol)
Glycine	0 (0)
Serine	1,67.3 (40)
Threonine	1,839.9 (440)
Histidine	2,090.8 (500)
Aspartic acid	2,258.1 (540)
Glutamic acid	2,299.9 (550)
Arginine	3,052.6 (730)
Alanine	3,052.6 (730)
Methionine	5,436.1 (1,300)
Lysine	6,272.4 (1,500)
Valine	7,066.9 (1,690)
Leucine	10,119.5 (2,420)
Proline	10,955.8 (2,620)
Phenylalanine	1,1081.2 (2,650)
Tyrosine	1,2001.2 (2,870)
Isoleucine	1,2419.4 (2,970)
Tryptophan	1,2544.8 (3,000)

Source: Srinivasan Damodaran and Kirk L. Parkin. Fennema's Food Chemistry, Fifth Edition. CRC Press. ISBN: 9781482208122.

Derived from the cleavage of α_{s1}-casein between residues 144–145 and 150–151, the peptide depicted in Figure 9.5 has a calculated Q-value of 9,576 based on J/mol (2,290 based on calories/mol). This peptide is extremely bitter and serves as an example of the readily derived, highly hydrophobic peptides from α_{s1}-casein. These peptides cause the bitterness that ages cheeses to acquire.

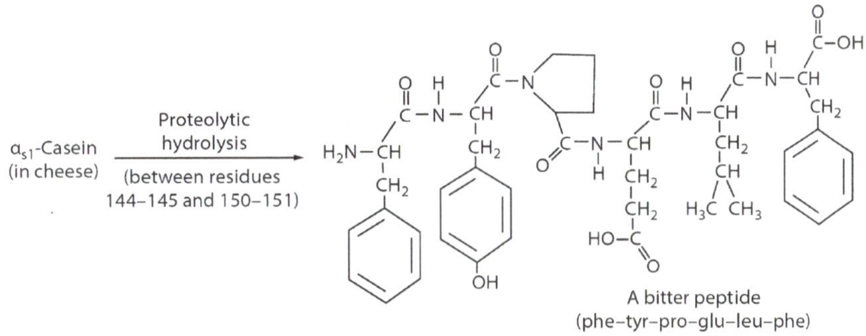

Figure 9.5. *Reaction depicting the hydrolysis of α_{s1}-casein to form a bitter peptide (phe–tyr–pro–glu–leu–phe) that exhibits strong overall nonpolar features.*

Source: Srinivasan Damodaran and Kirk L. Parkin. Fennema's Food Chemistry, Fifth Edition. CRC Press. ISBN: 9781482208122.

Humans differ in their ability to detect bitter substances due to genetic differences. Some substances, depending on the person, can be tasteless, bitter, or bitter–sweet at a certain concentration. While some people find saccharin to be exclusively sweet, others find it to be anywhere from mildly bitter and sweet to very bitter and sweet. Many other compounds also exhibit significant individual differences in how they are perceived, often tasting bitter or not being perceived at all.

One of the most well-known compounds in this category is phenylthiocarbamide (PTC) (VIII). It was found in the early 1930s that, when it comes to this compound, approximately 40% of the Caucasian-American population is taste blind to the bitterness taste attribute that is perceived by the other 60% of the population. The ability to detect PTC's bitterness was so obviously genetically controlled that it was quickly adopted as a marker to investigate behavioral and metabolic differences between bitter tasters and nontasters. More recently, researchers have mostly switched to 6-n-propyl-2-thio-uracil (PROP) (IX), which has no inherent odor, for further research because PTC also has a sulfurous odor. The bitter tastes of these two molecules are thought to be caused by the N–C=S grouping.

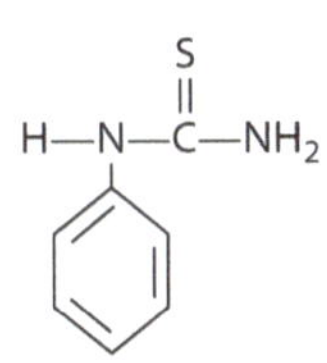

(VIII) Phenylthiocarbamide (PTC)

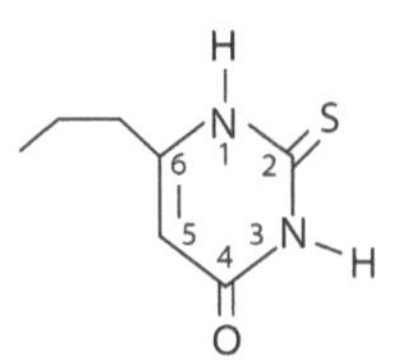

(IX) 6-*n*-Propyl-2-thiouracil (PROP or PRU)

According to recent research, people who find PROP to be extremely bitter are genetically predisposed to experience many flavors more intensely; these people are referred to as "supertasters." In an effort to identify underlying factors influencing food intake and preference as well as specific pathological conditions and health risks in the general population, researchers are currently looking into a wide range of physiological and psychological aspects of both PROP-insensitive and PROP-sensitive individuals.

While PTC and PROP are novel compounds not found in food, creatine (X) is a component of muscle foods that shares the same characteristics of varying population sensitivity to bitter tastes. Lean meats can contain up to mg/g of creatine, which is sufficient to give some soups a bitter taste to those who are sensitive to it.

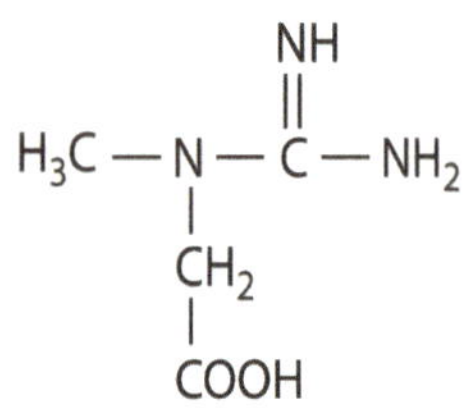

(X) Creatine

Salts have a bitter taste, and this sensory characteristic makes it difficult for consumers to replace sodium in foods that have had their composition changed to allow for a limited intake of sodium. It appears that the atomic (molecular) characteristics responsible for bitterness in salts differ significantly from those responsible for bitterness in organic compounds. The total ionic diameters of the anion and cation components that make up the salt in question appear to be correlated with the salt's bitterness. Salts with ionic diameters below 6.5 Å are purely salty in taste (LiCl = 4.98 Å, NaCl = 5.56 Å, KCl = 6.28 Å) although some individuals find KCl to be somewhat bitter. As the ionic diameter increases (CsCl = 6.96 Å, Csl = 7.74 Å), salts become increasingly bitter. Magnesium chloride (8.50 Å) is therefore quite bitter.

9.2.3. Salty Taste Substances

In addition to lithium chloride (LiCl), which is toxic and should not be used in food, sodium chloride (NaCl) provides a classic, clean salty taste. Generally speaking, salts have complex tastes that are typically characterized as psychological combinations of traditional sweet, bitter, sour, and salty perceptual components. Nonetheless, salts' tastes frequently defy conventional taste perceptions and are challenging to characterize using conventional flavor terminology. Other ambiguous terms, like "chemical" or "soapy," sometimes seem to capture the feelings that salts produce better than the traditional terminology.

In food applications, NaCl undoubtedly offers unique flavor-enhancing qualities, and its flavor effects seem to go well beyond those of the traditional tastes. It is simple to illustrate these effects by significantly lowering or eliminating NaCl from typical food formulations (e.g., bread and baked goods). It has been demonstrated that anions alter the fundamentally salty taste, while cations are responsible for the basic salty taste. Potassium and other alkaline earth cations produce a mixture of bitter and salty tastes, whereas sodium and lithium cations only produce a salty taste. Anions often add tastes of their own and also modify the tastes of salt by blocking the tastes of cations. The chloride anion is the least inhibitive to the salty taste among the anions that are frequently found in food, while the citrate anion is inhibitorier than the orthophosphate anions. Moreover, the orthophosphate anion imparts a stronger taste than the citrate anion, while the chloride anion has no taste at all.

Many foods, including processed cheese, are affected by anion taste effects because the citrate and phosphate anions in emulsifying salts both add and suppress the perceived saltiness caused by sodium ions. Similar to this, certain tastes elicited by the anions result in soapy tastes caused by sodium salts of long-chain fatty acids (XI) and detergents of long-chain sulfates (XII), which can completely mask the taste of the cation.

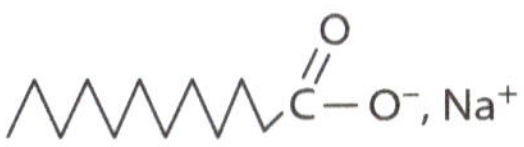

(XI) Sodium laurate

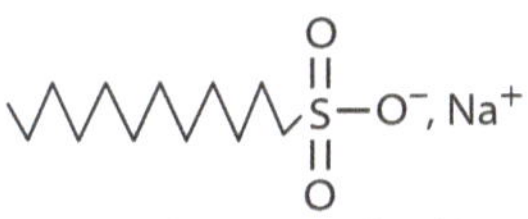

(XII) Sodium lauryl sulfate

National policies encouraging reduction in sodium consumption have stimulated interest in foods in which sodium salts have been replaced by alternative substances, particularly those containing potassium and ammonium ions. Since foods flavored with these alternatives taste different—and typically less appetizing—than foods flavored with NaCl, new efforts are being made to better understand the fundamental causes of the salty taste and to develop salt substitutes in the hopes of creating low-sodium goods that taste almost exactly like salt.

9.2.4. Sour Taste Substances

Because sour tastes are acidic by nature, they always contain one or more dissociable protons in aqueous solutions. While the binding of protons (H^+) to receptor cell membrane ion channels causes closure to Na^+ flow and depolarization, which is the first molecular-level event in the perception of acidic, sour, or tart flavors, the qualitative aspects of the sour taste response are not well understood. Contrary to common assumption, other poorly understood molecular features seem to be of primary importance and to determine the sour sensation rather than the acid strength in a solution (e.g., molecular weight, size, and overall polarity), as well as past practical experience, frequently dictate which acids are used in food applications.

9.2.5. Umami Taste Substances

Since the beginning of food preparation and cooking, people have used compounds that produce this taste sensation to enhance flavors. For a long time, scientists classified umami tastants—most notably, monosodium l-glutamate (MSG; XIII) and the 5′-ribonucleotides (inosine-5′-monophosphate [5′-IMP], XIV, and guanosine-5′-monophosphate)—as nonspecific taste responses because they had not yet been found to bind to specific taste receptors. Nonetheless, umami is now generally acknowledged as a fundamental taste response following the identification of taste receptors for these substances.

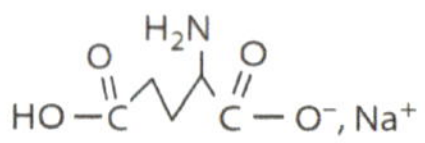

(XIII) Monosodium L-glutamate (MSG, umami)

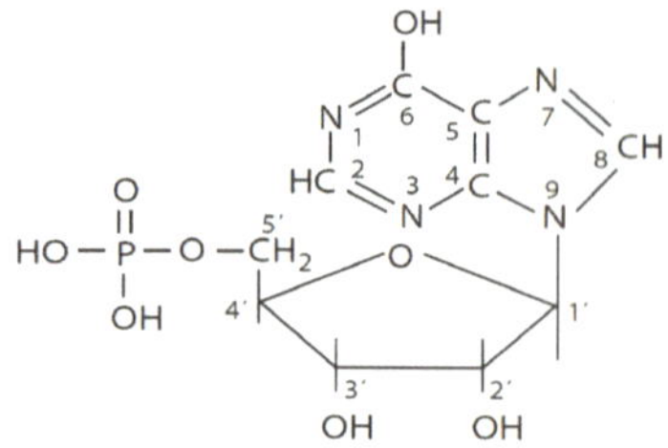

(XIV) Inosine-5′-monophosphate (IMP, umami)

When added to food at amounts above their independent detection threshold, umami compounds give it a delectable, mouthwatering taste; when added at levels below their independent detection thresholds, they alter and improve flavor. The flavors of vegetables, meats, poultry, fish, shellfish, and aged cheeses are significantly and favorably affected by them.

D-glutamate does not have any flavor-enhancing properties, nor do 2′- or 3′-ribonucleotides. A number of artificial 5′-ribonucleotide derivatives have potent flavor-enhancing qualities. These derivatives typically have substitutions on the 2-position purine moiety. MSG and 5′-ribonucleotides work in concert to enhance flavors and provide an umami taste, and commercial combinations of these compounds are commonly utilized. Some evidence suggests that the way umami compounds occupy receptor sites involved in the perception of sweet, sour, salty, and bitter tastes accounts for part of their flavor-enhancing effects. While the only commercially available flavor enhancers are MSG, 5′-IMP, and 5′-guanosine monophosphate, some natural amino acids, such as L-ibotenic acid and l-tricholomic acid, and 5′-xanthine monophosphate could also find application in the food industry. The 5′-ribonucleotides in yeast hydrolysates are largely responsible for the flavor that they add to food. A significant portion of the refined flavor enhancers used in the food industry come from microbiological sources, such as RNA-derived phosphorylated nucleosides that are produced *in vitro*.

9.2.6. Kokumi Taste Substances and Other Flavor Modifiers

Many food flavors can benefit greatly from the flavor-enhancing and flavor-modifying properties of common salt (NaCl). While specialized taste cells are responsible for detecting salt, there is a belief that salt can also enhance flavor by altering the functions of other basic taste receptor cells or by eliciting other sensations from other neural systems (e.g., trigeminal nerve) within the mouth. Therefore, salt probably shares some characteristics with other flavor-enhancing agents, the overall roles of which are still unclear.

To describe at least some of these compounds that do not elicit the umami response or any of the original four basic tastes, the Japanese have coined a new term, kokumi, to refer to them. These compounds improve food palatability by giving flavors more body, richness, complexity, continuity, and fullness. For instance, S-substituted cysteine sulfoxide amino acids (Figure 9.6) are the main precursors for the volatile aroma compounds that characterize garlic and onions. Both of these compounds are highly kokumi, which has a discernible impact on palatability, and they are easily soluble in water. Therefore, while the tastes of foods that contain garlic (e.g., pasta sauces, and sautéed meats) may not have overtly discernible volatile garlic flavor notes, but because S-(2-propenyl)-l-cysteine sulfoxide is present, their flavors are thought to be incredibly complex, full, and palatable.

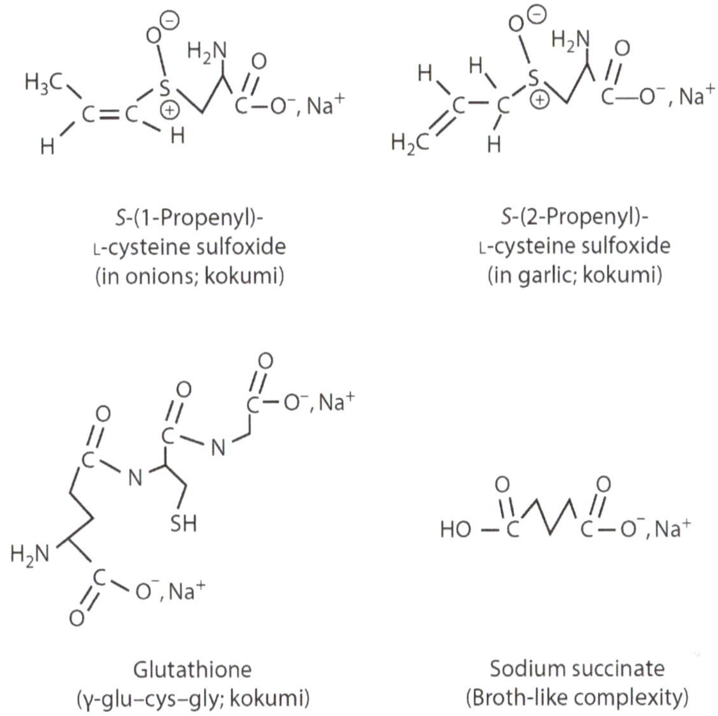

Figure 9.6. *Structures of some readily water-soluble, flavor-modifying compounds.*

Source: Srinivasan Damodaran and Kirk L. Parkin. Fennema's Food Chemistry, Fifth Edition. CRC Press. ISBN: 9781482208122.

While there aren't many easily soluble compounds that are said to give kokumi flavors yet, glutathione (Figure 9.6), another peptide that contains cysteines, is also kokumi active. A possibly related response is that succinic acid (and its soluble salts, Figure 9.6) tastes sour but also has a distinct broth-like flavor. While succinic acid's flavor isn't quite considered kokumi, it's used in commercial applications to give savory flavors—particularly those of meat-based sauces—a complex, broth-like quality.

Velvety, rich, creaminess, and juiciness are some other descriptive terms used to characterize a range of flavor modifications that seem to be influenced by kokumi. Such flavor modifications can be produced by a variety of substances, both synthetic and natural (Figure 9.7), some of which share some structural similarities. Vanillin-type flavors are among the most well-liked flavors globally, and most people find the aromas of vanillin and ethylvanillin to be very appealing. But vanillin-related compounds do more than just contribute to aroma; they also have flavor-modifying properties that improve the smoothness, richness, and creaminess of flavor sensations, especially in sweet, high-fat foods like ice cream.

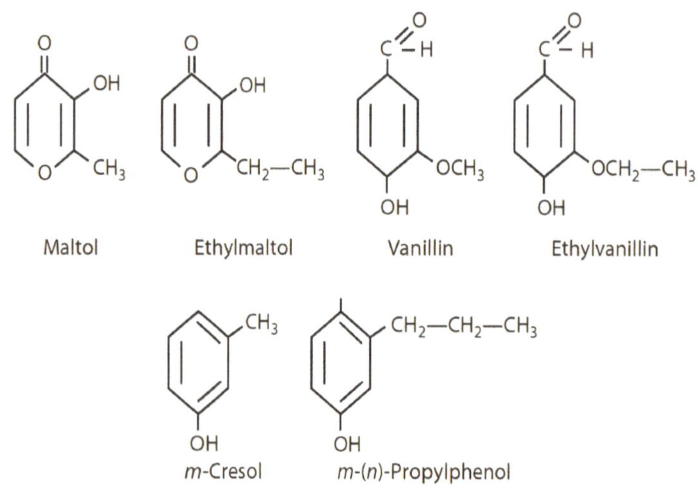

Figure 9.7. *Structures of some flavor-modifying compounds with limited water solubility.*

Source: Srinivasan Damodaran and Kirk L. Parkin. Fennema's Food Chemistry, Fifth Edition. CRC Press. ISBN: 9781482208122.

Maltol and ethylmaltol (Figure 9.7) have also been widely used as commercial flavor enhancers for fruit-based products and sweets. Both of these substances have pleasant, burnt caramel scents at high concentrations, but at relatively low concentrations (about 50 ppm), at which point the caramelic notes are not detectable, they are often marketed for the velvety, smooth feel they add to sweets and fruit juices. Maltol still typically reduces the detection threshold concentration for sucrose by a factor of two, but ethylmaltol is a more effective sweetness enhancer. Recently, it has been discovered that, even at very low concentrations (ng/g), some alkylphenols that naturally occur in ruminant milk and meat contribute to mouth coating, richness enhancing, and juiciness sensations. Among the members of this group, m-Alkyl substitution on the aromatic ring has the most significant flavor-modifying effects; m-cresol and m-(n)-propylphenol (Figure 9.7) are the most significant in ingredients and products derived from cows.

9.2.7. Pungent Substances

A variety of compounds found in spices and vegetables exhibit pungency, a chemesthetic property that produces distinct hot, sharp, and stinging sensations. Certain strong ingredients, like those in ginger, black pepper, and chili peppers, are not vaporized and primarily affect oral tissues unless they are released into the air through aerosol droplets. Other spices and vegetables have somewhat volatile pungent principles that cause pungency in the nasal and oral cavities as well as distinct aromas. These include onions, watercress, horseradish, mustards, and vegetable radishes. Clove is an aromatic spice that has eugenol as its active ingredient. All of these strong spices and veggies are added to food to give it a distinct flavor or to make it more palatable overall. When used in processed foods at low concentrations, it often gives flavors a lively quality by adding subtle notes that enhance the flavors that are perceived. Only the three main pungent spices—pepper, ginger, and chili peppers—are covered in this section of the chapter; other spices are covered later on in talks about flavor systems derived from plants (i.e., thiopropanal-S-oxide, eugenol, and isothiocyanates). Extensive analysis of aromatic compounds are also accessible.

A class of compounds called capsaicinoids, which are vanillylamides of monocarboxylic acids with variable chain lengths (C_8–C_{11}) and unsaturation, are found in chili peppers (Capsicum sp.). Capsaicin (XV) is an example of these strong concepts. A number of synthetic capsaicinoids that contain components of saturated straight-chain acids are used to replace natural chili extractives or oleoresins. The global content of capsaicinoid compounds in Capsicum sp. differs significantly; red pepper, for instance, has 0.06%; cayenne red pepper, 0.2 percent; Sannam, India, 0.3 percent; and Uganda, Africa, 0.85 percent. Sweet paprika is primarily used for its coloring effects and subtle flavors, which are partially derived from carotenoid oxidation, and has a very low concentration of pungent compounds. Additionally, chili peppers include a variety of volatile aroma compounds that enhance the flavor of food that is seasoned with them.

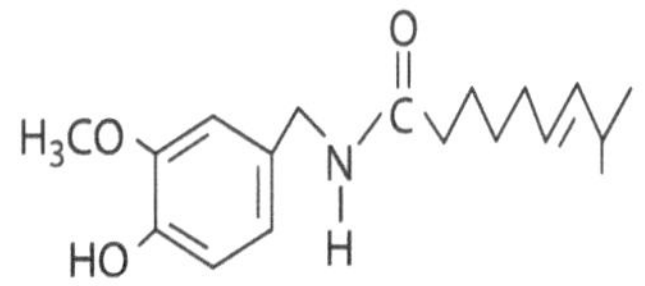

(XV) Capsaicin

The only difference between black and white pepper, which are both made from Piper nigrum berries, is that black pepper is made from young, green berries, while white pepper is made from more mature berries that are typically harvested just before they turn red. The main aromatic ingredient in pepper is the amide piperine (XVI). Strong pungency requires the transgeometry of the alkyl unsaturation, and isomerizations to cis forms of these double bonds are primarily responsible for pungency loss during light exposure and storage. L-formylpiperidine and piperonal (heliotropin), two volatile compounds found in pepper, also add flavor to food that has been seasoned with pepper or oleoresins. Additionally, piperine is synthesized to add flavor to food.

Zingiber officinale Roscoe is a tuberous perennial whose rhizome yields the spice known as ginger. The spice contains volatile aroma constituents along with pungent principles. Gingerols are a class of phenylalkyl ketones that give fresh ginger its pungent flavor; the most potent of these is [6]-gingerol (Figure 9.8).

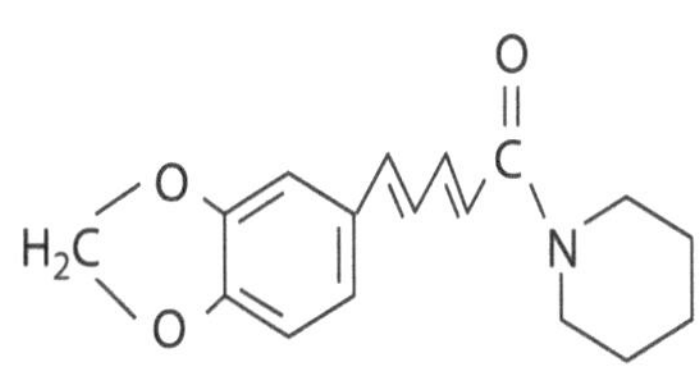

(XVI) Piperine

Gingerols differ in their external C-atom-substituted chain length (C5–C9).

Gingerols tend to lose moisture when they are dried and stored, which results in the formation of an external double bond that is conjugated with the keto group. Shogaols, a class of compounds produced by the reaction, are even more strongly aromatic than gingerols. When [6]-gingerol is exposed to significantly high temperatures, the alkyl chain outside of the keto group can cleave, producing the methyl ketone zingerone, which has a very mild pungent taste.

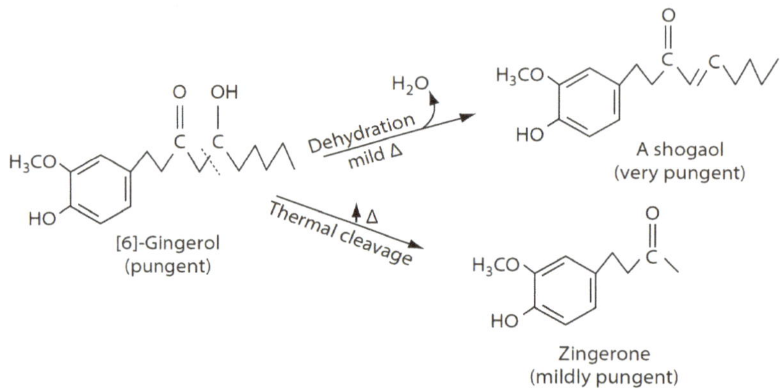

Figure 9.8. *Reactions altering gingerol that affect the relative pungency of ginger.*

Source: Srinivasan Damodaran and Kirk L. Parkin. Fennema's Food Chemistry, Fifth Edition. CRC Press. ISBN: 9781482208122.

9.2.8. Cooling Substances

Another chemesthetic sensation is cooling, which happens when specific chemicals come into contact with the tissues in the mouth or nose and activate the nonspecific neural systems (e.g., nerve (trigeminal). When naturally occurring substances cause these effects, flavors that resemble mint, such as peppermint, spearmint, and wintergreen, are most frequently linked to them. The feeling is produced by a number of compounds, but the most widely used form is (–)-menthol (XVII) in its natural form ($_L$-isomer). Numerous artificial cooling substances have been identified, and a concomitant camphoraceous scent is frequently produced by both natural and artificial substances. Because camphor (XVIII) has a distinct smell and a cooling effect, it is frequently used as a model for the camphoraceous group of compounds.

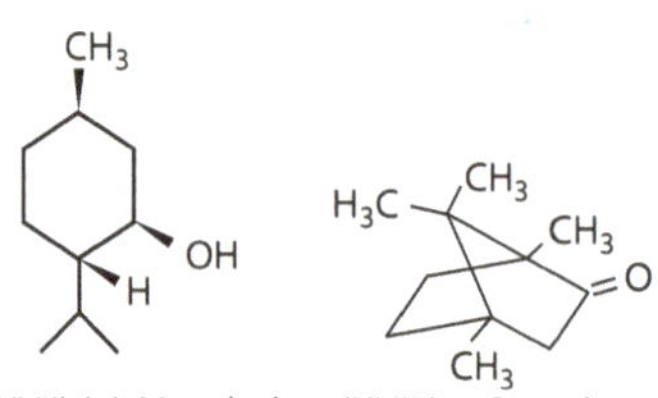

(XVII) (–)-Menthol (XVIII) D-Camphor

The slight cooling sensation experienced when polyol sweeteners, like xylitol, are tasted as crystalline materials is not the same as the cooling effect produced by the compounds related to mint. This is because of differences in their mechanisms. In the latter instance, the effect is thought to be caused by an endothermic dissolution of the materials.

9.2.9. Astringent Substances

Astringency is a sensation-related condition that is characterized by a rough puckering of the oral tissue and a dry mouth. Usually, astringency is caused by tannins or polyphenols binding to salivary proteins to form aggregates or precipitates. Furthermore, salivary proteins and mucopolysaccharides can react with sparingly soluble proteins, like those in some dry milk powders, to produce astringency. Because many people do not fully understand the difference between astringency and bitterness, many polyphenols or tannins, as in red wines, can produce both astringent and bitter sensations.

Condensed tannins from oxidative processes are frequently the more astringent tannins; these molecules have large cross-sectional areas (Figure 9.9) that are appropriate for hydrophobic interactions with proteins. Numerous phenolic groups found in tannins have the ability to change into quinoid structures, which then have the ability to chemically cross-link with proteins. It has been proposed that these cross-links may play a role in astringency activity.

In some cases, like with tea, astringency can be a desirable flavor attribute. However, by binding polyphenols with milk proteins, the practice of adding milk or cream to tea reduces its astringency. Red wine is a good example of a beverage that has tannins or polyphenols that cause both bitterness and astringency. Reducing polyphenol tannins, which are linked to the anthocyanin pigments, is common practice since too much astringency in wine is deemed sensory undesirable. Unwanted tastes in products containing unripe bananas can also result from the astringency caused by the polyphenols in unripe bananas.

9.3. VEGETABLE, FRUIT, AND SPICE FLAVORS

Fruit and vegetable flavors are not always amenable to logical groupings, making it difficult to categorize them into a manageable number of distinct groups. For instance, some flavor information derived from plants was covered in the section on pungency, while other flavor information was covered in the section on how "reaction" flavors are developed. This section focuses on the biogenesis and flavor development of significant fruits and vegetables. The reader is referred to the general references for more details on the flavors of other fruits and vegetables.

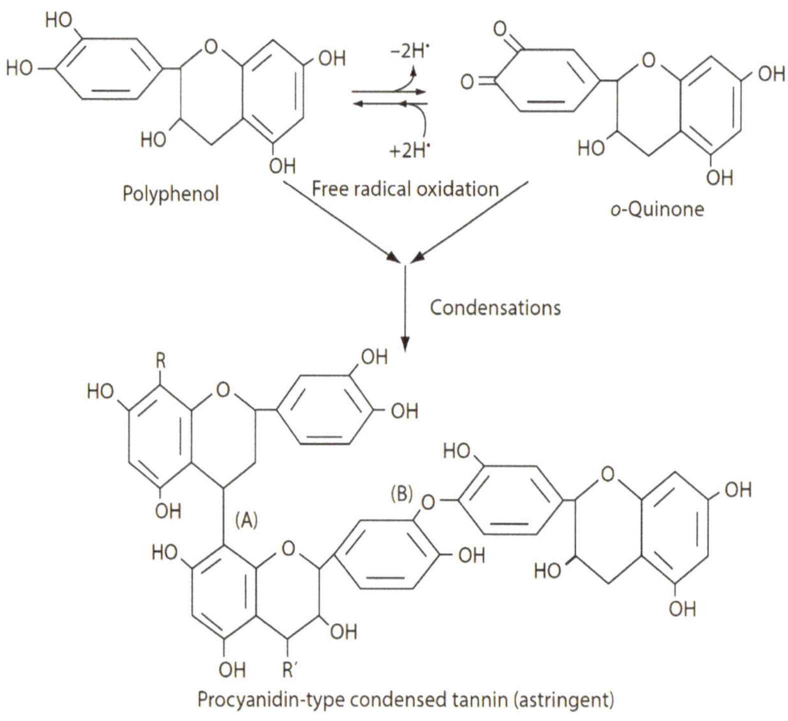

Figure 9.9. *Model reaction illustrating the formation of a procyanidin-type tannin with large planar hydrophobic areas capable of associating with proteins to cause astringency.*

Source: Srinivasan Damodaran and Kirk L. Parkin. Fennema's Food Chemistry, Fifth Edition. CRC Press. ISBN: 9781482208122.

9.3.1. Sulfur-Containing Volatiles in Allium sp.

Alliums are a group of plants that have powerful, pervasive scents. Key members of this genus include shallots, onions, garlic, leeks, and chives. Unless the tissue is injured and the enzymes are decompartmentalized to allow the flavor precursors to be transformed into odorous volatiles, these plants lack the potent aroma that makes them distinctive. Regarding onions (*A. cepa* L.), S-(1-propenyl)-$_L$-cysteine sulfoxide is the precursor of the sulfur compounds that give this vegetable its volatile flavor and aroma; additionally, this substance has kokumi flavor characteristics. Leek also contains this substituted cysteine sulfoxide precursor.

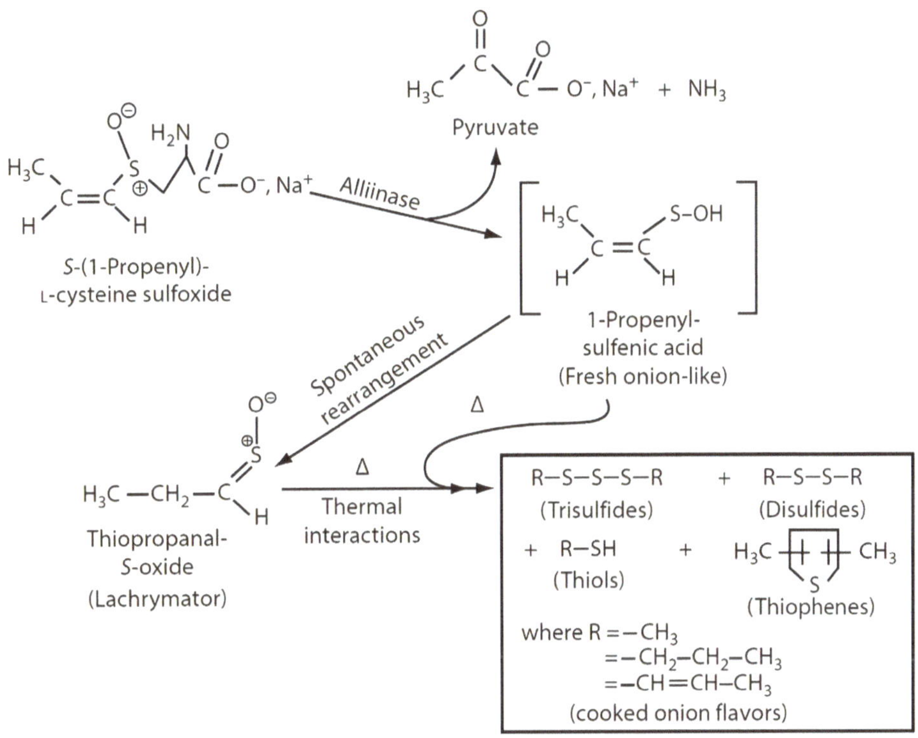

Figure 9.10. *Reactions involved in the formation of onion flavors.*

Source: Srinivasan Damodaran and Kirk L. Parkin. Fennema's Food Chemistry, Fifth Edition. CRC Press. ISBN: 9781482208122.

Alliinase in onions hydrolyzes S-(1-propenyl)-L-cysteine sulfoxide quickly, producing pyruvate, ammonia, and an unstable sulfenic acid intermediate (Figure 9.10). Additional rearrangements of the sulfenic acid result in the lachrymator thiopropanal-S-oxide, which is also linked to the overall scent of fresh onions. One stable byproduct of the reaction is pyruvic acid, which is generated by the enzymatic conversion of the precursor compound. This can be used as a good proxy for the flavor intensity of onion products. Additionally, a portion of the unstable sulfenic acid breaks down and rearranges into a sizable number of compounds belonging to the thiophene, disulfide, mercaptan, and trisulfide classes. The flavoring substances that give cooked onions their flavors are also formed more thoroughly from these compounds and other derivatives.

The general mechanism that forms the flavor of onions also forms the flavor of garlic (*Allium sativum* L.), with the exception that the precursor in garlic is S-(2-propenyl)-$_L$-cysteine sulfoxide. Garlic's flavor is influenced by diallyl thiosulfinate, or allicin (Figure 9.11), and it does not form an S-oxide lachrymator akin to that found in onions. Similar to how onion sulfenic acid is shown to break down and reorganize, garlic's thiosulfinate flavor compound does the same (Figure 9.10). This produces other principles in cooked garlic flavors and garlic oil, such as methyl, allyl, and diallyl disulfides.

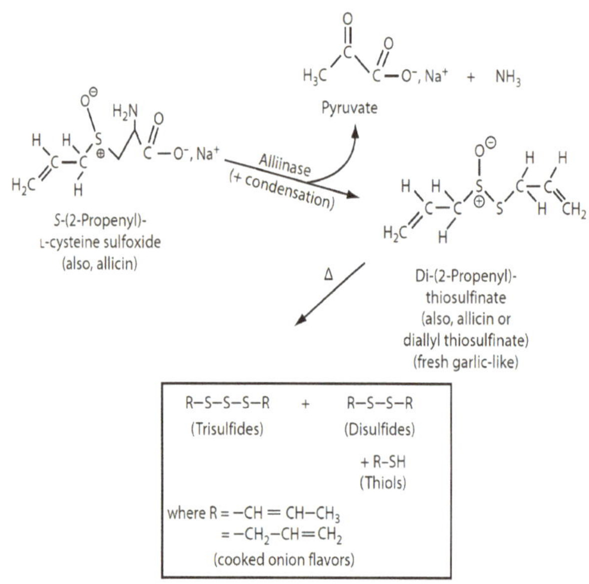

Figure 9.11. *Reactions involved in the formation of garlic flavor compounds.*

Source: Srinivasan Damodaran and Kirk L. Parkin. Fennema's Food Chemistry, Fifth Edition. CRC Press. ISBN: 9781482208122.

9.3.2. Sulfur-Containing Volatiles in the Cruciferae

Water cress (*Nasturtium officinale* R. Br.), radishes (*Raphanus sativus* L.), horseradish (*Armoracia lapathifolia* Gilib), Brussels sprouts (*B. oleracea* var. *gemmifera* L.), turnips (*Brassica Rapa* var. Rapa L.), brown mustard (*Brassica juncea* Coss.), and other Brassica plants are all members of the Cruciferae family. The active pungent principles in the Cruciferae are also volatile, as was mentioned in the discussion of pungent compounds, and as such, they contribute to distinctive aromas. Furthermore, the strong compounds in Cruciferae often have lachrymatory effects and produce strong irritation sensations, especially in the nasal cavity.

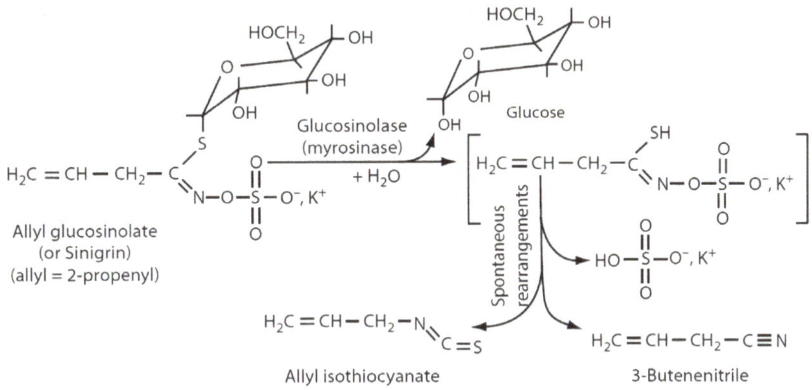

Figure 9.12. *Reactions involved in the formation of Cruciferae flavors.*

Source: Srinivasan Damodaran and Kirk L. Parkin. Fennema's Food Chemistry, Fifth Edition. CRC Press. ISBN: 9781482208122.

These plants produce their flavor compounds through cooking and enzymatic processes in damaged tissues.The primary source of fresh flavors in disrupted tissues is the action of glucosinolates on thioglycoside precursors, which produces isothiocyanates. The reaction depicted in Figure 9.12 serves as an example of how fresh Cruciferae form flavor. The primary source of pungent flavor and aroma in black mustard and horseradish is allyl isothiocyanate.

The Cruciferae contain a number of glucosinolates (S-glycosides), each of which results in distinct flavors. 4-methylthio-3-t-butenylisothiocyanate (XIX), an aroma compound, is responsible for the mild pungency of radishes. Glucosinolates yield thiocyanates and nitriles in addition to isothiocyanates.

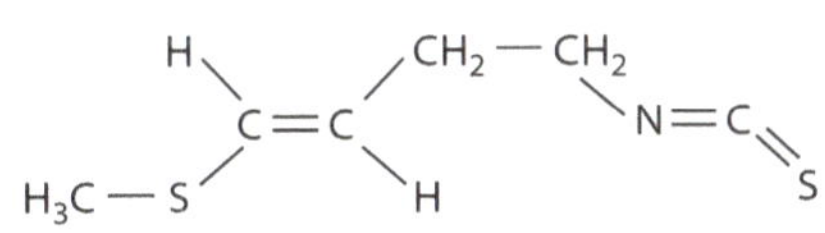

(XIX) 4-Methylthio-3-*t*-butenylisothiocyanate (radish)

Fresh cabbage and Brussels sprouts have the potential to contain both allyl isothiocyanate and allyl nitrile, though they are not usually overtly aromatic. The concentration of each depends on the stage of growth, where it is located in the edible portion, and how harshly they have been processed.

Processing (cooking and dehydrating) at temperatures significantly higher than room temperature tends to break down the isothiocyanates and increase the number of nitriles and other compounds that are involved in the breakdown and rearrangement of sulfur. Cruciferae contains a number of aromatic isothiocyanates; one of the primary aroma components of watercress is 2-phenylethyl isothiocyanate. Additionally, this compound adds a tingling, pungent sensation that affects the flavor of watercress salads.

9.3.3. Unique Sulfur Compound in Shiitake Mushrooms

The flavor of Shiitake mushrooms (Lentinus edodes), which are highly valued in Japan and other countries for their distinct and appealing flavor, is associated with a novel C-S lyase enzyme system.

The primary flavoring agent, lentinic acid, is derived from an S-substituted $_{L}$-cysteine sulfoxide that is bound as a γ-glutamyl peptide. γ-glutamyl transpeptidase is the first enzyme reaction in flavor development; it releases lentinic acid, a precursor of cysteine sulfoxide.

S-alkyl-L-cysteine sulfoxide lyase then attacks lentinic acid, producing products that eventually form lenthionine, the flavor compound responsible for the Shiitake flavor (Figure 9.13). The maximum flavor develops only after drying and rehydrating or after holding freshly macerated tissue for a while. These reactions are only triggered after the tissue is disturbed. In addition to lenthionine, other polythiepanes are also formed and add to the overall sulfurous flavor of flavor mushrooms.

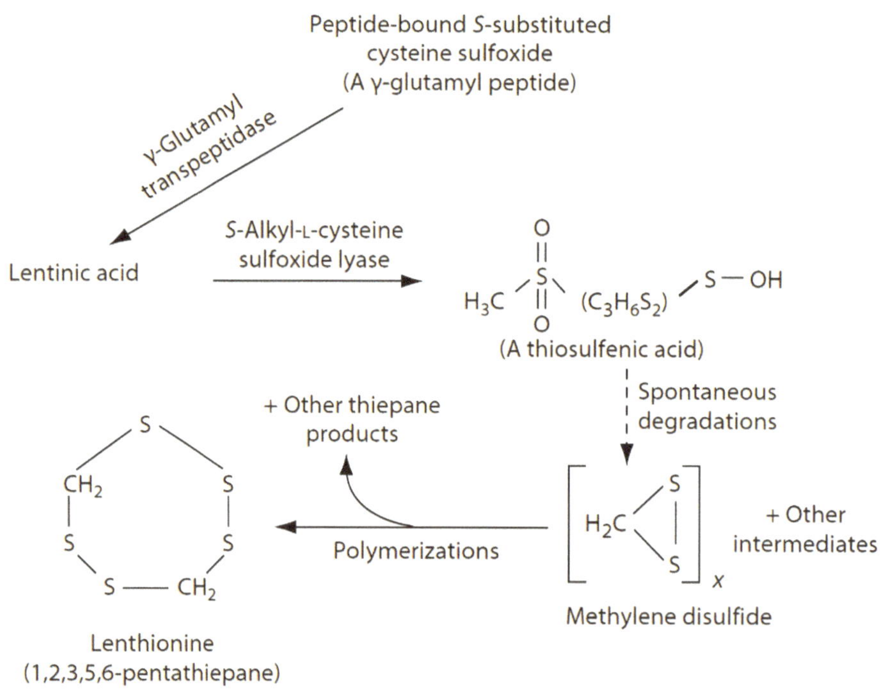

Figure 9.13. *Reactions involved in the formation of lenthionine in shiitake mushrooms.*

Source: Srinivasan Damodaran and Kirk L. Parkin. Fennema's Food Chemistry, Fifth Edition. CRC Press. ISBN: 9781482208122.

9.3.4. Methoxy Alkyl Pyrazine Volatiles in Vegetables

It has been discovered that the methoxy alkyl pyrazines are often responsible for the green-earthy aromas that many fresh vegetables have, which greatly contribute to their recognition.

Furthermore, methoxy alkyl pyrazines have lately been linked to the tastes of specific varietal grape wines. These compounds provide strong identifying aromas and have remarkably potent, penetrating odors.

The first of this class to be identified was 2-Methoxy-3-isobutylpyrazine, which has a strong bell pepper scent that can be detected at a threshold level of 0.002 ppb. 2-methoxy-3-isopropylpyrazine contributes significantly to the aroma of raw potatoes, green peas, and pea pods, while 2-methoxy-3-sec-butylpyrazine is responsible for the aroma of raw red beet roots. *Pseudomonas perolens* and *Pseudomonas tetrolens* are two strains of microorganisms that actively produce these special substances in addition to plants that biosynthesize these compounds. Methoxy alkyl pyrazine volatiles are derived from branched-chain amino acids, and the proposed mechanism is depicted in Figure 9.14.

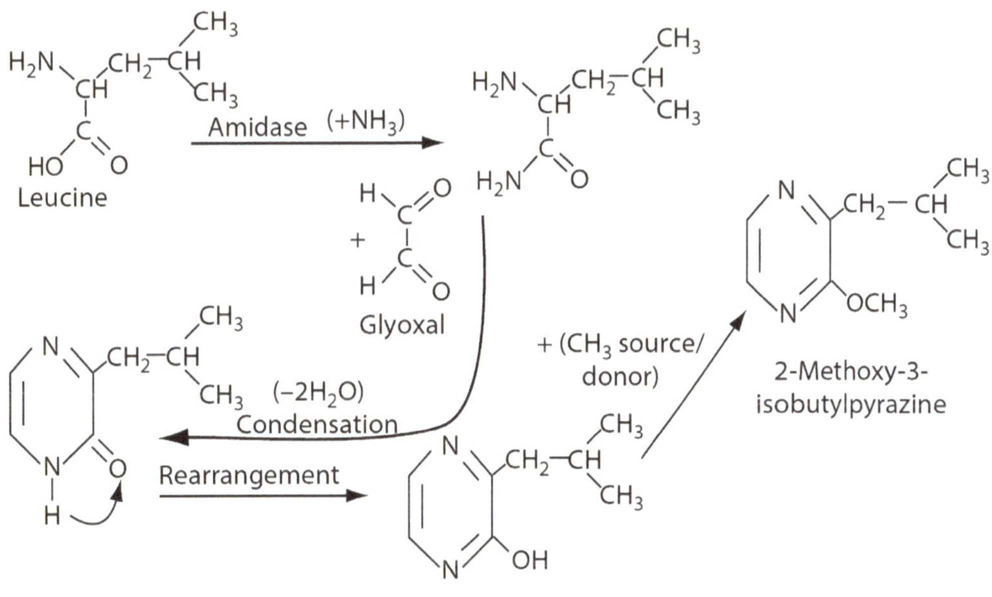

Figure 9.14. *Proposed enzymic scheme for the formation of methoxy alkyl pyrazines.*

Source: Srinivasan Damodaran and Kirk L. Parkin. Fennema's Food Chemistry, Fifth Edition. CRC Press. ISBN: 9781482208122.

9.3.5. Enzymatically Derived Volatiles from Fatty Acids

Fruits and vegetables' distinctive flavors are largely influenced by compounds that are produced by enzymes from long-chain fatty acids. Furthermore, these kinds of reactions may result in significant off-flavors, like those connected to processed soybean proteins. The talks regarding lipids and enzymes contain more details about these reactions.

9.3.5.1. Lipoxygenase-Derived Flavors in Plants

Enzyme-induced oxidative breakdown of unsaturated fatty acids takes place extensively in plant tissues, producing distinct aromas linked to certain ripening fruits and disrupted tissues. Enzyme-determined systems produce lipid-derived flavor compounds that are very distinctive, as opposed to purely autoxidizing systems that produce these compounds randomly. Figure 9.15 demonstrates the enzymatic specificity for generating flavor compounds. From unsaturated fatty acids, 1-octene3-one, t-2-c-6-nonadienal, and t-2-hexenal are formed, and these compounds give fresh mushrooms, cucumbers, and tomatoes, respectively, their distinctive flavors. Specific lipoxygenases and the ensuing lyase cleavage reactions control site-specific peroxidations of freed fatty acids. Oxo-acids are also produced during the cleavage of the fatty acid molecule, but they don't seem to affect flavor.

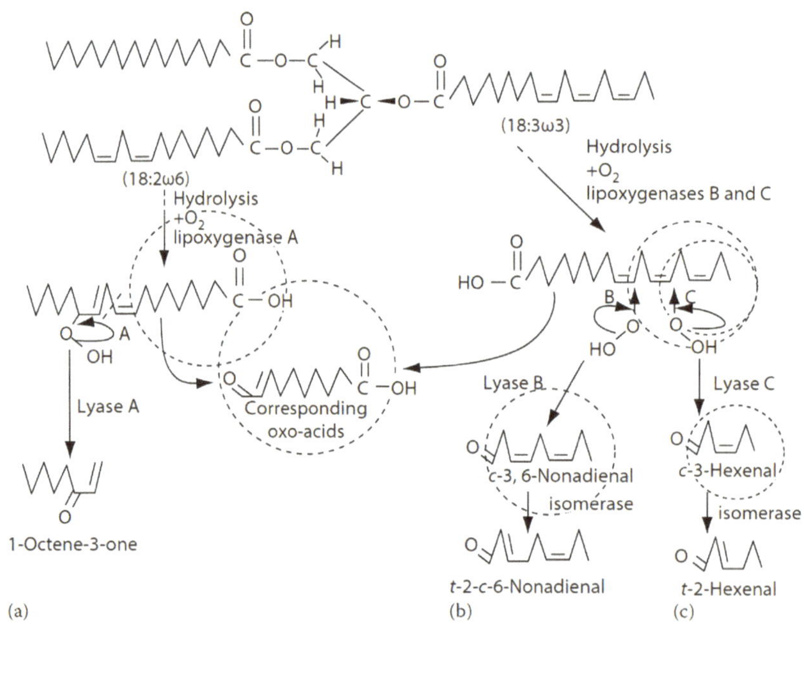

Figure 9.15. *Formation of lipoxygenase-directed carbonyl compounds from unsaturated fatty acids: (a) important in fresh mushrooms; (b) important in cucumbers; and (c) important in fresh tomatoes.*

Source: Srinivasan Damodaran and Kirk L. Parkin. Fennema's Food Chemistry, Fifth Edition. CRC Press. ISBN: 9781482208122.

Enzyme decompartmentalization is necessary to start this and other reactions, and since these reactions happen one after the other, flavors gradually shift over time. For instance, the aldehydes and ketones produced by lipoxygenase are changed into corresponding alcohols (Figure 9.16), which typically have stronger aromas and higher detection thresholds than the original carbonyl compounds. Furthermore, the presence of cis-trans isomerases converts aldehydes' cis-3 bonds to trans-2 isomers (Figure 9.15), which modifies the aldehydes' aroma quality. Green plant-like aromas, such as fresh-cut grass, are typically produced by C_6 compounds; cucumber and melons, by C_9 compounds; and mushrooms, or violet and geranium leaves, by C_8 compounds. While the C_8 compounds are secondary alcohols and ketones, the C_6 and C_9 compounds are primary alcohols and aldehydes.

Figure 9.16. *Conversion of an aldehyde to a corresponding alcohol resulting in a subtle flavor modification.*

Source: Srinivasan Damodaran and Kirk L. Parkin. Fennema's Food Chemistry, Fifth Edition. CRC Press. ISBN: 9781482208122.

9.3.5.2. Volatiles from β-Oxidation of Long-Chain Fatty Acids

When pears, peaches, apricots, and other fruits ripen, they release pleasant, fruity aromas. These aromas are often dominated by medium-chain-length (C^6–C^{12}) volatiles, which are produced by β-oxidation from long-chain fatty acids.

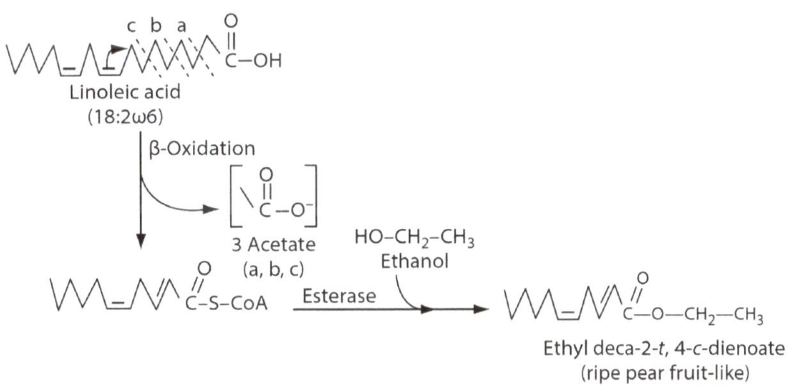

Figure 9.17. *Formation of a key aroma substance in ripened pears through β-oxidation of linoleic acid followed by esterification.*

Source: Srinivasan Damodaran and Kirk L. Parkin. Fennema's Food Chemistry, Fifth Edition. CRC Press. ISBN: 9781482208122.

Figure 9.17 shows how ethyldeca-2-t,4-c-dienoate is produced using this method. The Bartlett pear's impact or distinctive aroma compound is this ester. Hydroxy acids (C^8–C^{12}) can also be formed as part of the overall enzymatic process; however, they are not shown in the figure. They easily cyclize to yield γ- and δ-lactones. Similar processes take place when milk fat is metabolized and biosynthesized. The C^8–C^{12} lactones have distinct aromas that are reminiscent of peaches and coconuts, respectively.

9.3.6. Volatiles from Branched-Chain Amino Acids

Important flavor precursors for the biosynthesis of compounds linked to some ripening fruits are found in branched-chain amino acids. Because volatiles from amino acids contribute significantly to the ripe flavor of both bananas and apples, they are particularly good examples for this process.

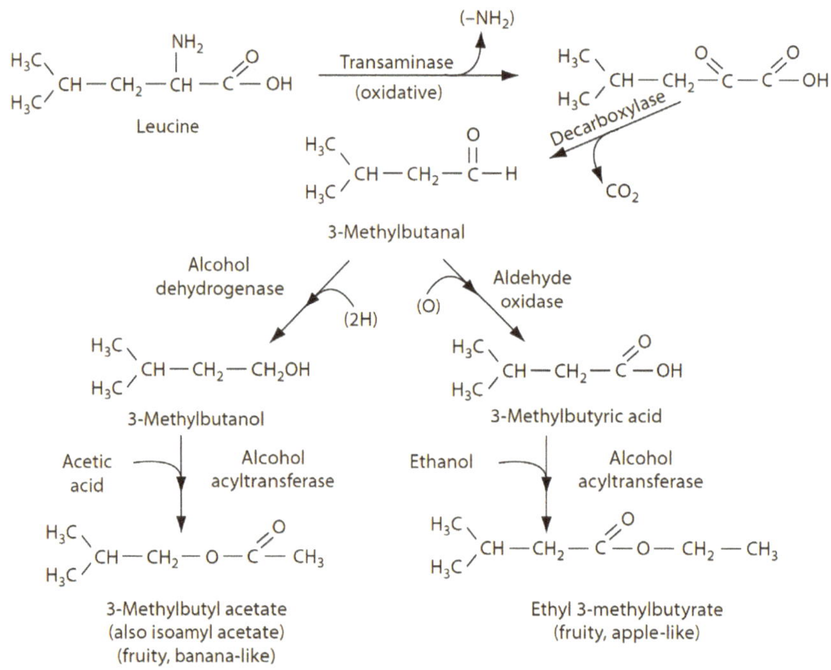

Figure 9.18. *Enzymatic conversion of leucine to volatiles illustrating the aroma compounds formed from amino acids in ripening fruits.*

Source: Srinivasan Damodaran and Kirk L. Parkin. Fennema's Food Chemistry, Fifth Edition. CRC Press. ISBN: 9781482208122.

Because the transamination and decarboxylation that take place during the initial phase of flavor formation (Figure 9.18) are similar to those that occur during nonenzymatic browning, this reaction is sometimes referred to as enzymic Strecker degradation. A number of microorganisms can also alter the majority of amino acids in a way that resembles the one depicted in Figure 9.18, including baker's yeast and strains of

Lactococcus lactis that produce malty flavors. Similar derivatives from amino acids other than leucine can also be produced by plants; this is why blooms sometimes contain 2-phenethanol that smells like roses or lilacs.

Character-impacting compounds that predominate are esters, although the aldehydes, alcohols, and acids from these reactions also directly contribute to the flavors of ripening fruits. Although isoamyl acetate (3-methylbutyl acetate) has long been recognized to play a significant role in banana flavor, other substances are also necessary to produce fully developed banana flavors. The primary note in the aroma of ripe, delicious variety apples is ethyl 2-methylbutyrate, which is even more akin to apples than ethyl 3-methylbutyrate.

9.3.7. Flavors Derived from the Shikimic Acid Pathway

The shikimic acid pathway, which is best known for its role in the synthesis of phenylalanine and other aromatic amino acids, supplies the aromatic portion of compounds related to shikimic acid in biosynthetic systems. The shikimic acid pathway produces additional volatile compounds that are commonly linked to essential oils, in addition to flavor compounds derived from aromatic amino acids (Figure 9.19). Additionally, it supplies the lignin polymers, which are the primary structural components of plants, with their phenylpropanoid skeleton. As seen in Figure 9.19, lignin undergoes pyrolysis to produce a large number of phenols, and compounds derived from precursors in the shikimic acid pathway are primarily responsible for the distinctive aroma of smokes used in food.

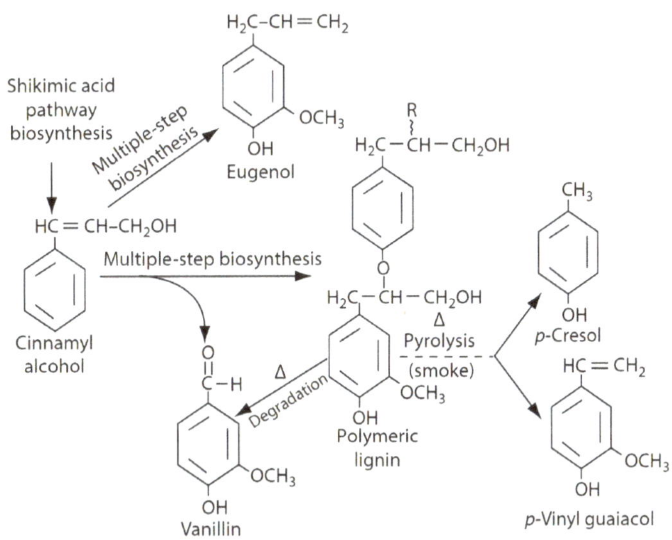

Figure 9.19. *Some important flavor compounds variously derived from shikimic acid pathway precursors.*

Source: Srinivasan Damodaran and Kirk L. Parkin. Fennema's Food Chemistry, Fifth Edition. CRC Press. ISBN: 9781482208122.

It is also evident from Figure 9.19 that this pathway can be used to naturally obtain vanillin, the most significant characteristic compound in vanilla extracts, or as a by-product of lignin during the processing of wood pulp and paper. In the vanilla bean, where it is first primarily present as vanillin glucoside until the glycoside linkage is hydrolyzed during fermentation, vanillin is also biochemically synthesized. Many of the key components of the compounds depicted in Figure 9.19 are also present in the methoxylated aromatic rings of the pungent principles found in ginger, pepper, and chili peppers. Eugenol is the main aroma and pungent component of cloves, while cinnamonyl alcohol is a component of cinnamon spice's aroma.

9.3.8. Volatile Terpenoids in Flavors

Terpenes are abundant in plant materials used in the essential oil and perfumery industries, which sometimes overvalues their significance in other flavors associated with plants. However, they are mostly in charge of the flavors found in many seasonings and herbs, as well as citrus fruits. Raw carrot roots get a lot of their flavor from volatile terpenes, which are found in various fruits in small amounts. The biosynthesis of terpenes occurs via the isoprenoid (C_5) pathway (Figure 9.20). Sesquiterpenes have 15 C atoms, while monoterpenes have 10 C atoms. Sesquiterpenes are also significant compounds that characterize aroma; two notable examples are β-sinensal (XX) and nootkatone (XXI), which give grapefruit and oranges, respectively, their distinctive flavors. Too big and nonvolatile to directly contribute to scents are the diterpenes (C_{20}).

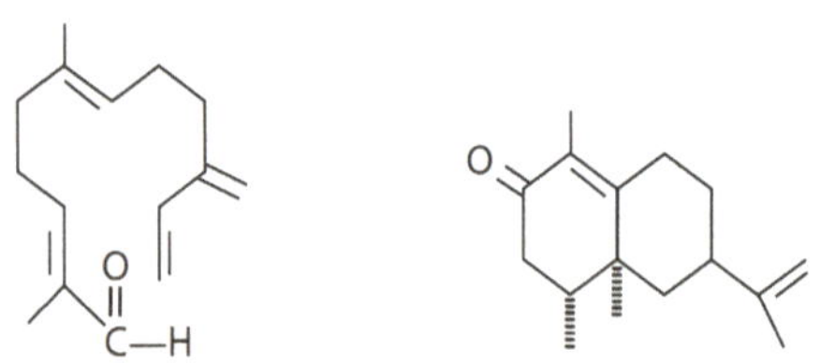

(XX) β-Sinensal (oranges) (XXI) Nootkatone (grapefruit)

Strong character-impact aroma properties are often associated with volatile terpenes, many of which are easily recognized by someone familiar with the scents of natural products.

Quantum isomers, or i.e., optical isomers of other nonterpenoid compounds, as well as structural mirror images) of terpenes, can have radically different odor characteristics. Regarding terpenes, a great deal of research has been done on the carvones. D-carvone [4S-(+)-carvone] (XXII) has a scent that is reminiscent of caraway spice, and l-carvone [4R-(–)-carvone] (XXIII) has a strong, distinct spearmint scent. Research on these pairs of compounds is particularly interesting because it sheds light on the structure–activity relationships of molecules as well as the basic mechanism of olfaction.

(XXII) 4S-(+)-Carvone (caraway-like) (XXIII) 4R-(−)-Carvone (spearmint-like)

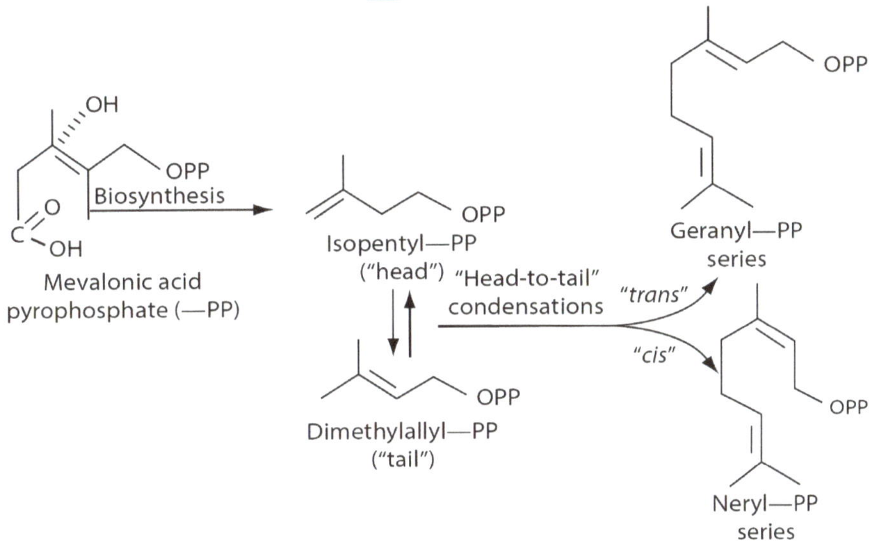

Figure 9.20. *Generalized isoprenoid scheme for the biosynthesis of monoterpenes.*

Source: Srinivasan Damodaran and Kirk L. Parkin. Fennema's Food Chemistry, Fifth Edition. CRC Press. ISBN: 9781482208122.

9.3.9. Citrus Flavors

The majority of knowledge regarding the flavor chemistry of natural citrus flavors comes from studies on processed juices, peel essential oils, essence oils, and aqueous essences used to flavor juice products. Citrus flavors are among the most popular flavors for both fresh fruits and beverages. Citrus flavors are primarily influenced by a number of flavor component classes, such as terpenes, alcohols, esters, and aldehydes. Numerous volatile compounds have also been found in the various citrus fruit extracts. Table 9.2 displays the flavor compounds that are important to some major citrus fruits. However, the important flavor compounds, including character-impact compounds, for citrus fruits are generally limited to relatively few.

Table 9.2. Some Volatile Compounds Considered Important in Citrus Flavors

Orange	Mandarin	Grapefruit	Lemon
Ethanol	Ethanol	Ethanol	Neral
Octanol	Octanol	Decanal	Geranial
Nonanal	Decanal	Ethyl acetate	β-Pinene
Citral	α-Sinensal	Methyl butanoate	Geraniol
Ethyl butanoate	γ-Terpinene	Ethyl butanoate	Geranyl acetate
d-Limonene	β-Pinene	*d*-Limonene	Neryl acetate
α-Pinene	Thymol	Nootkatone	Bergamotene
	Methyl-*N*-methyl-anthranilate	1-*p*-Methene-8-thiol	Caryophyllene
			Carvyl ethyl ether
			Linalyl ethyl ether
			Fenchyl ethyl ether
			Methyl epijasmonate

Source: Srinivasan Damodaran and Kirk L. Parkin. Fennema's Food Chemistry, Fifth Edition. CRC Press. ISBN: 9781482208122.

Orange and mandarin (tangerine is also used interchangeably with mandarin in US culinary applications) have subtle flavors that are frequently combined to varying degrees. Although many other compounds are present, Table 9.2 shows that only a small number of terpenes and aldehydes are thought to be necessary for these flavors. Orange and mandarin flavors contain both α- and β-sinensal (XX), with α-sinensal being particularly crucial for giving mandarin flavors a rich, orange citrus flavor. Nootkatone (XXI) and 1-p-methene-8-thiol, two character-impact compounds found in grapefruit, are largely responsible for the fruit's distinctive flavor. One of the reportedly few sulfur compounds that has an impact on citrus flavors is p-methene-8-thiol, and nootkatone is widely used to create artificial grapefruit flavor.

Many significant compounds must contribute to the flavor of lemons, and multiple terpene ethers are advantageous to lemon flavors. Similar to this, lime flavor has a high concentration of essential compounds and two varieties of lime oil are frequently used. Distilled Mexican lime oil, which is widely used in soft drinks like cola and lemon-lime, is the most common type of lime oil in commerce. It has a strong, harsh flavor. Due to their more authentic flavors, cold-pressed Persian and centrifuged Mexican lime oils are growing in popularity. Compared to distillation, cold-pressing and centrifugation involve less rigorous processing, which preserves some of the valuable but delicate fresh lime flavor compounds. Distilled lime oil gets its harsher flavors from the degradation of citral, which has a desirable fresh-like aroma, to coarse-flavored p-cymene and α-p-dimethylstyrene under acidic distillation conditions.

With the use of silicic acid chromatography and nonpolar and polar solvent elutions, citrus essential oils, or flavor extracts containing terpenes, can be divided into nonoxygenated (hydrocarbon) and oxygenated fractions, respectively. For instance, the oxygenated terpenes, aldehydes, and alcohols that have been extracted from orange oils make up the majority of terpeneless orange oil. Oxygenated terpene fractions are often preferred in flavors over nonoxygenated fractions due to their superior flavor quality.

9.3.10. Flavors of Herbs and Spices

Although there are some differences in definitions between industrial and the different domestic and international regulatory agencies, herbs and spices are recognized and controlled as natural vegetable products used to flavor, season, and add aroma to food. Alliaceous products, like onions and garlic, are not classified as spices by the U.S. Food and Drug Administration; however, these materials are frequently included in international and industrial spice classifications.

The regulatory definitions of the term "condiments" still refer to anything that adds flavor to food or a strong seasoning. Although some argue that the term should no longer be used, there are some circumstances in which it should still be used. Nonetheless, the term typically offers no foundation for aromatic plant material inclusion in classification schemes. Culinary herbs, or aromatic soft-stemmed plants like basil, marjoram, mints, oregano, rosemary, and thyme, as well as aromatic shrub (sage) and tree (laurel) leaves, are classified differently from spices in certain botanically based classification systems. All other aromatic plant materials used to flavor or season food are included in these classifications as spices.

These spices, which include rhizomes or roots (ginger), barks (cinnamon), flower buds (cloves), fruits (dill, pepper), and seeds (nutmeg, mustard), typically don't contain chlorophyll. Herbs and spices have been used since ancient times to give food and drink flavors character and to add savoriness, tanginess, and zest. A lot of them have also been used extensively in medicine and perfumery, and many of them have antibacterial or antioxidant properties. Around 70 herbs and spices are officially recognized as useful ingredients for foods, although there are many more in the world that are used for herbal medicines and perfumery. This group offers a vast variety of flavors for foods, though the flavor attributes of a particular spice frequently differ based on the place of growth and genetic variations. This list only includes herbs and spices that are frequently used as flavorings in food products. Herbs are typically derived from subtropical or temperate plants, whereas spices are typically derived from tropical plants. Additionally, phenylpropanoids from the shikimic acid pathway are typically found in high concentrations in spices (e.g., eugenol in cloves; refer to Figure 9.19); in general, herbs have greater p-menthanoids concentrations from terpene biosynthesis (e.g., peppermint, menthol in it).

Spices and herbs typically contain a high concentration of volatile compounds; however, in most cases, certain compounds—either abundant or minor volatile constituents—give the material its distinctive flavors and aromas. Tables 9.3 and 9.4, which list the main herbs and spices used in the food industry, provide an overview of significant flavor compounds. The key to successfully incorporating herbs and spices into food and drink recipes is to either have a thorough understanding of the ingredients or be aware of the prominent and nuances of flavor compounds. An analysis of the significant flavor compounds for spices and herbs listed in Tables 9.3 and 9.4 reveals which spices and herbs provide related flavors as well as the kinds of flavors that can be anticipated from their use. However, it's important to keep in mind that a variety of substances with less of an impact are also present and give herbs and spices their distinct flavors.

Table 9.3. Important Flavor Compounds Found in Some Culinary Herbs Commonly Used for Flavoring in the Food Industry

Herbs	Plant Part	Important Flavor Compounds
Basil, sweet	Leaves	Methyl chavicol, linalool, methyl eugenol
Bay laurel	Leaves	1,8-Cineole
Marjoram	Leaves, flowers	*c*- and *t*-Sabinene hydrates, terpinen-4-ol
Oregano	Leaves, flowers	Carvacrol, thymol
Origanum	Leaves	Thymol, carvacrol
Rosemary	Leaves	Verbenone, 1,8-cineole, camphor, linalool
Sage, clary	Leaves	Salvial-4(14)-en-1-one, linalool
Sage, Dalmatian	Leaves	Thujone, 1,8-cineole, camphor
Sage, Spanish	Leaves	*c*- and *t*-Sabinyl acetate, 1,8-cineole, camphor
Savory	Leaves	Carvacrol
Tarragon	Leaves	Methyl chavicol, anethole
Thyme	Leaves	Thymol, carvacrol
Peppermint	Leaves	*l*-Menthol, menthone, menthofuran
Spearmint	Leaves	*l*-Carvone, carvone derivatives

Source: Srinivasan Damodaran and Kirk L. Parkin. Fennema's Food Chemistry, Fifth Edition. CRC Press. ISBN: 9781482208122.

Table 9.4. Important Flavor Compounds Found in Some Spices Commonly Used for Flavoring in the Food Industry

Spice	Plant Part	Important Flavor Compounds
Allspice (pimento)	Berry, leaves	Eugenol, β-caryophyllene
Anise	Fruit	(*E*)-Anethole, methyl chavicol
Capsicum peppers	Fruit	Capsaicin, dihydrocapsaicin
Caraway	Fruit	*d*-Carvone, carvone derivatives
Cardamom	Fruit	α-Terpinylacetate, 1,8-cineole, linalool
Cinnamon, cassia	Bark, leaves	Cinnamaldehyde, eugenol
Clove	Flower bud	Eugenol, eugenylacetate
Coriander	Fruit	*d*-Linalool, C_{10}–C_{14} 2-alkenals
Cumin	Fruit	Cuminaldehyde, *p*-1,3-menthadienal
Dill	Fruit, leaves	Anethofuran, *d*-carvone
Fennel	Fennel	(*E*)-Anethole, fenchone
Ginger	Rhizome	Gingerol, shogaol, neral, geranial
Mace	Aril	α-Pinene, sabinene, 1-terpenin-4-ol
Mustard	Seed	Allyl isothiocyanate
Nutmeg	Seed	Sabinene, α-pinene, myristicin
Parsley	Leaves, seed	Apiol
Pepper	Fruit	Piperine, δ-3-carene, β-caryophyllene
Saffron	Stigma	Safranal
Turmeric	Rhizome	Turmerone, zingiberene, 1,8-cineole
Vanilla	Fruit, seed	Vanillin, *p*-OH-benzyl methyl ether

Source: Srinivasan Damodaran and Kirk L. Parkin. Fennema's Food Chemistry, Fifth Edition. CRC Press. ISBN: 9781482208122.

CASE STUDY

"DO US A FLAVOR"

Source: https://sites.psu.edu/cmsarosipsujourney/2019/07/17/case-study-do-us-a-flavor/.

In 2012, Lay's chip company launched its 'Do Us A Flavor' campaign to celebrate its 75th anniversary. By doing this, Lay's attempted to reach a new target audience of the younger generations using popular media platforms. The campaign's goal was to incorporate social media buzz to drum up business and have customers be able to directly impact the products they were buying. It was launched through Facebook's website where contestants could create a chip name, ingredients and a short description of the product they came up with. The winner of the competition would receive a $1 million cash prize. The first launch of the campaign was held in the UK in 2006, which had major success. After the accomplishments of the UK campaign, Lay's launched it again in the US, hoping for the same success.

Lay's used many different marketing techniques to make this campaign as successful as it was. Millennials, A.K.A the social media generation, were their target audience, so they decided to use Facebook as the source for contestants to submit their pitches. Lay's also incorporated the multi-platinum recording artist Nick Lachey into their campaign to surprise the four finalists and award the winner in 2015. (Get Your Pitch On!) Other celebrity endorsements included celebrity chef Michael Symon and actress Eva Longoria, who were on a panel of judges to help narrow down the submissions (Get Your Pitch On!). These celebrity appearances helped generate more buzz for the campaign and got more people talking about the brand. To help capture the interest of customers, Lay's allowed people to design their own bags, choose the ingredients that went into their product and allow them to share their bags all over Facebook to their friends and family.

Facebook was the catalyst used to expand its audience towards the younger generation. Lay's Facebook page allowed their interested audience to participate in the campaign and submit their pitches. Members of Lay's younger audience took part in the "Do Us A Flavor" campaign and pitched their flavor ideas partnered with a description and or story behind why they chose their new creation. Since the contest was previously held in the UK, it needed to be modified when it was redone in the US since Lay's was hoping for the new attempt to be much more advantageous. In order to come up with a new and improved strategy, Anindita Mukherjee (the North American branch- chief marketing officer) collected a team to generate a marketing idea that would most effectively address the largest audience possible through the medium of a social media campaign. At the time, Facebook was near its peak in popularity in regard to millennials, so Mukherjee and her team chose Facebook as their source.

PepsiCo's ultimate goal for the "Do Us A Flavor" campaign was to create social media buzz among the 18–34 age demographic. In order to do that, they made their submissions through Facebook and created buzz among millennials about their brand and the new flavors of chips. As the campaign in England received 1.2 million submissions, they were hoping the same would happen for the US campaign, and that this would result in a +3% uplift in sales (Brittany W). In the company's tenth campaign, they surpassed this goal, with 3 million submissions and 22.5 million Facebook page visits, they had an overall sales uplift of +12% (Brittany W). By 2015, Lay's had a net worth of 3.5 billion dollars after their lift in sales, and after the winning flavor of cheesy garlic bread was launched, sales rose 8% (Shultz). Ad recognition and purchase intent both increased during the campaign (Franklin), and their social media following had more activity than ever by the end of the campaign. Not only did they attract attention to their products, but they also strived for a more emotional connection with consumers. They wanted their audience to feel heard, and to be interactive with the people who buy their products, in order to gain more consumers. Therefore, Lay's achieved their goal of being the talk of the town (Frito-Lay).

Source: https://sites.psu.edu/cmsarosipsujourney/2019/07/17/case-study-do-us-a-flavor/.

Frito-Lay created a marvelous marketing technique that works for their company and now many others. By incorporating the audience into their products, they built an emotional connection with their customers. The campaign was effective by utilizing audience participation and various social media platforms such as Facebook, Twitter, Instagram, and Youtube to reach the millennials and the younger demographics Lay's was looking for. As a result, the campaign reached record numbers, much higher than what Lay's had projected. The massive food products supplier, Frito-Lay, greatly grew their quantity of consumers by their "Do Us A Flavor" campaign, therefore accomplishing their mission. The company renewed customer consumption and introduced an extremely effective method of marketing, so significant that it permanently changed the marketing world forever.

Source: https://sites.psu.edu/cmsarosipsujourney/2019/07/17/case-study-do-us-a-flavor/

CLASS ACTIVITY

To demonstrate the way that our senses of touch and smell combine with taste, conduct this activity. Before class, cut each fruit or vegetable into small cubes (approximately one centimeter in width). Place two samples of each food on a plate or a paper towel, but don't let the students see the food samples. Organize students into pairs. Within each pair, one person will be the "taster" and one will be the "tester." Give each of the testers a copy of the "Taste Survey" student handout. Blindfold the tasters and ask them to plug their noses. Distribute the food samples to the testers and have them give each sample to the tasters. Discuss the results. Which foods were most difficult to identify? Which taste sensations were strongest? What texture clues helped students analyze each food sample?

SUMMARY

- Flavor in food is important in how food smells and tastes to the consumer, especially in sensory analysis. Some of these products occur naturally like salt and sugar, but flavor chemists (called a "flavorist") develop many of these flavors for food products. Such artificial flavors include methyl salicylate which creates the wintergreen odor and lactic acid which gives milk a tart taste.
- The appreciation of flavor is a universal, everyday experience and is the integrated response to a complex mixture of stimuli primarily on the senses of smell and taste, but also on those associated with sight (color and appearance), tactile sensations (texture and mouth-feel) and pain (pungency). As a consequence, flavor chemistry in its broadest context is a specialized branch of organic chemistry that impinges on many other disciplines, including botany, zoology, anatomy, physiology, histology, cytology, etc., for its complete understanding.
- Flavors may be classified in many ways, but from the point of view of their chemistry, the most realistic classification is that based on their mode of formation, either naturally by biosynthetic pathways from known precursors, or by processing in which biological, chemical or physical conditions are imposed on natural or artificial starting materials.
- Natural flavors are mostly metabolites so that their formation in living tissues is complex and dependent on genetic factors and is influenced by environmental conditions during the natural growth cycle.
- Sensory assessments are used for the documentation of qualitative flavor characteristics of flavor and aroma chemicals, and both qualitative and quantitative flavor or aroma sensations for single or combinations of chemicals vary with the concentration.
- Sweet substances have been the focus of much attention because of interest in sugar alternatives and the desire to find suitable replacements for certain low-calorie sweeteners, including saccharin and cyclamate.
- The bitterness sensation appears to be closely related to sweetness from a molecular structure– receptor relationship. Bitterness resembles sweetness because of its dependence on the stereochemistry of stimulus molecules, and the two sensations are triggered by similar features in molecules, causing some molecules to yield both bitter and sweet sensations. Although sweet molecules must contain two polar groups that may be supplemented with a nonpolar group, bitter molecules appear to have a requirement for only one polar group and a hydrophobic group.
- Salts exhibit complex tastes usually described as consisting of psychological mixtures of classic sweet, bitter, sour, and salty perceptual components. However, the tastes of salts often fall outside the traditional taste sensations and are difficult to describe in the classic flavor terms.
- Sour taste substances are acidic in nature and thus contain at least one proton that is dissociable in aqueous systems. Although the initial molecular-level event in the perception of acidic, sour, or tart flavors involves the binding of

protons (H^+) to receptor cell membrane ion channels that results in closure to Na^+ flow and depolarization, the qualitative aspects of the sour taste response are poorly understood.

- Umami substances contribute a delicious, mouthwatering taste to foods when used at levels in excess of their independent detection threshold, and they modify and enhance flavors at levels below their independent detection thresholds. Their effects are prominent and desirable in the flavors of vegetables, meats, poultry, fish, shellfish, and aged cheeses.
- Common salt (NaCl) provides profound flavor-enhancing and flavor modifying effects to many food flavors. Although salt is specifically sensed by specialized taste cells, many believe that it also may provide flavor enhancing by modifying the functions of other basic taste receptor cells or through other sensations emanating from other neural systems (e.g., trigeminal nerve) in the oral cavity.
- Pungency is a chemesthetic property exhibited by a number of compounds found in spices and vegetables that cause characteristic hot, sharp, and stinging sensations. Some pungent principles, such as those found in chili peppers, black pepper, and ginger, are not volatile and, unless made airborne via aerosol droplets, exert their effects principally on oral tissues.
- Cooling is another chemesthetic sensation that occurs when certain chemicals contact the nasal or oral tissues and stimulate the nonspecific neural systems (e.g., trigeminal nerve). These effects when caused by naturally occurring substances are most commonly associated with mint-like flavors, including peppermint, spearmint, and wintergreen.
- Astringency is a feeling-related phenomenon perceived as a dryness in the mouth along with a coarse puckering of the oral tissue. Astringency usually results from the association of tannins or polyphenols with proteins in the saliva to form precipitates or aggregates.

REVIEW QUESTIONS

1. What is flavor?
2. What are natural flavors?
3. What is a sensory assessment of flavor?
4. Are all sour substances acidic in nature?
5. What are citrus flavors?

REFERENCES

1. Acree, T. E., Barnard, J., & Cunningham, D. G. (1984). The analysis of odoractive volatiles in gas chromatographic effluents. In P. Schreier (Ed.), *Analysis of Volatiles: Methods and Applications* (pp. 251–267). New York: Walter de Gruyter.
2. Angelino, S. A. G. F. (1991). Volatiles in Beer. In H. Maarse (Ed.), *Volatile Components in Foods and Beverages* (pp. 581–616). New York: Marcel Dekker, Inc.

3. Arctander, S. (1969). *Perfume and Flavor Materials, Vol. 1 and 2*. Montclair, NJ: S. Arctander.
4. Bedoukian, P. A. (1991). Bedoukain's 47th Annual Review: Perfumery and flavor materials. *Perfumery & Flavorist, 16*(3), 1–27.
5. Beets, M. G. J. (1978). The sour and salty modalities. In *Structure–Activity Relationships in Human Chemoreception*. London, U.K.: Applied Science Publishers, pp. 348–362.
6. Birch, G. G. (1981). Basic tastes of sugar molecules. In J. Solms & R. L. Hall (Eds.), *Criteria of Food Acceptance*. Zurich, Switzerland: Forster Verlag, pp. 282–291.
7. Hasegawa, S., Patel, M. N., & Snyder, R. C. (1982). Reduction of limonin bitterness in navel orange juice serum with bacterial cells immobilized in acrylamide gel. *Journal of Agricultural and Food Chemistry, 30*, 509–511.
8. Maarse, H., & Visscher, C. A. (1989). *Volatile Compounds in Food Volumes I, II, and III*. Netherlands: Grafische.
9. Ney, K. H. (1990). Aromagrams-A New Approach to Flavor Classification and Description. In G. Charalambous (Ed.), *Flavors and Off-Flavors '89* (pp. 561–576). Amsterdam: Elsevier.
10. Shallenberger, R. S., & Acree, T. E. (1967). Molecular theory of sweet taste. *Nature (London), 216*, 480–482.
11. Ulrich, F., & Grosch, W. (1987). Identification of the most intense volatile flavor compounds formed during autoxidation of linoleic acid. *Zeitschrift für Lebensmittel-Untersuchung und -Forschung, 184*, 277–282.

CHAPTER

10 Food Additive

LEARNING OBJECTIVES

After studying this chapter, you will be able to:

- Understand the overview of food additive
- Define the classification of food additives
- Discuss on risks and benefits of food additives

KEY TERMS FROM THIS CHAPTER

Acceptable daily intake (ADI)
Antioxidants
Bacterial contamination
Cholesterol
Citric acid
Dietary intakes
Flavor enhancers
Food additives
Food preservatives
Anticaking agents
Attention deficit hyperactivity
Butylated hydroxyanisole (BHA)
Chromosomal aberrations (CAS)
Clastogenic properties
Disodium phosphate
Flavorings
Food coloring
Food quality protection act (FQPA)

INTRODUCTION

Materials with minimal or no nutritional value that are nonetheless utilized, particularly in industrialized nations, in the preparation or preservation of food or animal feed include food coloring, flavoring, antioxidants, food preservatives, vehicles, excipients, and other materials of a similar nature. Many of the same ingredients work better as pharmaceutics when they are added to drugs as opposed to food. Food additives are compounds that are added to food in order to improve its flavor or appearance. For centuries, people have preserved food by pickling it in vinegar, salting it like bacon, preserving sweets, or adding sulfur dioxide to some wines, among other methods. Many more additives, both natural and artificial, have been added since the second half of the 20th century with the introduction of processed foods. Food additives are sometimes mistakenly believed to be a relatively new concept, although there has undoubtedly been a rise in public awareness of the issue.

Not all of this has been fully informed, and there are indications that consumer pressure and food producers using deceptive marketing strategies have affected business interests. Several labeling laws have been implemented to guarantee that consumers are aware of the ingredients in processed foods and that food is fresh—a crucial factor in unprocessed foods that is likely to remain significant even in the presence of preservatives.

In order to keep the food from spoiling, we also need to add some preservatives. Foods that have direct additives added to them on purpose do so for specific reasons. Food is infused with indirect additives while it is being processed, packaged, and stored. Food preservatives are additives used to stop bacteria, molds, and yeasts from growing in food. While some additives are synthetic and created by humans, others are made from natural sources like soybeans, corn, and beets. Instead of cooking their own food at home, most people prefer to consume the prepared food that is sold in stores. These foods have preservatives and additives of some kind to preserve their flavor and quality and prevent bacteria and yeast from ruining them.

There are over 3,000 preservatives and additives on the market that function as antimicrobials and antioxidants. Among the most often used additives are sugar and salt. Aluminum silicate, amino acid compounds, ammonium carbonates, sodium nitrate, propyl gallate, butylated hydrozyttoluene (BHT), butylated hydroxyanisole (BHA), monosodium glutamate, white sugar, potassium bromate, potassium sorbate, sodium benzoate, etc., are a few of the frequently used food additives and preservatives. The foods are also given some artificial coloring to give them a visually appealing appearance.

Erythrosine (red), cantaxanthin (orange), amaranth (Azoic red), tartrazine (Azoic yellow), and annatto bixine (yellow orange) are these coloring agents. In order to preserve the food's quality and flavor during long-term storage, additives and preservatives must be used. Foods containing too much water may encourage the growth of bacteria, fungi, and yeasts. The use of preservatives and additives stops food from spoiling because they stop bacteria and fungi from growing.

Preservatives and additives keep food consistent and of high quality. Additionally, they preserve the food's wholesomeness and palatability, boost or preserve its nutritional value, regulate pH levels appropriately, add color and leavening, and improve flavor. Certain food products are composed solely of chemicals.

Candies, coffee creamers, and sugar replacements are almost entirely made of artificial ingredients. Such tampering with our food can significantly impact the distinct biochemical equilibrium within our bodies. Any food that needs to be kept for a longer period of time needs to be properly processed. The food is given certain materials and chemicals, referred to as additives, during this processing. Foods continue to remain of a high caliber thanks to additives.

Figure 10.1. *Showing classification of food additives.*

Source: https://portal.abuad.edu.ng/lecturer/documents/1587976825Food_additives.pdf.

10.1. OVERVIEW OF FOOD ADDITIVE

The term "food additive" refers to any substance that is not typically consumed as food on its own or used as a typical ingredient in food, regardless of its nutritional value. It also does not include contaminants or substances added to food for nutritional qualities maintenance or improvement.

Instead, food additives are intentionally added for technological (including organoleptic) purposes during the manufacture, processing, preparation, treatment, packaging, transport, or holding of food, or it may reasonably be expected to result (directly or indirectly) in the food or its by-products becoming a component of or otherwise affecting the characteristics of such food. To put it more simply, food additives are the things that food manufacturers add to food to make processing easier or to enhance the food's texture, flavor, appearance, and shelf life. The term excludes substances added to food to preserve or enhance its nutritional value, as well as accidental contaminants that may unintentionally find their way into our food. Its application is limited to ingredients that are purposefully added to food. These materials include vitamins, propionate sorbate, flavors, and oxidizing agents, among others.

10.1.1. Additives Used in Foods

Food additives serve many beneficial purposes that are sometimes overlooked. Additives help keep food wholesome and appealing while it is being transported to markets, sometimes thousands of miles away from where it is grown or manufactured, as the majority of people no longer live on farms.

In addition to increasing a food's nutritional value, additives can enhance a food's flavor, texture, consistency, or color and increase its appeal. We can better comprehend this by examining the ingredients that go into making high-quality bread.

To achieve a desired taste and flavor, the flour is treated with a mild oxidizing agent to make it whiter; vitamins can be added to improve nutritional quality; glycerol monostearate is added to provide a soft texture; propionates or sorbates are added to improve the quality of the product for long-distance marketing and transportation; and finally, salt, sugar, and flavorings are added. The ingredient needed to produce high-quality bread. Every extra ingredient added during the bread-making process improves the final product's desirable quality, which is crucial for its marketability and consumer acceptance.

Having looked at the role of additives in making good quality bread, the different uses of additives in foods, in general, can be summarized as under:

- **To Maintain Product Consistency:** Emulsifiers give products a consistent texture and prevent them from separating. Stabilizers and thickeners give a smooth uniform texture. Anti-caking agents help substances such as salt to flow freely.
- **To Improve or Maintain Nutritional Value:** Vitamins and minerals are added to many common foods such as milk, flour, cereal and margarine to make up for those likely to be lacking in a person's diet or lost in processing.
- **To Maintain Palatability and Wholesomeness:** Preservatives retard product spoilage caused by mold, air, bacteria, fungi or yeast. Bacterial contamination can cause food borne illnesses which could be life-threatening. Antioxidants are preservatives that prevent fats and oils in baked goods and other foods from becoming rancid or developing an off-flavor. They also prevent cut fresh fruits such as apples from turning brown when exposed to air.
- **To Provide Leavening or Control Acidity/Alkalinity:** Leavening agents that release acids when heated can react with baking soda to help cakes, biscuits and other baked goods to rise during baking. Other additives help modify the acidity and alkalinity of foods for proper flavor, taste and color.
- **To Enhance Flavor or Impart Desired Color:** Many spices and natural and synthetic flavors enhance the taste of foods. Colors, likewise, enhance the appearance of certain foods to meet consumer expectations.
- To Enhance the Keeping Quality or Stability of a Food: Use of certain preservatives, stabilizers, anti-caking agents etc., increases the shelf-life of food products.

10.1.2. Dangers of Food Additives and Preservatives

While preservatives and additives are necessary for food storage, they can also cause a number of health issues.

They can trigger a variety of allergies and ailments in certain individuals who are chemically sensitive, including attention deficit disorder and hyperactivity. Additive-containing foods can aggravate eczema and cause hay fever, asthma, and other reactions like rashes, vomiting, headaches, tight chests, and hives. Some of the known dangers of food additives and preservatives are as follows:

- Benzoates can trigger allergies such as skin rashes and asthma as well as believed to be causing brain damage.
- Bromates destroy the nutrients in foods. It can give rise to nausea and diarrhea.
- Butylates are responsible for high blood cholesterol levels as well as impaired liver and kidney function.
- Caffeine is a colorant and flavorant that has diuretic, stimulant properties. It can cause nervousness, heart palpitations and occasionally heart defects.

- Saccharin causes toxic reactions and allergic response, affecting skin, gastrointestinal tract and heart. It may also cause tumors and bladder cancer.
- Red Dye 40 is suspected to cause certain birth defects and possibly cancer.
- Mono and di-glycerides can cause birth defects, genetic changes and cancer.
- Caramel is a famous flavoring and coloring agent that can cause vitamin B6 deficiencies. It can cause certain genetic defects and even cancer.
- Sodium chloride can lead to high blood pressure, kidney failure, stroke and heart attack.

You should stay away from foods that contain additives and preservatives to reduce your chance of experiencing health issues as a result of them. You need to review the ingredients of the canned food before buying it. Organic foods are free of artificial additives, so you should purchase them. Eat as many freshly prepared foods as you can instead of processed or canned ones.

10.1.3. Effects of Food Additives

One of the most important steps to improving your health and reducing your risk of disease is to minimize or avoid toxins in your diet. These toxins can come from food, cosmetics, and medications, among other things. Food additives can have negative long-term effects if you are exposed to them often, or they can have immediate negative effects. Headaches, changes in energy levels, and adjustments to behavior, immune response, or mental focus are examples of immediate effects. Long-term consequences could raise your chance of developing heart disease, cancer, and other degenerative diseases. While it might seem challenging to break old habits and find healthy alternatives for foods you enjoy, keep in mind that you will be incorporating some new, wholesome foods into your diet that you will grow to love even more. Instead of consuming packaged and processed foods, take advantage of the abundance of fresh fruits, vegetables, grains, beans, nuts, and seeds that nature has to offer. Look for foods that bear some resemblance to their original appearance during cultivation.

10.1.4. Cytotoxic Effects of Food Additives

Behavioral disorders like attention deficit hyperactivity disorder (ADHD) have been linked to the onset and severity of exposure to non-nutritional food additives during the critical development window. The combined effects of food additives are still unknown, despite the fact that individual additives used at controlled concentrations are thought to be reasonably safe for neuronal development. Four common food additives have been evaluated for possible interactions in terms of their neurotoxic effects when combined with two other ingredients (quinoline yellow and aspartame, brilliant blue and L-glutamic acid). Additives were used to stimulate the differentiation and growth of neurites in mouse NB2a neuroblastoma cells. Cells were fixed, stained, and the length of the neurites was measured using light microscopy and computerized image analysis after a 24-hour period. Neurite outgrowth inhibition was used to quantify neurotoxicity. The combination effects were examined using two separate models: dose additivity and effect additivity. In both models, there was a noticeable synergy between the combinations of Quinoline Yellow and aspartame and

Brilliant Blue and Lglutamic acid. Using the NMDA antagonist CNS-1102, the role of N-Methyl-D-aspartate (NMDA) receptors in neurite inhibition induced by food additives was investigated. The presence of CNS-1102 decreased the neurotoxicity caused by L-glutamic acid and aspartame, but it had no effect on the neurotoxicity caused by food color. Using estimated percentage absorption from the gut and content analysis of food, theoretical exposure to additives was computed. At additive concentrations that are theoretically attainable in plasma through consumption of a typical snack and drink, inhibition of neurite outgrowth was observed. Additionally, the cellular toxicity of food additives on the viability of NB2a cells was assessed using Trypan Blue dye exclusion; both combinations had a clear additive effect on cytotoxicity. These findings have consequences for how common chemicals, both taken alone and in combination, affect cells (Lau et al., 2006).

Gallic acid is produced by hydrolyzing tannic acid, which is present in tea, coffee, red wine, and immature fruits. It is added to food to stop oxygen-induced lipid peroxidation. Additionally, tannic acid has been applied as a food additive. Gallic acid's effects on mouse spermatogonia, mouse spermatocytes, and mouse Sertoli cells were studied *in vitro*. To examine the effects of gallic acid on the *in vitro* growth of testicular cells, each cell line was first cultivated for three hours at predetermined concentrations. Cell viability was then measured using the MTT cytotoxicity assay. Second, dihydrorhodamine 123 was used as a probe to measure the intracellular levels of hydrogen peroxide in mouse spermatogonia, mouse spermatocytes, and mouse Sertoli cells treated with gallic acid in order to assess the pro-oxidative property of gallic acid. Gallic acid raises the intracellular level of hydrogen peroxide in mouse spermatogonia significantly ($p < 0.05$) and in a dose-dependent manner inhibits the growth and proliferation of testicular cells, according to the obtained results. It is possible that the pro-oxidative activity of gallic acid causes testicular cells to undergo cytotoxicity. In light of the potential toxicity of gallic acid to the testicles, the cytotoxicity that gallic acid induces *in vitro* in mouse spermatogonia, mouse spermatocytes, and mouse Sertoli cells may be of interest for toxicological research. (Park et al., 2008).

In Iran, the practice of using Carthami flos, or the flowers of Carthamus tinctorius, as a food coloring and flavoring has become more popular. To assess its safety, researchers looked into the teratogenic effects of carthami flos on the development of mice's central nervous systems. To conclude safety assessments, its cytotoxic effect on a culture of rat nerve cells was investigated. Mice that had naturally mated were split into test and control groups for teratogenic investigations. From the first to the eighth day of pregnancy, the groups received varying dosages of aqueous carthami flos extract. The 13th day of gestation saw the isolation of embryos, which were subsequently examined for morphometric, microscopic, and macroscopic traits. Results indicated that the embryos were absorbed at higher doses (1.6 and 2 mg/kg/day), while alterations in cellular degeneration, open neuropores, changes in cellular orientation, and external, internal, and longitudinal diameters were seen at lower doses (1.2 mg/kg/day). Carthami flos extract has a cytotoxic effect that is concentration-dependent, according to the results of the cytotoxic assay. It is concluded that a reevaluation of carthami flos as a food additive is warranted (Nobakht et al., 2000).

On cultured fetal rat hepatocytes, the cytotoxicity of 11 dyes—used as food dyes in Japan—was investigated. Phloxin, rose bengal, and erythrosine are examples of xanthene dyes that have halogen atoms in their molecules and are therefore more toxic than other classes of food dyes. Additionally, the impact of food dyes on hepatocyte cell growth was investigated. When the dye was added three days after plating, a dose-response relationship was seen between the concentration of phloxin and the hepatocytes' growth. Phloxin was particularly toxic to cell growth (Sako et al., 1980).

Using primary cultured rat hepatocytes, the synergistic effect of food additives or food colors on the toxicity of 3-amino-1,4-dimethyl-5H-pyrido[4,3-b]indole (Trp-P-1) was examined. Rat hepatocytes given a typical diet were subjected to a combination of four common food additives (sorbitol, sodium 1. Using either a combination of six common artificial food colors (erythrosine, allura red, new coccine, brilliant blue, tertrazine, and fast green) or (+)-glutamate, benzoic acid, and propylene glycol, the *in vitro* treated food-color mixture itself demonstrated cytotoxicity: a decrease in cell viability and a decrease in the activities of gluconeogenesis and ureogenesis. The food-color combination clearly increased Trp-P-1's cytotoxicity. The impact of food coloring or additives dosed *in vivo* on toxicity caused by Trp-P-1. Rats given a diet containing a combination of food additives or a combination of food colors in half the permitted daily intake for four weeks were used to harvest and culture hepatocytes. The hepatocytes received different doses of Trp-P-1 for a duration of 12 hours. Food coloring and food additives administered *in vivo* did not show any synergistic effects on Trp-P-1-induced cytotoxicity, as measured by decreased DNA and protein synthesis and decreased cell viability. Conversely, it was found that food colors given *in vivo* together with Trp-P-1 treatment facilitated the reduction of gluconeogenesis and ureogenesis activities in hepatocytes. These findings imply that when dietary carcinogens are exposed to liver cells on a daily basis, the consumption of artificial food coloring may affect hepatic functions like gluconeogenesis and ureogenesis (Ashida et al., 2000). Study was conducted to investigate the impact of food additives like boric acid, citric acid and sodium metabisulphite individually, in different concentrations, on root tips of Vicia faba L. Cytological studies revealed significant decrease in mitotic index with an increase in concentration of the food additives. Most frequent cytological abnormalities observed are fragments, disturbed metaphase, C-mitosis, laggards, bridges, stickiness, precocious movement of chromosomes, unequal and late separation of chromosomes. Bridges and fragments were more frequent at anaphase. The percentage of chromosomal aberrations at mitosis increased with an increase in concentration of the food additives (Pandey & Upadhyay, 2007).

Citric acid, a food additive that is frequently used as an antioxidant, was examined for its clastogenic qualities in human peripheral blood lymphocytes.

Ester gum (EG) serves as a coloring or weighting agent in beverage flavorings that contain citrus oils. The genotoxic and clastogenic potential of the food additive was assessed in this study using sister chromatid exchange and chromosomal aberration as the cytogenetic endpoints. The male Swiss albino mice were given oral doses of 50, 100, and 150 mg/kg body weight. In spite of its weak clastogenicity and ability to slightly raise sister chromatid exchange frequencies, EG did not pose a risk to human health at the doses that were investigated (Mukherjee et al., 1992).

Acesulfame-K, a sweetening agent, was evaluated *in vivo* for its genotoxic and clastogenic potentials. Swiss albino male mice were exposed to the compound by gavage. Bone marrow cells isolated from femora were analyzed for chromosome aberrations. Doses of 15, 30, 60, 450, 1,500 and 2,250 mg of acesulfame-K/kg body weight induced a positive dose-dependent significant clastogenicity (trend test $\alpha < 0.05$). These doses were within the no-toxic-effect levels (1.5–3 g/kg body weight in rats) reported by the Joint Expert Committee for Food Additives of the World Health Organization and the Food and Agriculture Organization of the United Nations. In view of the present significant *in vivo* mammalian genotoxicity data, acesulfame-K should be used with caution. (Mukherjee & Chakrabarti, 1997).

In diets and drinks, aspartame and acesulfame-K, two non-nutritive sweeteners, are acceptable in separate amounts. When these sweeteners of various classes are combined, it has been discovered that they have a synergistic sweetening effect. It's unclear if they also have a synergistic genotoxic effect. Mixtures of aspartame (3.5, 35, and 350 mg/kg body weight) and acesulfame-K (1.5, 15, and 150 mg/kg body weight) were given intraperitoneally to male Swiss Albino mice. We examined the bone marrow cells taken from the femora for chromosomal abnormalities. The results of the statistical analysis indicate that there is no discernible genotoxicity when aspartame and acesulfame-K are combined (Mukhopadhyay et al., 2000).

The purpose of this work was to investigate how saccharin affected certain rat blood indices and how vitamins C and E could be therapeutic. In the current study, adult female Rattus norvegicus albino rats weighing between 100 and 120 g were used. Saccharin was administered at a dose of 35 mg kg^{-1} body weight (b. What?) day^{-1} for 35 days dramatically reduced the levels of total protein, albumin, triglycerides, cholesterol, and serum glucose. In comparison to the control level (p value < 0.01), these deviations were 20.16%, 22.76%, 44.92%, 20.16%, and 40.44%, respectively. However, it raised kidney function index levels. Saccharin had a more noticeable effect on creatinine. Rats treated with saccharin showed a significant increase in the activities of alkaline phosphatase (ALP), aspartate aminotransferase (AST), and alanine aminotransferase (ALT). In terms of hematoligical parameters, the most notable alterations were seen in the rise in platelets (PLT), mean corpuscular volume (MCV), and white blood cells (WBC), as well as the fall in hematocrit, hemoglobin (Hb), and red blood cell count in reaction to saccharin administration. Generally speaking, 150 mg kg^{-1} b. What? day^{-1} for 35 days) was able to mitigate the impact of consuming saccharin. Nonetheless, both vitamins often help to lessen the variations in the parameters under study (Abdelaziz et al., 2011).

Monosodium phosphate (MSP – E339i), disodium phosphate (DSP – E339ii), and trisodium phosphate (TSPE339iii) are used as antimicrobials, pH control agents (buffers), boiler water additives, cleaners, coagulants, dispersants, leavening agents, stabilizers, emulsifiers, sequestrants, texturizers, nutrients, and dietary supplements. The effects of these have been studied on root tips of Allium cepa L. Roots of A. cepa were treated with a series of concentrations, ranging from 300 to 500 ppm for 24, 48 and 72 h.

The results indicated that these food preservatives reduced mitotic division in A. cepa when compared with the respective control. Mitotic index values were generally decreased with increasing concentrations and longer treatment times. Additionally, variations in the percentage of mitotic stages were observed. The total percentage of aberrations generally increased with increasing concentrations of these chemicals and longer period of treatment. Different abnormal mitotic figures were observed in all mitotic phases. Among these abnormalities were stickiness, anaphase bridges, C-mitosis and micronuclei. These food additives remarkably depressed the DNA content in the root meristems of A. cepa. The interphase nuclear volume (INV) also varied between the untreated (controls) and the treated plants (Turkoglu Sifa, 2009).

Citric acid, a food additive that is frequently used as an antioxidant, has been shown to have clastogenic properties in human peripheral blood lymphocytes. Citric acid is used as an acidulant, pH regulator, flavor enhancer, preservative, and antioxidant synergist in many foods, including soft drinks, jelly sweets, baked nutrients, jam, marmalade, candy, and tinned vegetable and fruit food. Chromosomal aberrations (CAs) increased significantly when treated with citric acid at all tested concentrations and times. At 24 and 48 hours, the mitotic index (MI) was significantly reduced by citric acid at concentrations of 100 and 200 lg ml^{-1}, respectively. It did not, however, considerably lower the replication index (RI). Sister chromatid exchanges (SCEs) at 100 and 200 lg ml^{-1} concentrations at 24 h and in all concentrations at 48 h were also significantly increased by citric acid. When this chemical was used instead of the negative control, the micronuclei frequency (MN) increased significantly. It had a non-statistically significant effect on the cytokinesis-block proliferation index (CBPI), as well (Serkan et al., 2008; Serkan Yılmaz et al., 2008).

10.2. CLASSIFICATION OF FOOD ADDITIVES

Out of all the classification schemes used for food additives, the functional classification is the most widely accepted. This means that food additives are categorized according to how they work in food, i.e., the reason behind the food's incorporation of the additive.

10.2.1. Antioxidants

Antioxidants lessen the oxidative damage that causes food to go rancid, lose its flavor, color, and nutritional value. When exposed to air, fats, oils, flavorings, vitamins, and colors can all oxidize naturally with oxygen. The amount of natural antioxidants and other components, the availability of oxygen, and the substance's sensitivity to oxidation, temperature, and light, among other factors, can all have a significant impact on the rate of deterioration. There are several ways to prevent or delay oxidation, including using sequestering agents, cooling, replacing air with inert packaging gases, removing oxygen with glucose oxidase, and adding UV-absorbing materials to transparent packaging materials. These might not always be feasible or sufficient to give some foods a suitable shelf life. Antioxidants are therefore used to prolong shelf life and slow down oxidative deterioration. Certain antioxidants deplete oxygen through the process of self-oxidation ascorbic acid, while others, such as others, obstruct the oxidation mechanism. BHA, BHT, gallic acid esters, and tocopherols. Each has unique qualities that make them better suited for certain uses than others. Because of their synergistic effects, combining two or more antioxidants can often yield greater results than using just one. Because sequestering agents like citric acid decrease the availability of metallic ions that could catalyze oxidation reactions, they may also have a synergistic effect. The strong synthetic antioxidants BHA, BHT, and gallic acid esters have very limited applications. Tocopherols, which come in natural or synthetic forms, are less regulated but less successful in shielding processed foods from harm. Antioxidants only slow down the oxidation process; they cannot reverse the effects of oxidation on food. Oxidation must be taken into consideration as soon as possible because it is a chain reaction process. Thus, the fats and oils used in the manufacturing process are where antioxidants are most useful.

10.2.2. Emulsifiers and Stabilizers

Emulsifiers and stabilizers are used to help substances that wouldn't normally mix together, like fat and water, mix more easily. Stabilizers then keep the aqueous and lipid phases from mixing. For example, these additives are necessary for the manufacturing of chocolate products, mayonnaise, and fat spreads. Emulsifiers and stabilizers are necessary for the production of fat spreads, which are lower-fat alternatives to butter and margarine that have significantly influenced consumer choice and dietary modification. This technology is also necessary for other reduced- and low-fat versions of several products. The advantages of this technology will be appreciated by anyone who has ever made an emulsified sauce, such as mayonnaise or hollandaise, but even more so by those who have failed miserably at the technique and ended up with a costly mess of curdled ingredients. Aside from this use, the term "stabilizer" can also refer to substances that help food pieces bind together to form reconstituted food, as well as substances that help preserve, enhance, or stabilize a food's natural color. Food labels now need to list the source of some emulsifiers due to growing awareness of issues with food allergies and intolerances. For instance, lecithin derived from soy is unsuitable for someone with a soy allergy; hence, it is essential to clearly label the ingredient's source to enable consumers to select products that are safe for people with particular dietary needs.

10.2.3. Colors

Foods' aesthetic qualities are improved by the use of color. Their use is especially contentious, in part because some people believe that color can be used to mislead consumers about the nutritional value of food, but also because some of the products with the brightest colors are meant for younger consumers. Like with all additives, their use is strictly regulated and only allowed in situations where it can be demonstrated that it is necessary, such as to ensure color consistency, to repair color lost during processing (like canning or heat treatment), or to serve as a decorative element. In the UK, there is a long and honorable tradition of using color in food. The chefs of the Middle Ages especially enjoyed it. In addition to green spinach and parsley juice, saffron (the name of Saffron Walden) and the reddish-hued Saunders (powdered sandalwood) were used to color soups in stripes or create a marbleized effect (1). Therefore, it is undeniable that the careful application of color improves the appearance of many foods, even though some people may view food coloring as an unnecessary cosmetic that serves no purpose for consumers. A few retailers attempted to offer a variety of canned fruits and vegetables, like peas and strawberries, without restoring the color that was destroyed during heat processing. Many years later, they were still attempting to get rid of the unsold returns! Color plays a significant role in how consumers perceive food and frequently indicates a particular flavor. Orange flavor is therefore anticipated to be orange in color, and strawberry flavor to be red. Adding color is therefore justified by consumer expectations.

In the popular press, food colorings in particular have long been blamed for behavioral issues in kids. Feingold first proposed that artificial food coloring and preservatives had a negative impact on children's behavior more than 30 years ago. Since then, studies on the impact of food colors and preservatives on kids' behavior have fueled the negative perception that consumers have of these additives, especially when it comes to products targeted at kids. Children's hyperactive behavior significantly changed when colorants and preservatives were cut out of their diets. This result was gender-neutral, and the decrease in hyperactivity happened regardless of the child's initial level of hyperactivity—from extremely hyperactive to nonexistent. More recently, in 2007, research was conducted on the effects of two combinations of specific artificial food colors and the preservative sodium benzoate. The results indicated that the two mixtures, as well as the two age groups, had a negative effect on children's hyperactive behavior when compared to a placebo, though the increases in hyperactive behavior levels were not consistently significant for either combination. The results of an earlier study conducted in 2004 on preschoolers are both replicated and expanded upon in this new study. The European Food Safety Authority's (EFSA) work on the reevaluation of colors already includes the colors used in this study.

10.2.4. Coloring Foodstuffs

It has become common practice to refer to colorings that are made from approved foods and processed so as to preserve the fundamental qualities of the original food as "coloring foodstuffs." This is not the case with natural colors, which are thought of as additives and in which the pigment is concentrated and extracted in a specific way. Food coloring can be listed as an ingredient on the label without having to explain its purpose, as the law only requires this for additives.

These coloring foods include vibrant yellow hues from safflower, turmeric, and oleoresin; golden yellow to naturally orange hues from carrots and paprika; caramelized sugar syrup gives a toffee brown color; green hues from stinging nettles and spinach leaves, which are high in chlorophylls; and red, blue, and purple hues from concentrated red and blue fruits, red cabbage, and beetroot, which are high in anthocyanins. While it is evident that food coloring can produce the entire color spectrum, creators must make sure the food coloring has the same stability and vibrancy of color in the finished product as traditional food coloring.

10.2.5. Sweeteners

Sweeteners serve a clear purpose. They are allowed in foods that are either low in energy or don't contain added sugar. They are classified into two basic categories: "bulk" and "intense." Additionally, they are offered directly to customers as "table-top" sweeteners, which diabetics and dieters are familiar with. For instance, Splenda contains sucralose, but Sunette, a tabletop sweetener, contains acesulfame-K. As their name implies, intense sweeteners like aspartame, saccharin, acesulfame-K, and sucralose have a very high sweetening property that varies depending on the type but is typically several orders of magnitude higher than that of sucrose.

Aspartame has a sweetness ratio of about 200 times sugar for every unit of weight; saccharin has a ratio of 300–500; and acesulfame-K has a ratio of 130–200.) Bulk sweeteners, such as erythritol, sorbitol, isomalt, and lactitol, are less sweet but give mouthfeel due to their volume and mostly polyol content. Maltitol is one of the sweetest polyols, and xylitol, the sweetest of them all, has a sweetness intensity that is identical to sucrose. Since most polyols have lower sweetness levels, you can increase their sweetness and taste quality by blending them with other polyols or strong sweeteners. We call this quality sweetness synergy. The ability to cover up the unpleasant bitter metallic aftertaste of some strong sweeteners is another advantage. Combinations such as saccharin and cyclamate, erythritol and acesulfame-K, and aspartame and acesulfame-K are frequently utilized. Both kinds of sweetener—bulk and intense—are helpful in low-calorie goods, which are becoming more and more popular among consumers, as well as in products for specific diets, like those meant for people with diabetes. The product's cariogenic qualities are also reduced by the absence of sucrose.

10.2.6. Flavor Enhancers

This particular set of additives, monosodium glutamate in particular, has drawn negative attention because it is commonly linked to an intolerance reaction that has come to be known as "Chinese Restaurant Syndrome." Flavor enhancers are substances that, when added to food, bring out and enhance the flavors of the food they are combined with. They do not have a distinct flavor or taste of their own. Salt is the most commonly used flavor enhancer, despite not being considered a food additive due to its unique taste. The next most well-known is glutamic acid and its salts, which are most frequently found as monosodium glutamate and have been used as a condiment in savory goods for several centuries in the Far East. It is a necessary amino acid that is found in the body and a typical component of all proteins. Reputable scientific studies never confirmed the supposed intolerance reaction. Anybody who reacts to MSG added as an additive will inevitably react to foods like cheese and tomatoes that naturally contain large amounts of MSG. Certain sweeteners have been approved for use because it has been discovered that they enhance flavor. For instance, acesulfame K, aspartame, and thaumatin are used to improve the flavor of chewing gum and desserts, while neohesperidine DC can improve the flavor of meat products and margarine.

10.2.7. Flavorings

While flavor enhancers fall under the additive category, flavorings are regulated differently and have a different technological background, despite the public's common misconception that they are the same. The definition of flavorings is the addition of aroma and/or taste to food. Typically, flavorings are used in combinations of various preparations and specific chemicals. They do not include edible materials and goods meant to be eaten as such, nor do they include materials with only a sweet, sour, or salty flavor, such as common culinary ingredients like sugar, vinegar, lemon juice, or salt. The most recent draft of the proposed new EC Regulation on Flavorings would also exclude raw and non-compound foods, as well as tea and spice mixtures, from the definition of flavorings, provided that they are not used as food ingredients. Apart from various flavoring varieties like smoke or process flavors, there exist three unique categories of flavoring substances: natural (like citral); nature identical (like vanillin); and artificial (like ethyl vanillin).

Then there are preparations for flavoring, like extract from vanilla. A lot of flavorings are marketed as intricate blends of distinct preparations and flavoring ingredients that are typically proprietary to the flavoring's manufacturer. Laws have been created to safeguard business privacy when adding recently found flavoring substances to the EC list. Food labels typically list "flavorings" in the ingredients list due to the complexity of the flavoring used in the product. This is all that is needed by law, though the manufacturer is free to be more detailed if he so chooses. Listing every single substance would frequently be very long and nearly impossible for the consumer to understand. Flavorings that are marketed as "natural" have to adhere to the legal guidelines. There are guidelines for using the term "natural" in product labeling that have been released by the Food Standards Agency. In the future, there will probably be stricter regulations for the labeling of natural flavorings due to the new proposal for an EC Regulation on flavorings and certain food ingredients with flavoring properties for use in and on foods. Similar to additives, certain flavorings are supplied directly to consumers for use in domestic cooking. Among the most well-known are peppermint and vanilla, along with the widely used brandy and rum essences. The tiny amounts in which they are used will be appreciated by anyone who has ever overflashed a batch of peppermint creams or a homemade cake. Similar to this, flavoring is used in minuscule amounts in commercial manufacturing when compared to other ingredients. The majority of flavorings are made from ingredients that are found in food naturally. For example, one of the most popular natural sources of materials used to flavor substances and preparations is orange and citrus oils.

10.2.8. Other Additives

Due to the nature of their functions, colors and sweeteners fall under very specific and well-defined classes of additives that are governed by particular laws. The term "miscellaneous" now encompasses all other additive classes. This more general grouping includes additional categories in addition to the larger ones previously mentioned. These include thickeners, acids, acidity regulators, anticaking agents, anti-foaming agents, bulking agents, glazing agents, humectants, raising agents, and sequestrants.

With the possible exception of sequestrants, the majority of these have clear functions just by looking at their names. These are materials that combine with metallic ions to form chemical complexes. They are not commonly used, and food labels hardly ever list this class of additives. Conversely, thickeners are among the most widely utilized additives because they modify the viscosity and texture of food and beverage products. Similar to how different kinds of flour are frequently used in cooking to thicken sauces, soups, stews, and other meals that contain a lot of liquid, the majority of commercial thickeners are based on starch or gum and accomplish the same thing. Packaging gases are one class of additive that does not have a domestic equivalent. These are the natural atmospheric gases that are currently widely used in a variety of pre-packaged goods, including fresh pastas, meat, fish, and shellfish, as well as ready-made vegetables that are found in sealed containers on chilled food counters. The phrase "packaged in a protective atmosphere" refers to the process of replacing the air within the pack and altering the atmosphere by adding one or more gases, depending on the product, to the "headspace" of the container. This helps delay the bacteriological deterioration that would otherwise occur under normal atmospheric conditions. Although it could be argued that the gases serve only to keep food fresher for longer in its packaged state and are not present in the food itself, they were considered additives for regulatory purposes and must be labeled as such. Naturally, carbon dioxide is also a common ingredient in a lot of fizzy drinks, which serves as an example of the wide range of purposes and applications for additives.

10.3. Risks and Benefits of Food Additives

A comprehensive analysis of food additives and colorants would be lacking if it did not address the advantages and disadvantages of their application. Any risks connected to the use of food and color additives should be weighed against other risks related to food, as well as against the risks connected to the use of substitute additives or particular formulations without additives. Food additives must be functional in order to be covered by the FD and C Act; however, potential benefits are not taken into consideration. Nonetheless, the advantages of food additives are closely linked to their uses. Food additives are required by the FD&C Act to pose a minimal or negligible risk to human health. Therefore, by all measures of comparison, the advantages of food and color additives, GRAS ingredients, and previously approved ingredients usually exceed the risks. However, U.S. law does not require comparisons. S. food regulations. Furthermore, evaluations of risk alone are used to determine whether or not a food ingredient is acceptable. Therefore, even a moderate amount of risk associated with a beneficial additive would not be permitted. The Federal Food, Drug and Cosmetic Act's Delaney clause mandates the prohibition of any food additive with a demonstrable risk of cancer. Although this provision has not often been used, the zero tolerance policy for carcinogens has drawn a lot of attention to the health dangers associated with food additives and mildly carcinogenic foods.

In the United States, there is very little risk involved with eating food. The majority of the world's developed nations share this trait. That does not preclude the possibility of lowering the risk level even more. There is little doubt that significant progress could be made to increase the global food supply's wholesomeness, but the majority of this improvement must come from developing nations. It is accurate to say that there has never been a safer food supply, particularly in developed nations like the US. Currently held beliefs that the food supply poses serious risks appear to stem from a number of factors. First, even minute amounts of potentially dangerous substances can now be found in food thanks to advancements in analytical technology. Such levels of detection were unattainable just a few decades ago. Second, although naturally occurring substances can occasionally be food additives, food additives receive a lot more attention under US food laws than naturally occurring chemicals in foods. Although food additives should undergo extensive safety testing, the relative risks of food additives have become less clear as a result of the insufficient testing of naturally occurring chemicals.

10.3.1. Functions and Usage of Food Additives

Numerous chemical additives are either directly added to food or seep in through the environment or packaging materials. A material whose intended use will result in its incorporation into the food or will alter the food's properties can also be used to define a food additive. The food producer, processor, or consumer typically benefits in some way from these additives. Additives can enhance the nutritional value, facilitate meal preparation, and improve the organoleptic properties of food for the consumer. Adding additives can help food processors and producers by enhancing product variety, safety, and quality.

Additives can be found in foods in different amounts, serve different purposes in foods and ingredients, and work in concert with other additives to create unique effects. Generally speaking, their roles fall into one of four categories: (1) preserving or enhancing nutritional quality; (2) preserving or enhancing product safety or quality; (3) supporting processing or preparation; and (4) enhancing sensory attributes (FDA, 1979, 1992). Vitamins and minerals are the main additives that have an impact on nutritional quality. Some foods may contain these added to enhance the flavor or replenish nutrients lost during processing. To supplement nutrients that are frequently deficient in human diets, vitamins and minerals are added for fortification to other foods. To stop bacteria and fungi from growing in food, preservatives and antimicrobial agents are employed. These additives can prolong the finished product's shelf life or postpone spoilage. By postponing rancidity or lipid oxidation, antioxidants are additives that can also increase the shelf life of food. Additives that preserve product quality may also guarantee consumer safety when it comes to food products. For instance, adding acids may also stop the growth of bacteria that can lead to foodborne illnesses by preventing the growth of bacteria that cause spoiling.

The texture of ingredients and prepared foods is typically impacted by additives used in processing or preparation. Emulsifiers, stabilizers, thickeners, leavening agents, humectants, and anti-caking agents are a few of these classifications. This class of chemicals is also used to modify the volume, stability, and homogeneity of food products. Enhancing a food's flavor or color to increase consumer appeal is the fourth main purpose of food additives. The original flavor or aroma of food ingredients can be enhanced or flavors lost during processing can be restored with the use of flavoring chemicals. Foods can have artificial and natural coloring added to them to improve their appearance, enhance their flavor, bring back color that has been lost during processing, or intensify their natural color.

10.3.1.1. Usage

Food additives can be synthetic or naturally occurring, originating from plants or animals. Furthermore, food additives that are functionally or chemically identical to those that may come from natural sources can be created. Certifiable colors also include synthetic colors in the case of color additives. Each batch of these coloring chemicals is tested by the manufacturer and the U.S. Food and Drug Administration to ensure the safety, quality, consistency, and strength of the color additive prior to its use in foods.

There are currently nine artificial color additives that can be used in food. Pigments obtained from natural sources, such as plants, minerals, or animals, as well as synthetic substitutes for naturally occurring colors, are exempt from this certification process. Turmeric, beet powder, paprika, and annatto extract are a few examples. Food additives are substances that are added artificially to certain foods or that naturally occur in certain foods and ingredients. A food's package label may indicate that an additive that comes from a natural source is "artificially added" if it is added to something that isn't typically present in the food. The way that additives are added to food determines another category for them. As with the preservatives BHA and BHT inside breakfast cereal bags, certain additives are added directly to foods and ingredients, while others may be added indirectly through contact with packaging materials. When added during the manufacturing process, water is typically regarded as a food ingredient, but it can also be thought of as a direct additive. For instance, trimmed and cleaned carcasses are submerged in a cold water "chiller tank" for one to two hours in poultry processing facilities in the United States in order to lower the temperature of the carcasses. The carcasses may put on weight during this period as a result of absorbing tank water. Regulators are currently discussing whether or not to require processed poultry to have a label indicating that water (and weight) has been added. Foods may inadvertently contain some unwanted food additives. For instance, during the growing, harvesting, and storage of produce and grains, pesticides and fumigants may come into contact with them. While using these chemicals on certain crops may be legal, doing so on others may not be. Occasionally pesticides are applied inadvertently or mistakenly at the incorrect concentration or too soon before harvest. Occasionally, pesticides have been discovered in amounts and on specific crops that are prohibited by the Environmental Protection Agency. Food chemicals that inadvertently find their way into food items can also come from inadequate storage procedures and materials used for packaging, particularly plastics. Foods can become contaminated by a variety of chemicals from the environment. Dioxin, polychlorinated biphenyl compounds (PCBs), and heavy metals like lead and mercury are among the often mentioned examples.

Generally speaking, food additives are meant to offer significant advantages to food producers or consumers. However, occasionally these additives have unfavorable or unhealthy effects. Food additives can have unfavorable effects when they are used excessively, unintentionally, at the wrong stage of production, processing, or storage, or when their purity or quality is compromised. Numerous additives have been the subject of extensive research to demonstrate their safety for ingestion and efficacy in fulfilling their intended purpose. Food processors have proven the safety and effectiveness of other additives through extensive and continuous use. The federal government of the United States has determined the source, recommended use, and allowable quantity of most additives used in food products. The usage of food additives is regulated, and this section explores how these regulations are influenced by the advantages and disadvantages of additive consumption.

10.3.2. Regulation of Food Additive Usage

Food additives are defined as compounds that have the potential to either directly or indirectly, through their intended uses, become components of food or otherwise alter the properties of food. The term encompasses substances meant for use in food production, manufacturing, packing, processing, preparation, treatment, packaging, transportation, or holding, as well as sources of radiation intended for any of these purposes (FDA, 1998a). Certain additive classes—such as pesticide chemicals for unprocessed agricultural products, novel animal medications, certified colors, and colors exempt from certification—are not covered by this definition. Though different laws or acts govern them, these additives are similarly regulated. Furthermore, food additives that have a track record of safe use are not subject to the Food Additives Amendment's regulatory procedures. These additives fall into a category known as "generally recognized as safe" (e.g., vitamins, salt, etc.). Because of their long history of use in foods prior to 1958 or because of published scientific evidence, they have been widely acknowledged by experts as safe. Prior approved additives are a different category of exempted substances. These are substances that the FDA or U.S. Department of Agriculture (USDA) had determined were safe for use in specific foods prior to 1958 (e.g., sodium nitrite to preserve meat).

The 1990 Nutrition Labeling and Education Act, the 1996 Food Quality Protection Act, and the 1960 Color Additive Amendment to the FD and C Act all provide additional significant federal regulation of food and color additives. The amount of food additives and colorants used in production may be restricted by current good manufacturing practice (CGMP) regulations. Only the amount of an additive required to produce the intended result may be used by manufacturers. In the case of pesticides for foods and animal feeds, the U.S. Environmental Protection Agency and the FDA have regulatory responsibilities for the approval and monitoring of pesticide use.

The EPA and FDA's authority to regulate pesticide chemicals is outlined in three significant legislative acts (EPA, 1998). The EPA is in charge of registering or granting licenses for the use of pesticide products in the US, according to the Federal Insecticide, Fungicide, and Rodenticide Act. The Federal Food, Drug and Cosmetic Act establishes the upper limit of pesticide residues that can be present in or on particular foods for humans and animals. The Food Quality Protection Act (FQPA) of 1996 made various changes to these two laws. For instance, the FQPA mandates that the combined exposure from dietary and other non-occupational sources be taken into account when assessing the risks of pesticide residues in foods or feeds. In addition, the EPA must take into account any unique risks to young children and infants when establishing new or modified pesticide tolerance criteria.

10.3.2.1. Monitoring

To ascertain whether an additive is safe and effective for the purposes for which it is intended, federal regulations are in place. Regulations may also specify the amounts of additives that can be used in food or their approved uses. However, analytical testing and label review of final products or ingredients can be used to monitor the use and labeling of food additives. Better chemical analytical techniques have made it possible to extract and identify food additives more quickly. Nowadays, a lot of testing procedures can identify extremely low chemical concentrations (e.g, parts per billion) and with a rise in prejudice. Federal authorities inspect imported goods more often than domestic products to ensure they comply with food additive regulations. It's possible that processed foods imported from other nations contain additives that are prohibited in the US. Alternatively, additives may not be properly labeled or may use descriptions that are foreign to consumers in the United States.

10.3.2.2. Labeling

Customers must be made fully aware of the advantages and risks associated with food additives and ingredients. Almost always, the FD and C Act mandates a detailed list of every ingredient in food. The Nutrition Labeling and Education Act, which modified the Food and Drug Administration and Control Act, mandates that the majority of foods have nutrition labels. Additionally, food labels that include claims about nutrient content and certain health messages must adhere to certain regulations. In order to safeguard consumers' health, two exemptions from the requirement to label ingredients on products have led to specific product labeling initiatives. First, the act allows flavorings, colorings, and spices to be declared as a group without individual names. A tiny portion of people may be allergic to or sensitive to this additive, so it needs to be mentioned expressly in the finished food's ingredients list. Second, incidental additives (like processing aids) that are present in food at negligible amounts and do not affect the finished product's functionality or technical aspects are exempt from ingredient declaration requirements under FDA regulations. Peanuts are a significant illustration of an incidental additive. There are more and more products that have been identified as potentially containing peanuts. Although peanuts or ingredients derived from peanuts were not purposefully added to these products, peanut residues from prior production runs or from the processing of peanuts on adjacent equipment may have contaminated these products. Many companies have decided to label some products with "may contain peanuts" because peanuts are one of the main foods that trigger allergic reactions. Certain processors have made the decision to process their peanut-containing products in a different area than their non-peanut product lines.

Certain foods may have additives listed on the label that can enhance one's own or the public's health. Numerous breakfast cereal products that are ready to eat are enhanced with multiple vitamins and minerals. These ingredients are added to the cereal in a quantity that allows a consumer to anticipate consuming 25–100% of the daily recommended intake of that nutrient from a serving of a specific size. On the packaging of finished products, advertisements for additives that can enhance human health can appear in places other than the ingredients list and nutritional label. Advertising claims regarding specific additives, ingredients, or natural compounds that suggest consuming the product will improve health are allowed. For instance, it is possible to promote consistent folic acid intake as a means of shielding newborns from neural tube defects.

10.3.3. Regulatory Assessment of the Risks of Food Additives

10.3.3.1. Determination of Safe and Effective Quantities for Foods

The amount of certain additives — particularly vitamins and minerals — added to a final food product might not be enough to produce the desired health effects. Certain foods that are frequently eaten by the majority of people might have extra vitamins or minerals added to them. The health benefits of regularly consuming a fortified food may outweigh the benefits of just one serving. For instance, iodine is frequently added to table salt to offer a consistent supply of this crucial mineral. Most Americans regularly eat table salt, and typical diets frequently lack iodine. The average consumer's lifetime salt consumption must be taken into account when determining the optimal, yet safe, amount of iodine to add to table salt. When used as directed, a food additive can benefit both the processor and the consumer; however, if used in excess, it can harm either the food or the consumer. When an additive is processed into a product in excess, it may not have the desired effect or the product may have an unfavorable quality defect. For instance, if a beverage contains a high concentration of artificial sweetener, like aspartame, the result could be an overly sweet product that customers reject. Alternatively, a consumer may suffer an acute illness or injury as a result of an overly high concentration of a chemical additive, as is the case with sulfiting agents. Individuals who have asthma or are otherwise sensitive to sulfur-containing additives are more likely to experience an allergic reaction when exposed to high sulfite levels. When determining the safety of using an additive, it is important to take into account the consequences of consuming an additive in excess of its recommended use level, whether by accident or on purpose. Food additive safety has been evaluated using toxicological principles (FDA, 1993).

Additionally, if an additive is taken over an extended period of time in moderate doses, it may be harmful to health. Estimating the long-term or lifetime consumption of the additive is necessary when weighing the advantages or disadvantages of using food additives. These drugs may interact with other biological or chemical substances in the body, have cumulative effects on health, or cause different reactions in users based on their age or health.

10.3.3.2. New Additive Approval

The U.S. Food and Drug Administration determines if new additives will be permitted for food use, or if existing additives can be used in additional food products or for new functions (FDA, 1998b). A commercial company typically submits a new additive petition to the FDA in order to request approval of a new additive or additive use. For instance, the FDA approved the new additive olestra for use in food in 1996 as a result of a petition filed by the Proctor and Gamble Company. Olestra, which is approved to replace fats and oils in some foods, is made of sucrose esterified with medium and long chain fatty acids (FDA, 1998c). The FDA keeps a database for the assessment and tracking of these substances and reviews new additive petitions to ascertain whether substances are deemed safe to be added to the food supply. All relevant information about the additive, such as its chemical identity and composition, physical, chemical, and biological properties, and any potential byproducts or impurities, must be included in this petition. Additional details that need to be provided include the preparation process, the manufacturer's identity, the stability assessment, suggested applications or dosages, and techniques for analysis or recovery. The most crucial information to include is proof that the additive will have the desired effects and that the studies have shown it to be safe for ingestion by humans.

10.3.3.3. Risk Assessment

Determining which applications and amounts of additives to permit in food can be a difficult task. Additives that are directly or indirectly added may be beneficial to food processors and consumers, but when ingested in excess or by vulnerable populations, they may have detrimental toxicological effects. The intrinsic toxicity of a food additive and the average consumer consumption or exposure must be estimated in order to assess whether the additive poses a risk to the consumer's health. It is important to take into account both the lifetime cumulative exposure and the short-term exposure since certain additive uses can have negative acute or long-term health effects. In order to safeguard people from these substances' potential negative effects, studies evaluating the risk of chemical residue exposure ought to be carried out. Hazard identification, exposure assessment, dose-response assessment, and risk characterization are the fundamental elements of a risk assessment (NRC, 1980, 1983; Winter & Francis, 1997).

The process of determining which chemicals are causally related to the manifestation of particular health effects is known as "hazard identification." These could consist of diseases, birth and developmental defects, and anomalies related to reproduction. It needs to be ascertained whether ingesting a chemical can cause cancer as well. Most often, long-term exposure to a carcinogenic substance results in the development of cancer. After a high-level or brief exposure to an additive, additional harmful health effects could be noticed. Stated differently, certain additives might only be dangerous when used in certain amounts.

The relationship between the likelihood of negative effects and human exposure to the chemical is predicted using a dose-response assessment. It is believed that there may be no threshold level of exposure for carcinogens, meaning that any amount of the substance could be harmful. Toxic effects from noncarcinogenic hazards usually don't show up until a threshold dose is reached. Toxicology studies can be planned to determine the dose that is just above the threshold at which effects are observed (lowest observed effect level, or LOEL) and the dose that is just below the threshold at which no effects are observed (no observed adverse effect level, or NOAEL). The NOEL has frequently been adjusted for uncertainty to produce a number known as the acceptable daily intake (ADI). The acceptable chemical exposure per unit of body weight per day can be used to describe this term. When the NOAEL is derived from animal studies, it is typically divided by 100; when it is derived from human data, it is typically divided by 10. These calculations yield the ADI (Renwick, 1996; WHO, 1987). The choice of whether to add a given amount of an additive should take into account both the ADI level and the minimal quantity thought to be required to produce the intended technical effect.

In order to estimate the probable level of human exposure to an additive, exposure assessment is required. It can be challenging to estimate how much a given population may consume and how individual consumption differs for a variety of foods and food additives. Different amounts of food additives may be exposed to different population subgroups. Due to factors like consumer age, gender, ethnicity, socioeconomic status, health, and so forth, food consumption patterns differ significantly. Furthermore, the amount of additives in some foods may be known, unknown, or vary widely in quantity. The amount of indirect additives or pesticide residues on food at the time of consumption can vary greatly and be unpredictable, even though the concentration of many food additives may be known or comparatively constant. The origin, scope, and uncertainties of estimates of the health risk are described in the risk characterization. A food additive's potential for harm should be estimated based on exposure, its specific hazardous properties should be taken into account, and the amount of exposure should be estimated when assessing the overall risk of using it. Risk characterization for noncarcinogens usually connects the estimated exposure to the allowable daily intake. Exposures at the ADI level carry a very small risk. Exposure to chemicals above the recommended daily allowance (ADI) would raise the likelihood of a negative health outcome. The process of multiplying exposure estimates by cancer potency factors yields the estimated cancer risks associated with carcinogens. This procedure frequently yields numerical cancer risks that can be used to express the proportion of individuals (e.g., 1 in 1 million) who are anticipated to experience a new cancer following prolonged exposure to the food additive. A risk characterization should include qualitative assessments of risk because of the significant uncertainties and variety of data used to estimate risks (Winters & Francis, 1997).

10.3.4. Categorization of Risks and Benefits

10.3.4.1. Categories of Risk

Because we can't measure the risks associated with individual components well, the question of what the risks are associated with our food supply can be very confusing. Food additives are not linked to the majority of foodborne illnesses. The main food-related risks, including additives, have been divided into five groups by Roberts (1981) and Wodicka (1977), and these groups are arranged according to significance: (1) foodborne risks of microbial origin; (2) nutritional hazards; (3) environmental contaminant hazards; (4) foodborne risks of natural origin; and (5) food and color additive hazards. The order of the above list is frequently not how the public perceives the risks connected to particular foods (Oser, 1978). To better comprehend the entire food safety issue, it is crucial to look into each of these areas. Although food additives are the main topic of this book, in order to give readers a perspective on the relative risks associated with food additives, it is necessary to examine all of the risks related to the food supply. Among food-related hazards, foodborne diseases of microbial origin are the most common. Inadequate hygienic measures during food preparation and/or storage in a home, restaurant, or processing facility may lead to microbial contamination. According to March (1981), there are four main causes of foodborne illness outbreaks: improper cooking, using contaminated equipment when handling food, keeping food at the incorrect temperature (which includes inadequate cooling), and poor personal hygiene on the part of the food handler. Because of their great diversity, foodborne diseases of microbial origin are important for food safety. These microbial infections can range in severity from less serious conditions like staphylococcal food poisoning to more serious ones like botulism. The most significant risk associated with foodborne illnesses is posed to young children, the elderly, and people with disabilities. Listeriosis, which is linked to an infection with Listeria monocytogenes, is a prime illustration of an opportunistic foodborne illness that primarily affects vulnerable people. As long as a sample of the contaminated food is still present to verify its contamination with the suspected organism, foodborne illnesses of microbial origin are easily identifiable and frequently treatable. This foodborne illness category's ascent to the top of our ranking has been largely attributed to the development of distinct cause-and-effect relationships.

Nutritional hazards are linked to food as the second major risk. Due to the fact that nutritional hazards can result from excesses or deficiencies in nutrient intake, they have earned a high ranking (Stults, 1981). Most nutritional risks result from an imbalance in the amount of food consumed in the diet. The most well-known risks associated with inadequate nutrient intakes are likely diseases like scurvy (vitamin C), pellagra (niacin), rickets (vitamin D), beriberi (thiamin), and goiter (iodine). These illnesses were common in the United States in the early 20th century, but they have all but disappeared thanks to nutrient-fortified foods like milk and table salt, increased dietary intake, and better distribution and preservation of perishable goods. On the other end of the spectrum, there are risks related to consuming large amounts of some trace elements and fat-soluble vitamins. Differentiating between the water-soluble (vitamin C) and fat-soluble (vitamins A, D, E, and K) is crucial when talking about the toxicity of vitamins. Due to their storage in body fat, fat-soluble vitamins—particularly vitamins A and D—may accumulate and have harmful side effects if taken in excess. Conversely, excess water-soluble vitamin excretion is typically eliminated through sweat and urine (Stults, 1981), though mild cases of toxicity are occasionally documented. The interactions between the trace elements have a significant impact on the degree of toxicity. For example, it is well known that harmful levels of iron can obstruct the uptake and utilization of copper, zinc, and manganese, and that high manganese levels can obstruct the uptake of vitamin B-12 (Davies, 1978). Consuming a sufficient and varied diet poses no appreciable risk for these nutritional excesses and deficiencies for the majority of healthy individuals. Thus, one could contend that food safety issues shouldn't come in second on the list of priorities for nutritional hazards. However, there is a case to be made for moving microbial hazards to the second rank and moving nutritional hazards to the top due to the prevalence of heart disease, cancer, and stroke as well as the potential role that diet plays in these illnesses. Furthermore, a lot of people around the world are unable to access a sufficient and varied diet due to subpar food distribution networks, ineffective manufacturing practices, and inadequate storage facilities. Saturated fats and cholesterol in the diet may play a role in the development of coronary heart disease, though opinions on this matter vary. More significant risk factors than dietary cholesterol appear to be smoking, genetics or family health history, obesity, low economic status, lack of exercise, and stress.

Environmental contaminants rank third on the list of food-related hazards. Environmental pollutants can enter the food chain through natural sources or through the release of industrial chemicals. There are certain common traits among the chemical substances in this category, despite their wide variety. For instance, these pollutants frequently withstand degradation and continue to exist in the environment. Due to their slow rates of metabolism and excretion, these substances may accumulate in specific bodily tissues. Additionally, some environmental pollutants, like mercury in swordfish and sharks or neurotoxins in shellfish, can build up in the food chain. The food supply is at risk from a number of environmental contaminants, including lead, mercury, dioxins, and polychlorinated biphenyls (PCBs). Human illness has been linked to PCB and mercury exposure through ingestion of contaminated fish (Munro & Charbonneau, 1981). Natural sources of contamination typically originate from soils that have high concentrations of specific substances by nature or from the erosion of rock formations. The main naturally occurring contaminants are tin, arsenic, mercury, selenium, and cadmium. This category also includes residues of drugs and pesticides in animals raised for food. It is important to keep in mind that toxicity is a function of dose when evaluating regulations for the management of environmental pollutants. As a result, it's important to understand the amount of contaminated food that is typically consumed as well as its contamination level. Regulatory action levels for certain chemical residues in food have been established, though this can get pretty complicated. These allowed levels seem to reduce the amount of certain contaminants that people may come into contact with when eating. However, action levels for some environmental contaminants, like mercury, have been set at the level of natural occurrence, minimizing exposure and eliminating any possibility of contamination or dose impact resulting from industrial processes. Other cases (PCBs) allow for greater flexibility in determining hazardous levels because action levels are established based on risk assessment and toxicity knowledge. According to Rodricks and Pohland (1981), naturally occurring toxicants are ranked fourth among the main causes of foodborne illness. These contaminants are ranked relatively low because it appears that they only cause issues in specific extreme situations. Additionally, they are ranked poorly because, according to Rogers (1983), the public is more concerned about man-made pollutants than "natural" hazards. The naturally occurring toxicants found in food include mycotoxins, marine toxins, oxalates in spinach, glycoalkaloids in potatoes, mercury in swordfish, and mushrooms. Food hazards can also be caused by other substances that can be created during the processing, preparation, or storage of food, such as nitrosamines and biologically active amines.

The toxicity of a chemical is unaffected by whether it is synthetic or "natural." Therefore, more research is required to ascertain whether there is actually little risk from naturally occurring contaminants in the area of food safety. Acute foodborne illness is frequently caused by naturally occurring seafood toxins and mushroom toxins (Hughes et al., 1977). Studying the effects of exposure to naturally occurring toxins on humans is challenging because the consumption of these toxins is often unpredictable and varies widely. Furthermore, consuming too many natural toxins usually leads to chronic or long-term illness, the origin of which can be hard to find. The risk associated with food additives is ranked lower than the other four food hazards. This classification would include the GRAS ingredients. Legally speaking, GRAS ingredients are not food additives, but the general public does not recognize this. There are thousands of substances in this class. The level at which a food additive is consumed and its toxicity will determine any possible risk to humans from that additive. The four most commonly used direct food additives are dextrose, corn syrup, sucrose, and salt. Together, they make up 93% of the total weight of direct food additives (Clydesdale, 1982; Roberts, 1981). It is challenging to quantify the amount of indirect additives that humans are exposed to, but it is very low. Most direct food ingredients are used because it has been determined that they are GRAS or previously approved. While only a small portion of the GRAS list's items have undergone extensive evaluation, a review of some of them has revealed that the majority pose no significant risk when used normally (Roberts, 1981). The FDA regulates the use of the other direct food additives that are used in food products and has approved their use.

So why is the public's perception of food additives and some GRAS ingredients so negative? Roberts (1981) and Oser (1978) hypothesized that the reason for the public's negative perception of food ingredients is that these foodborne substances need to be demonstrated to be "safe." It is impossible to guarantee a substance's total safety for every human being under every set of usage circumstances. As a result, any doubt regarding a food additive's safety may lead to the public believing that a far higher risk exists. One example of an ingredient whose use in food is still up for debate is the noncaloric fat replacer olestra (brand name Olean), which was just approved (ACSH, 1998). Olestra's molecules are too big for the body to absorb or digest, despite being made of sugar and vegetable oil. Similar to fats and oils, it can enhance the flavor and texture of savory snack foods without adding any calories. Numerous customers have complained of experiencing gastrointestinal distress after consuming products containing olestra, and some are worried that this product will hinder the body's ability to absorb carotenoids and fat-soluble vitamins. Regardless of the kind of chip they were actually eating, individuals in one study who thought they were eating Olestrĭ snack chips reported gastrointestinal symptoms about 50% more frequently than those who thought they were eating regular chips. The participants' perceptions of what they were eating and their awareness of the product labels that state that olestra may cause loose stools and abdominal cramps may have had an impact on how many symptoms they reported (Sandler et al., 1999). Numerous scientific studies have demonstrated that eating standard diets that include the fat replacer Olestrat does not cause gastrointestinal distress and that the body would only experience a loss of vitamin absorption when eating a meal that included Olestrat-formulated products. FDA regulators have determined that when taken as part of a varied and well-balanced diet, olestra is intrinsically safe. However, there will likely be ongoing debate regarding the efficacy and safety of olestra.

The public's perception of food additives and specific GRAS ingredients as unneeded, involuntary sources of risk is equally significant. Few people are aware of the health advantages of these culinary ingredients. Additionally, some customers are more suspicious of these food ingredients because they are not thought of as natural or typical food ingredients. Additives will always be utilized because they can raise the quantity and quality of food. In order to reassure the public that food additives are safe, the FDA must thus keep an eye on how they are used.

10.3.4.2. Categories of Benefits

The advantages that arise from the food supply can be broadly classified into four groups: (1) health advantages that lower certain health risks or offer health advantages like better nutrition; (2) supply advantages that are related to quantity, variety, and affordability; (3) hedonistic advantages that offer sensory gratification; and (4) advantages that result in greater convenience (Darby, 1980; Food Safety Council, 1980). Food additives can contribute significantly to each of these benefit categories by boosting supply, boosting appeal, boosting convenience, or improving health. The most important of these benefits should be health-related, followed by supply-related and finally, less important benefits like enhanced appeal and convenience.

Food additives and other food components may offer two types of health benefits: those that improve nutrition or prevent or lower the incidence of particular diseases. Because nitrites have anti-botulinal properties, they may lessen the possibility of botulism in cured meats. The main source of nutritional advantages is the inclusion or presence of nutrients. The addition of vitamins and minerals to bread, milk, and salt, among other staple foods, increases their nutritional wholesomeness. One way to think of fortifying food is as a way to prevent deficiency illnesses like scurvy, beriberi, or goiter. Naturally, as was already mentioned, over-fortification of foods with specific nutrients can raise risks. The use of food ingredients that stop food from spoiling, boost processing technique yield, or offer new sources for desired functions also improves supply benefits. Preservatives increase supply and decrease costs by preventing food spoilage. By preserving nutrients, stopping the growth of dangerous bacteria, and assisting in ensuring that a plentiful and nutrient-dense food supply is available, preservatives also indirectly contribute to good health. In 1999 the U.S. The Food and Drug Administration announced that it would provide an expedited review of a new food additive petition if the additive is intended to significantly reduce foodborne human pathogens or their toxins.

Improved color, flavor, and texture are examples of hedonic benefits that increase consumer appeal. Benefits related to convenience arise from food ingredients that save preparation time. Rich societies typically place more value on these advantages.

10.3.4.3. Striking a Balance

It is rarely an exact science to quantify the level of risk or benefit associated with a food additive. Benefits are not taken into account and risk assessment is the only focus of our current regulatory statutes. Even though risk-benefit analyses have never been performed, it might occasionally be able to calculate the net risk by contrasting the risks associated with using and not using an additive. Because risk is easier to quantify than benefit, it is easier to balance one risk against another than it is to balance risks against benefits, and risk assessment is a widely recognized regulatory concept, the risk–risk approach is more widely accepted. Additionally, the hazards associated with additive use are typically weighed against the harm they do to human health. While improving health or preventing disease is one advantage of some food additives, the majority of benefits are related to food processors' bottom lines as well as consumer convenience and sensory aspects. Therefore, it can be challenging to conduct and quantify a sufficient risk-benefit analysis for each food additive (IFT, 1988).

The idea of a defined, socially acceptable risk level is crucial to both the risk-benefit and risk-risk approaches to decision making (Food Safety Council, 1980). Every chemical carries some inherent risk. However, opinions on how much to worry about health risks from chemical consumption are frequently divided. The definition of a socially acceptable risk level is especially needed when reviewing carcinogenicity data from animal experiments and the extrapolation of these data to human experiences. Which risks are less tolerable—ones that could cause acute illness versus chronic illness? or risks that may lead to many cases of temporary illness versus a few cases of mortality? or risks that affect children more than adults? The FD&C Act's Delaney clause sets zero as the acceptable risk level, but past experience has shown that consumers are willing to take on greater risk in exchange for a significant perceived benefit. With regard to pesticide residues in food, the Delaney Clause was eliminated in 1996 by the Food Quality and Protection Act. A "reasonable certainty of no harm" standard was established by this legislation, which takes into account the risks associated with various exposures, the risks to various population subgroups, and the numerous toxicological effects of pesticides on human health (EPA, 1998; Winter & Francis, 1997).

When a chemical is consumed at typical levels over the course of a lifetime, the FDA has operationally defined acceptable risk from chemical consumption as up to one additional case of cancer per million cases, or 10^{-6}. On the other hand, the Food Safety Council (1980) examined risk acceptability in greater detail from a theoretical standpoint. Four scenarios were identified by the Food Safety Council as potential outcomes of risk-benefit analyzes: (1) no discernible risk for the chemical; (2) a level of risk for the chemical that is unacceptable; (3) a measured risk level for the chemical that is below the acceptability threshold; and (4) a measured risk level for the chemical that exceeds the acceptability threshold. In each of these scenarios, the potential advantages of utilizing an additive are weighed differently.

Any quantifiable benefit should permit the use of the substance in the food supply if there is no discernible risk. That being said, it shouldn't be unduly consoling that there is no discernible risk. Every chemical has some level of intrinsic toxicity. It's possible that the lack of evidence of any toxicity is just the result of improper testing procedures. Some risk may be identified by additional testing, testing in different species, or testing at higher dose levels. In these kinds of situations, a later review of the risk-benefit analysis may be required. These evaluations might be required for any approved substance in light of recent findings on toxicology.

Under no circumstances should substances carrying blatantly unacceptable risks be permitted in food products. The risks in this case are much greater than the rewards. This idea is widely applied in the decision tree approach to toxicological assessment. When a substance exhibits lethal effects at relatively low dose levels in an acute toxicity test, it is eliminated from consideration without additional testing. Usually, before a petition for a food additive is created, substances falling into this category are found and rejected. Any toxicologist should be able to identify risks that are obviously unacceptable even though there isn't a set level of acceptability for the majority of toxic effects. Theoretically, a situation could occur where a substance has some known risk, but not to the extent that society considers acceptable. In such circumstances, the Food Safety Council's logic would permit use of the substance only insofar as a significant net benefit was demonstrated (Food Safety Council, 1980). This scenario presumes that a risk threshold that is socially acceptable has been established for this specific kind of toxic reaction. For the majority of toxic reactions, such definitions do not exist. According to the Food Safety Council (1980), compounds in this category need to be carefully assessed to make sure the advantages outweigh the risks. Benefits to nutrition and health would be most relevant in this hypothetical comparison. Larger benefits would be required to balance the known risks if the benefits were primarily related to supply, appeal, or convenience.

The last scenario concerns substances where the measured risk level is higher than what is considered socially acceptable. The issue remains that, aside from cancer, no level of acceptability has been determined for any other toxicological response. If such definitions were possible, a substance falling into this category would still need to offer enough advantages to make the socially acceptable risk threshold for that specific instance of use reevaluate itself. Large health benefits would presumably be one potential counterbalance. It may be necessary to impose extra regulatory restrictions in order to warn consumers about potential risks. Saccharin might be a good illustration. Saccharin is more carcinogenic than is considered to be acceptable. Those who wish to manage their weight or diabetes conditions are said to benefit from saccharin. Warning labels are required to inform consumers about the potential for cancer. Nonetheless, the label requirement was removed in 2000 as a result of additional safety analysis of saccharin. The kind and extent of risks as well as the benefits must be carefully considered when making risk-benefit decisions. The inability to accurately extrapolate risks from animal experiments to human situations, the lack of appropriate methods for quantifying benefits, and the challenges associated with comparing health risks with non-health risks all complicate these decisions. Two of the three issues mentioned above can be mitigated by approaching these choices using the risk-risk approach. Health risks and health benefits are compared in risk-risk approaches, with little attention paid to other kinds of benefits. Put differently, the risk associated with using an additive is contrasted with the risk of not using it. Food additives are permitted to be used if they provide a net improvement in health. Although this method is praiseworthy, it is not easily applicable to the majority of food additives because some additives have no health benefits or only minimal risks associated with not using them. It can, however, be applied in certain circumstances and offer an intriguing viewpoint. It is important to stress once more that the regulations as they stand do not require such an approach.

There are some situations where the relative risks are clear. For instance, one must weigh the possibility of using nitrites and developing cancer from nitrosamine exposure against the possibility of not using nitrites and contracting Clostridium botulinum toxin from cured meat (IFT, 1988). Although there is very little chance of contracting botulism, the disease is frequently fatal. In other situations, the relative risks are harder to define or measure. For instance, the minor chance of developing bladder cancer from consuming saccharin needs to be compared to the other risks. In theory, there would be two alternative risks: a higher chance of all the diseases linked to obesity and a higher intake of sucrose. There are, nevertheless, other nonnutritive sweeteners available, and substituting highly caloric sweeteners for these sweeteners does not always result in an equivalent sweetness intake. Using saccharin doesn't stop obesity. The risk-risk approach can be even harder to implement in certain situations. There may be some evidence of risk associated with the use of the food colorant, so how could it be applied to tartrazine or another food coloring, for example? However, it would be extremely challenging to pinpoint any risks connected to the substance's unavailability. Maybe people would start choosing fewer safe foods, or maybe the food industry would choose a food coloring that is more dangerous as a substitute. These alternative risks, however, would be challenging to predict or measure. Therefore, it is likely that the risk-risk approach will only work with specific food ingredients.

CASE STUDY

FOOD INDUSTRY COLOR CONTROL

When it comes to food, color and appearance are the most important first impressions, even before one's olfactory sense is tickled with a pleasing aroma. This point is illustrated in the famous Dr. Seuss book, Green Eggs and Ham, loved by children and adults alike. The relentless character, Sam-I-Am, pesters a nameless skeptic to taste his green eggs and ham dish. It also shows the reality of how resistant consumers are to trying "off colored" foods. In today's retail world of behind-glass, chilled, frozen, boxed, dried, and plastic wrapped foods, eye appeal is far more important than nose appeal. Both fresh and processed food producers know this well and are increasingly adopting instrumental color measurement technologies and practices to control color better across a wide range of applications.

1. Two Principal Color Measurement Techniques

In current food industry practices, two principal color measurement techniques are used:

Colorimetry is the technique that quantifies color by measuring three primary color components of light seen by the human eye, specifically red, green, and blue. This "tristimulus" color measurement provides data on how much of these three components are present in the light reflected (solids) or transmitted (typically liquids) by a food product. Such data may be used, for example, to adjust the color components in a prepared food or beverage recipe to improve "eye appeal," gauge "doneness" in a baked product, and, in fresh foods, to determine factors such as degrees of ripeness and spoilage during shipping, storage, shelf-life, palatability, and disposal cycles. Although there is no strict demarcation line where the benefits of colorimetry in foods end, it should be recognized that it measures color much the same as the human eye. That is, secondary and tertiary colors such as orange, yellow, violet, tans, browns, etc., are not individually quantifiable. This leaves a variability factor that can hamper the consistent reproduction of a desired color in prepared food products formulated for a specific, consistently produced look.

Spectrophotometry, a scientific "step up" so to speak, is presently the most precise and accurate technique for the measurement, formulation, and quality control of desired colors in prepared food products. Spectrophotometer instruments measure the spectral reflectance or transmittance of an object across the full spectrum of human visible light wavelengths, 400 nm to 700 nm (nanometers), enabling precise specification of any desired color. Spectrophotometers offer greater specificity, making them the instruments of choice for food product color formulation, specification of standards and tolerances, inter-plant color communication, and color quality control in processing operations.

Over the last decade, the use of spectrophotometers in the food industry has significantly increased for color standardization and QC inspection of ingredients, specification of final product color (particularly in jams, jellies, preserves, beverages, etc.), research and development of new food and beverage products, and potential food screening and sorting techniques of factors ranging from carcass natural fat content to quality grading of meats, poultry, and fish.

2. Software and Methodologies

One indispensable, key factor in the food industry's growing use of colorimetry and spectrophotometry is the availability of easy-to-understand and easy-to-adapt color data software for color measurement, color matching, color formulation, and color quality control. When correlated with all other food laboratory and quality control data, color can become an integral clinical data component of factors ranging from product appeal to shelf-life to possible spoilage and/or contamination risks.

Color formulation software and color matching software packages are available with portable and bench-top spectrophotometers for use in laboratory color control systems. Color quality control software with sample pass/ fail comparison testing and database update functions is a standard component of those systems and is also an integral onboard QC inspection function of our Konica Minolta portable colorimeters and spectrophotometers. The art in the science of integrating color as a food quality component lies in the methodology developed and used by food producers and the food processors. In food science, technologies are developed in response to needs vocalized by food industry scientists and technologists. In Konica Minolta's Instrument Systems division, we view our role as being technologically responsive: that is, scientists and technologists foresee their color measurement and methodology goals and, in turn, we attempt to create and apply technologies to fulfill those needs. Over the last 20 years, we have moved through nearly four generations of technologies. The following examples demonstrate the continuing advancement of food color measurement, as well as provide hints as to how this technology may progress in the future.

3. Baking "Goodies" at Pepperidge Farm

Judging baked goods by eye is not as easy as many home bakers might think, according to Mike Davis, Bakery Technologist, at Pepperidge Farm. Until Pepperidge Farm installed BC-10 Baking Contrast Meters, evaluating the acceptable color of finished baked goods on plant floors was a real challenge. Previously, "acceptable color" was established by comparing a production sample against a color photograph of how the product should look when it exits the baking oven. Despite pictures being taken in a special room with carefully controlled lighting conditions, the color on the line was still problematic.

Visual color judgment is simply too subjective. Pepperidge Farm recognized the influence of color and appearance on consumer perception and the flavor profile of baked and fried products. Because of this, a strong focus was placed on incorporating objective standards and measurement techniques into their operations. Portable, hand-held, and battery-powered, the Konica Minolta BC-10 colorimeter eliminated the problem of subjectivity by establishing acceptable color contrast standards (the brightness of a baked or fried product) to communicate numerically to their production plants. Now, color standard and tolerance measurements at Pepperidge Farm are expressed in terms of Baking Contrast Units (BCUs). The Baking Contrast scale is calibrated so that a difference of one-tenth of a BCU corresponds to one perceptible difference in shade as perceived by a normal observer. The BC-10 colorimeter can also report in the commonly used CIE L*a*b* color space and scale units, measuring lightness and darkness in units from the darkest, 0, to the lightest, 100.

4. **"Preserving" the Right Color at Welch's**

A few years ago, Welch Foods, Inc., the famous cooperative of some 1,400 grape growers, was seeking a more pragmatic way of formulating color in recipes for their many bottled and canned juices, jellies, jams, and frozen products. The challenge, their laboratory explained, was the annual, seasonal, and regional variability of the natural grapes used, including Concord Grapes and Niagra Grapes. Grown in five states and a Canadian province, and processed in six plants, the grapes varied in a number of factors. Color, of course, is important because many of Welch's products are in clear glass and plastic bottles. Testing samples of each season's harvest against nutritional and recipe standards, an on-going process at the laboratory, was very precise both qualitatively and quantitatively. Color testing, however, was not precise because this was performed subjectively. Working with Konica Minolta's application engineers, Welch's food technologists determined that spectrophotometry was the optimum color measurement and evaluation methodology for them. Experimenting with portable and bench-top spectrophotometers, the laboratory found that it could work backward from the ideal final product and establish specifications and tolerances for color at each stage of the product manufacturing process. Color variations from the standard could then be adjusted to better assure color consistency in final products. Spectrophotometry now plays an important part in Welch's product development.

5. A Flour by Any Other Name

One would think that in manufacturing white pasta flour, the color wouldn't be much of an issue. Well, with every kitchen becoming a home gourmet's retreat, the surge in specialty flours and the homegrown sprouting of epicurean eyes and palates, it is now an issue – and one to be avoided. This challenge was presented by a semolina flour manufacturer hoping to determine the best method for standardizing the color of their semolina products for consumer lines as well as commercial products. The problem was natural: semolina wheat grains vary in color. The manufacturer was already combining both grain and flour lots to average the color mixtures but felt numerical standards and tolerances were needed to assist operations in averaging the ingredients. The solution was first to standardize an ideal color of semolina flour using a Konica Minolta colorimeter system and then measure individual flour lots for their color characteristics. By sample averaging, color variability within each lot was reduced, and computer color matching of lots for successful color mixture potential in a final flour blend became possible.

6. Fresh Mangos in Frozen Maine

In the 1990's, Trans FRESH Corporation, a leader in the development of modified and controlled atmosphere systems for the transport of perishable foods, initiated an advanced program to gage the maturation and spoilage rates of specific fresh produce items. Creating a controlled atmosphere to keep produce fresh during shipment requires extensive testing. Because of the direct correlation between the ripeness of various fruits and their color, Trans FRESH implemented a program using a Konica Minolta colorimeter to chart the ripening process under various controlled atmospheric conditions over typical durations of shipment. For example, the measured color of an avocado rind provided data about its point of ripeness. Similarly, the rate of discoloration of the rind allowed plotting of the maturation process. Today, Trans FRESH can precisely determine the right atmosphere needed for a particular product shipment to arrive fresh at its retailer.

7. Surf, Turf, and Barnyard

Since the mid-1990s, Konica Minolta has been working with companies in the seafood, meat, and poultry industries to help them develop and improve color testing and measurement methodologies within their operations. This has included working on a barge in the Pacific Northwest to help a fisheries association establish quality and nutritive content standards of Pacific salmon. To do this, the pink flesh of the salmon was measured using portable and bench-top spectrophotometer systems. Similarly in the North American beef industry, both colorimetry and spectrophotometry are being tested and used in various ways to classify the fat content of carcasses and cuts both on and off processing lines. This is done by quantifying the relative "marbling" of meat through precise reference to overall color. In the poultry industry, our portable spectrophotometers are now widely used on processing lines where the skin "yellowness" of carcasses is measured to determine the fat content. From ice cream, yogurt, and cheeses in the dairy section to lunch meats, sausages, and sauces, Konica Minolta Sensing Americas, Inc. provides food technologists with the color measurement solutions to ensure favorable "eye appeal" of their final product – a key ingredient to retail success.

CLASS ACTIVITY

Protecting the Quality of Food

Answer the Following Questions

1. What are food additives?
2. Why are food additives added to food?
3. List 3 types of food additives..
4. Draw a line to link the examples of food additives in the right-hand column to the correct group of food additives in the left-hand column.

Preservatives	MSG
	Tatrazine
Flavourants	Benzoates
	Nitrates
Colourants	Sulphites

SUMMARY

- Food additive, any of various chemical substances added to foods to produce specific desirable effects. Additives such as salt, spices, and sulfites have been used since ancient times to preserve foods and make them more palatable.
- Food additives are substances not normally consumed as food by themselves and not normally used as typical ingredients in foods. Most minimally processed and unprocessed foods do not contain food additives.
- Authoritative bodies at the national, regional and international levels are responsible for evaluating the safety of food additives.
- Nutritional additives are used for the purpose of restoring nutrients lost or degraded during production, fortifying or enriching certain foods to correct dietary deficiencies, or adding nutrients to food substitutes.
- Emulsifiers are used to maintain a uniform dispersion of one liquid in another, such as oil in water. The basic structure of an emulsifying agent includes a hydrophobic portion, usually a long-chain fatty acid, and a hydrophilic portion that may be either charged or uncharged.
- Stabilizers and thickeners have many functions in foods. Most stabilizing and thickening agents are polysaccharides, such as starches or gums, or proteins, such as gelatin. The primary function of these compounds is to act as thickening or gelling agents that increase the viscosity of the final product.
- The autoxidation of unsaturated fatty acids involves a reaction between the carbon-carbon double bonds and molecular oxygen (O_2). The products of autoxidation, called free radicals, are highly reactive, producing compounds that cause the off-flavors and off-odors characteristic of oxidative rancidity.
- Antimicrobials are most often used with other preservation techniques, such as refrigeration, in order to inhibit the growth of spoilage and pathogenic microorganisms. Sodium chloride ($NaCl$), or common salt, is probably the oldest known antimicrobial agent.
- Color is an extremely important sensory characteristic of foods; it directly influences the perception of both the flavor and quality of a product. The processing of food can cause degradation or loss of natural pigments in the raw materials. In addition, some formulated products, such as soft drinks, confections, ice cream, and snack foods, require the addition of coloring agents.
- A flavor additive is a single chemical or blend of chemicals of natural or synthetic origin that provides all or part of the flavor impact of a particular food. These chemicals are added in order to replace flavor lost in processing and to develop new products.
- Sucrose, or table sugar, is the standard on which the relative sweetness of all other sweeteners is based. Because sucrose provides energy in the form of carbohydrates, it is considered a nutritive sweetener. Other nutritive sweeteners include glucose, fructose, corn syrup, high-fructose corn syrup, and sugar alcohols (e.g., sorbitol, mannitol, and xylitol).

REVIEW QUESTION

1. What do you understand by the term 'functional role' of additives? Give any five functional roles of food additive.
2. Why are artificial sweeteners also referred to as intense sweeteners?
3. Which food additive can be added to salt to make it "free flowing"?
4. Describe how flavor enhancers and artificial sweeteners affect the taste of food.
5. Discuss the importance of reading the label of a processed food item before consuming it.
6. What role do antioxidants play in food preservation?
7. Explain the role of thickeners and gelling agents in food texture modification.
8. How do regulatory agencies determine the acceptable daily intake (ADI) of food additives?

REFERENCES

1. Anon. (2006). Coloring Foodstuffs–The Clean Label Colorful Alternative to Additives? *Innovations in Food Technology, 30*, 76–77.
2. Ashida, H., Hashimoto, T., Tsuji, S., Kanazawa, K., & Dasno, G. I. (2000). Synergistic effects of food colors on the toxicity of 3-amino-1,4-dimethyl-5h-pyrido[4,3-b]indole (Trp-P-1) in primary cultured rat hepatocytes. *Journal of Nutritional Science and Vitaminology, 46*(3), 130–136.
3. Bateman, B., et al. (2004). The effects of a double-blind, placebo-controlled, artificial food colorings and benzoate preservative challenge on hyperactivity in a general population sample of preschool children. *Archives of Disease in Childhood, 89*, 506–511.
4. Darby, W. J. (1980). The nature of benefits. *Nutrition Reviews, 38*, 37–44.
5. Dean, T. (2002). Do food additives cause hyperactivity? In V. Emerton (Ed.), *Food Allergy and Intolerance, Current Issues and Concerns*. Leatherhead: Leatherhead Food International, 93–101.
6. Directive 89/107/EEC on the approximation of the laws of the Member States concerning food additives authorized for use in foodstuffs intended for human consumption, as amended. (1989). *The Official Journal of the European Communities, 32*(L40), 27–33.
7. Dubois, G. E. (1992). Sweeteners, non-nutritive. In Y. H. Hui (Ed.), *Encyclopedia of Food Science and Technology*. New York: Wiley, pp. 2470–2487.
8. FDA. (1992). *Food Additives*. FDA/IFIC Brochure.
9. FDA. (1993). Toxicological principles for the safety assessment of direct food additives and color additives used in food (Redbook II). Springfield, VA: National Technical Information Services.

10. FDA. (1998). Food additives permitted for direct addition to food for human consumption. *Code of Federal Regulations, Title 21, Part 172*. Washington, D.C.: Government Printing Office.

11. Feingold, B. F. (1975). Hyperkinesis and learning disabilities linked to artificial food flavors and colors. *American Journal of Nursing, 75*, 797–803.

12. Hughes, J. M., Horwitz, M. A., Merson, M. H., Barker, W. H., Jr., & Gangarosa, E. J. (1977). Foodborne disease outbreaks of chemical etiology in the United States, 1970–1974. *American Journal of Epidemiology, 105*, 233–244.

13. Kendrick, A. (2005). Clean up your label with the colorful alternatives to additives. *Confectionery Production, 71*(4), 14–15.

14. Lau, K., McLean, W. G., Williams, D. P., & Howard, C. V. (2006). Synergistic interactions between commonly used food additives in a developmental neurotoxicity test. *Toxicological Sciences, 90*(1), 178–187.

15. Marth, E. H. (1981). Foodborne hazards of microbial origin. In H. R. Roberts (Ed.), *Food Safety*. New York: Wiley, pp. 15–65.

16. McCann, D., et al. (2007). Food additives and hyperactive behavior in 3-year-old and 8/9-year-old children in the community: A randomized, double-blinded, placebo-controlled trial. *The Lancet, 370*, 1560–1567.

17. McKendry, M. (1973). *Seven Hundred Years of English Cooking*. London: Treasure Press.

18. Mukherjee, A., & Chakrabarti, J. (1997). In vivo cytogenetic studies on mice exposed to acesulfame-K—A non-nutritive sweetener. *Food and Chemical Toxicology, 35*(12), 1177–1179.

19. Mukherjee, A., Agarwal, K., & Chakrabarti, J. (1992). Genotoxicity studies of the food additive ester gum. *Food and Chemical Toxicology, 30*(7), 627–630.

20. Mukhopadhyay, K. M., Mukherjee, A., & Chakrabarti, J. (2000). In vivo cytogenetic studies on blends of aspartame and acesulfame. *Food and Chemical Toxicology, 38*(1), 75–77.

21. Munro, I. C., & Charbonneau, S. M. (1981). Environmental contaminants. In H. R. Roberts (Ed.), *Food Safety*. New York: Wiley, pp. 141–180.

22. Nobakht, M., Fattahi, M., Hoormand, M., Milanian, I., Rahbar, N., & Mahmoudian, M. (2000). A study on the teratogenic and cytotoxic effects of safflower extract. *Journal of Ethnopharmacology, 73*(3), 453–459.

23. Opinion of the Scientific Panel on food additives, flavorings, processing aids and materials in contact with food (AFC) related to para hydroxybenzoates (E 214–219) (Adopted on 13 July 2004).

24. Renwick, A. G. (1996). Needs and methods for priority setting for estimating the intake of food additives. *Food Additives and Contaminants, 13*(4), 467–475.

25. Rodricks, J. R., & Pohland, A. E. (1981). Food hazards of a natural origin. In H. R. Roberts (Ed.), *Food Safety*. New York: Wiley, pp. 181–237.

26. Sandler, R. S., Zorich, N. L., Filloon, T. G., Wiseman, H. B., Lietz, D. J., Brock,

M. H., Royer, M. G., & Miday, R. K. (1999). Gastrointestinal symptoms in 3181 volunteers ingesting snack foods containing olestra or triglycerides. *Annals of Internal Medicine, 130*, 253–261.

27. Türkoğlu, S. F. (2009). Genotoxic effects of mono-, di-, and trisodium phosphate on mitotic activity, DNA content, and nuclear volume in *Allium cepa* L. *Caryologia, 62*(3), 171–179.
28. Wilson, R. (2007). *Ingredients Handbook–Sweeteners* (3rd ed.). Leatherhead: Leatherhead Food International.
29. Winter, C. K., & Francis, F. J. (1997). Assessing, managing, and communicating chemical food risks. *Food Technology, 51*(5), 85–92.
30. Yılmaz, S., Ünal, F., Yüzbaşıoğlu, D., & Aksoy, H. (2008). Clastogenic effects of food additive citric acid in human peripheral lymphocytes. *Cytotechnology, 56*, 137–144.

Index

D

E

F

G

R

S

T

U

V

W

FSC
www.fsc.org
MIX
Papier | Fördert
gute Waldnutzung
FSC® C083411

Kolibri 360 GmbH
Ferdinand-Jühlke-Straße 7
99095 Erfurt, Deutschland
produktsicherheit@kolibri360.de